Recent Advances in Geochemistry

Recent Advances in Geochemistry

Guest Editors

Ningqiang Liu
Qingjie Gong

MDPI

Basel • Beijing • Wuhan • Barcelona • Belgrade • Novi Sad • Cluj • Manchester

Guest Editors

Ningqiang Liu
School of Earth Sciences
and Resources
China University of
Geosciences (Beijing)
Beijing
China

Qingjie Gong
School of Earth Sciences
and Resources
China University of
Geosciences (Beijing)
Beijing
China

Editorial Office
MDPI AG
Grosspeteranlage 5
4052 Basel, Switzerland

This is a reprint of the Special Issue, published open access by the journal *Applied Sciences* (ISSN 2076-3417), freely accessible at: https://www.mdpi.com/journal/applsci/special_issues/85A6N78TS2.

For citation purposes, cite each article independently as indicated on the article page online and as indicated below:

Lastname, A.A.; Lastname, B.B. Article Title. *Journal Name* **Year**, *Volume Number*, Page Range.

ISBN 978-3-7258-8313-4 (Hbk)
ISBN 978-3-7258-8314-1 (PDF)
https://doi.org/10.3390/books978-3-7258-8314-1

Contents

applied sciences

Editorial

Recent Advances in Geochemistry: Risk Assessment of Soils and Provenance of Geological Materials

Qingjie Gong * and Ningqiang Liu

School of Earth Sciences and Resources, China University of Geosciences, Beijing 100083, China
* Correspondence: qjiegong@cugb.edu.cn

1. Introduction

This Special Issue, 'Recent Advances in Geochemistry: Risk Assessment of Soils and Provenance of Geological Materials', was organized to present papers from the 10th National Conference on Applied Geochemistry in China held in Kunming, Yunnan Province, in November 2024, which aimed to facilitate the academic exchange and presentation of new ideas in the field of applied geochemistry. This Special Issue is edited by the Committee of Applied Geochemistry of the Chinese Society for Mineralogy, Petrology, and Geochemistry (CSMPG).

This Special Issue contains nine scientific papers that present various recent advances, challenges, and illustrations in applied geochemistry. These contributions cover broad fields, including environmental geochemistry on pollution risk assessment of soils, provenance of geological materials, exploration and analysis, etc.

2. Pollution Risk Assessment of Soils

Heavy metal pollution in soils is an environmental challenge that has attracted global concern [1,2]. Three papers in this Special Issue discuss the pollution risk of metals in soils, crops, and coals. Chen et al. (contribution 1) discuss the risk assessment methods of heavy metals in soils, Li et al. (contribution 2) assess the risk of Selenium in soils and crops, and Junussov and Mustapayeva (contribution 3) discuss the harmful risk of potentially toxic elements in coal ash samples.

Methods for assessing heavy metal pollution are generally categorized into three types: pollution indices, released standards, and the integration of these two methods (contribution 1) on single or individual indices. In addition, integrated indices that combine these single indices are also commonly used.

2.1. Methods on Pollution Indices

Pollution indices are usually used on heavy metals without released standards. They include the contamination factor (Cf) [3,4], ecological risk factor (Er) [3,5], enrichment factor (EF) [6–9], and index of geo-accumulation (I_{geo}) [10,11]. Here, the index of geo-accumulation is calculated as

$$I_{geo} = \log_2[C_i/(1.5C_{ri})] \quad (1)$$

where C_i is the concentration of the element i in a sample and C_{ri} is the background concentration or reference value of the element i such as the values in reference [12]. Seven

 https://doi.org/10.3390/app16125739

classes of pollution risk are commonly used [13–15] and were primarily suggested by Muller [10] on I_{geo} values as

$$\begin{array}{lll} I_{geo} \leq 0, & \text{class 0,} & \text{unpolluted} \\ 0 < I_{geo} \leq 1, & \text{class 1,} & \text{from unpolluted to moderately polluted} \\ 1 < I_{geo} \leq 2, & \text{class 2,} & \text{moderately polluted} \\ 2 < I_{geo} \leq 3, & \text{class 3,} & \text{fromm oderately to strongly polluted} \\ 3 < I_{geo} \leq 4, & \text{class 4,} & \text{strongly polluted} \\ 4 < I_{geo} \leq 5, & \text{class 5,} & \text{from strongly to extremely polluted} \\ I_{geo} > 5, & \text{class 6,} & \text{extremely polluted} \end{array} \quad (2)$$

Specifically, seven classes are divided on the I_{geo} classification, while four classes on the contamination factor [3], five classes on the ecological risk factor [3], and six classes on the enrichment factor [6]. Although these classifications are widely used in the assessment of environment risk [16–18], the risk levels are not consistent on different indices because of their different classifications.

2.2. Methods on Released Standards

Different methods are usually used on heavy metals with released standards from different countries and organizations [19–21]. In China, the national standard GB15618-2018 [22] lists the different thresholds of heavy metals for different pH values and agricultural land types (contribution 1). Pollution risk is described in three levels: background level, screening level, and intervention level [22]. These three levels can also be described by pollution indices as

$$PI = \begin{cases} 0 & C < C_{RS}, \quad \text{the background level} \\ 1 & C_{RS} \leq C < C_{RI}, \quad \text{the screening level} \\ 10 & C_{RI} \leq C, \quad \text{the intervention level} \end{cases} \quad (3)$$

where C represents the concentration of heavy metals in soil samples, and C_{RS} and C_{RI} are their risk screening values and risk intervention values, as specified in GB15618-2018, on different land use types and pH levels (contribution 1).

Based on Equation (3), metals can be evaluated for their released standard values. However, metals cannot be assessed using Equation (3) if they are not specified in these standards.

2.3. Integration Methods on Pollution Indices and Released Standards

In order to consistently assess the pollution indices and released standards, Chen et al. (contribution 1) and Huang et al. [1] studied the assessment results of both pollution indices and released standards for heavy metals specified in GB15618-2018. They suggest using the integration method to assess heavy metals:

$$PI = \begin{cases} 0 & C < C_{RS} \text{ or } I_{geo} \leq 1.4, \quad \text{the background level} \\ 1 & C_{RS} \leq C < C_{RI} \text{ or } 1.4 < I_{geo} \leq 3.7, \quad \text{the screening level} \\ 10 & C \geq C_{RI} \text{ or } I_{geo} > 3.7, \quad \text{the intervention level} \end{cases} \quad (4)$$

where C, C_{RS}, and C_{RI} are those described in Equation (3), and I_{geo} is calculated by Equation (1). Notably, the dividing values of I_{geo} are 1 and 4 in Chen et al. (contribution 1) while the values are 1.4 and 3.7 in Huang et al. [1] as listed in Equation (4). This difference resulted from the different values of C_{ri} in Equation (1); therefore, consistent values of C_{ri} in Equation (1) should be derived in China to keep assessment results that are consistent with those of released standard GB15618-2018.

https://doi.org/10.3390/app16125739

2.4. Integrated Indices That Combine Single Indices

Based on the aforementioned single indices such as I_{geo}, PI, Cf, Er, and EF, an integrated index can by formed using their sum, average, weighted average when the single indices are in a normal distribution, or on their product, root of product, and weighted power product when the single indices are in a log-normal distribution [23].

In addition, the Nemerow index is also commonly used, regardless of the distribution pattern of the single indices [24–28].

3. Provenance of Geological Materials

Provenance of geological materials is a hot topic in current scholarship on paleoenvironment [29,30], mineralization [31], and diagenesis [32–34]. Two papers in this Special Issue discuss the provenance of sediments and rocks. Cui et al. (contribution 4) discuss the provenance of sediments in the Beibu gulf of the South China Sea, and Liu et al. (contribution 5) discuss the parent sources of volcanic rocks in the Fuyun-Qinghe area of the southern margin of Chinese Altay.

With respect to these provenance methods, a promising method called geochemical genes is introduce here to trace the geological materials of soil–sediment–sedimentary-rock in lithology and of magmatic rocks in their parent source, respectively [35–37].

3.1. Provenance on Lithology of Soil–Sediment–Sedimentary Rock

Geochemical genes were introduced and reviewed by Gong et al. [35], and geochemical spectral lines are similar to those of spider diagrams and REE patterns with quantitative descriptions [38–40]. On the provenance of geological materials from soil–sediment–sedimentary-rock systems, LG01 and LG03 lithogenes are commonly used [41,42], and they are listed in Table 1.

Table 1. Elemental sequences and their reference values for geochemical genes.

Gene	Sequence No.	1	2	3	4	5	6	7	8	9	10	11	References
LG01	Gene elements	Zr	Ti	Al_2O_3	TFe_2O_3	SiO_2	P	Pb	Mn	Th	Nb	U	[35,41]
	Reference Values	147	4016	14.8	6.4	60	756	19	940	5.72	14.5	1.2	
LG03	Gene elements	Nb	Ti	Zr	Cr	La	V	Pb	Co	U	Ni	Th	[35,41]
	Reference Values	14.5	4016	147	81	35	130	19	24	1.2	32	5.72	
CRG01	Gene elements	Gd	La	Dy	Ce	Ho	Pr	Er	Nd	Yb	Sm	Lu	[37,43]
	Reference Values	0.207	0.241	0.256	0.619	0.056	0.094	0.165	0.474	0.169	0.154	0.025	
CRG02	Gene elements	Cr	Pb	TFe_2O_3	Th	Mn	U	V	La	Ti	Nb	Ni	[37]
	Reference Values	250	0.5	10	0.2	1200	0.07	240	3.5	5700	2.5	90	

Notes: Reference values of major oxides as %; others as μg/g.

LG01 and LG03 lithogenes can be viewed as a tool to classify geological materials compositionally because of their stability during weathering and alteration processes [36,40]. The sample with the acidic-like composition has the same gene code on LG01 and LG03 as 12020202020, while the sample with the basic-like composition has the gene code 10202020202 [35]. Acidic similarity (or R_{Acidic}) can be used to classify or trace parent materials to three types: acidic-like composition with $R_{Acidic} \geq 80\%$, intermediate-like composition with $20\% < R_{Acidic} < 80\%$, and basic-like composition with $R_{Acidic} \leq 20\%$ [36,41]. This classification or provenance is applicable to soil–sediment–sedimentary-rock systems.

3.2. Provenance on Source of Mgamatic Rocks

In order to trace the origin of magmatic rocks in the earth's crust, Li [37] proposed two geochemical genes on the continental crust, CRG01 and CRG02, which are also listed in Table 1. CRG01 was constructed on rare earth elements, and CRG02 was constructed on major and trace elements. The reference values of CRG01 are from CI carbonaceous chondrite abundance [43], while the values of CRG02 are determined by the elemental content data of lower ocean crust samples and continental crust samples [37]. These two continental crust genes are used to distinguish rocks sourced from the continental crust or lower ocean crust [37].

If the rock is sourced from the continental crust, its gene codes, CRG01 and CRG02, are the same as 12020202020. However, if the rock is derived from the lower ocean crust, its CRG01 gene code is 10202020211 and its CRG02 gene code is 10202020202. In geochemical gene research, gene similarity between a sample and continental crust with the code of 12020202020 is called continental crust similarity (R_{Crust}). R_{Crust} can be viewed as a tool to distinguish the sources of magmatic rocks as three types: CC-like (continental-crust-like) rock with $R_{Crust} \geq 80\%$, LOC-like (lower-ocean-crust-like) rock with $R_{Crust} \leq 30\%$ on CRG01 or $R_{Crust} \leq 20\%$ on CRG02, and TC-like (transitional-crust-like) rock with $30\% < R_{Acidic} < 80\%$ on CRG01 or $20\% < R_{Acidic} < 80\%$ on CRG02 [37]. This classification or provenance is applicable to magmatic rocks.

4. Exploration and Analysis

Geochemical exploration is the main branch of applied geochemistry [44,45], and analysis of geological materials is a fundamental technique used to consolidate geochemical exploration [46,47]. Three papers in this Special Issue discuss geochemical exploration and analytical methods of geological materials.

Zhang et al. (contribution 6) introduce a non-traditional geochemical exploration approach called geo-electrochemical-integrated technology [48] in a case study on a uranium deposit in the Guangzitian area of northern Guangxi, China. This geo-electrochemical-integrated technology includes three exploration approaches: geo-electrochemical extraction measurement [49], soil ionic conductivity measurement [50], and thermally released soil mercury measurement [51]. Zhang et al. (contribution 7) report a clean extraction processes to investigate Li concentrations in a clay-type Li deposit; they found that Li is concentrated in fine-grained fractions (<45 μm) in this case deposit, which can be utilized on other clay-type Li deposits [52].

Despite these field measurements and lab analyses, information extraction from survey data is also a key step for geochemical exploration [53–56]. Regarding a core principle of geoscience big data, "to uncover correlations within data to address geological issues", Zhu et al. (contribution 8) report a suite of prospecting models by using multisource data to identify potential gold deposits in Gansu's Beishan area, China. Specifically, Feng et al. (contribution 9) report a thermodynamic model to calculate the solubility of CO_2-SO_2 mixtures in aqueous electrolyte solutions, which may have potential applications in exploration and mineralization.

5. Perspective

Although only nine papers in this Special Issue are published, recent advances in geochemistry are clearly illustrated in the fields of environmental geochemistry, geochemical exploration, and geochemical principles on provenance. The innovative studies presented here have great potential, and more advances in geochemistry, especially in applied geochemistry, will emerge in the near future. We thank the authors and reviewers for their contributions to this Special Issue, and anticipate that more in-depth research in applied geochemistry will be showcased at the 11th National Conference on Applied Geochemistry in China.

https://doi.org/10.3390/app16125739

Author Contributions: Conceptualization, Q.G. and N.L.; writing—original draft preparation, Q.G. and N.L.; writing—review and editing, Q.G. and N.L. All authors have read and agreed to the published version of the manuscript.

Acknowledgments: We are grateful to all of the contributors who made this Special Issue a success. Our thanks and congratulations are extended to all the authors for submitting their work. Our sincere gratefulness is also given to the reviewers and editors for the time and effort they spent helping the authors to improve their papers.

Conflicts of Interest: The authors declare no conflicts of interest.

List of Contributions

1. Chen, X.; Zhao, P.; Huang, J.; Liu, J.; Cao, X.; Che, J.; Liao, H.; Zhu, X.; Gong, Q. Risk Assessment of Soil Heavy Metals in the Jiahe River Basin of Yantai City, China. *Appl. Sci.* **2025**, *15*, 70.
2. Li, J.; Xie, S.; Yang, W.; Zhou, W.; Carranza, E.J.M.; Wen, W.; Shi, H. Prediction of Selenium-Enriched Crop Zones in Xiaoyan Town Using Fuzzy Logic and Machine Learning Ap-proaches. *Appl. Sci.* **2025**, *15*, 4943.
3. Junussov, M.; Mustapayeva, S. Preliminary XRF Analysis of Coal Ash from Jurassic and Carboniferous Coals at Five Kazakh Mines: Industrial and Environmental Comparisons. *Appl. Sci.* **2024**, *14*, 10586.
4. Cui, Z.; Hou, Y. Impacts of Holocene Sea Level Rise and the Opening of the Qiongzhou Strait on the Provenance of Sediments in the Beibu Gulf, South China Sea. *Appl. Sci.* **2025**, *15*, 4224.
5. Liu, J.; Liu, C.; Liu, Q.; Luo, Z.; Liu, Y.; Zhou, C.; Guo, X.; Yu, X.; Wang, M. Paleo-Asian Ocean Ridge Subduction: Evidence from Volcanic Rocks in the Fuyun–Qinghe Area, Southern Margin of the Chinese Altay. *Appl. Sci.* **2025**, *15*, 3736.
6. Zhang, X.; Wen, M.; Luo, Q.; Ma, Y.; Jiang, Y.; Jiang, Y.; Ye, W.; Zhang, J. Research on the Prediction of Concealed Uranium Deposits Using Geo-Electrochemical Integrated Technology in the Guangzitian Area, Northern Guangxi, China. *Appl. Sci.* **2025**, *15*, 7426.
7. Zhang, H.; Li, P.; Zhang, W.; Li, J.; Chen, Z.; Yin, J.; Fang, Y.; Liu, S.; Kang, J.; Zhu, D. Occurrence State and Extraction of Lithium from Jinyinshan Clay-Type Lithium Deposit, Southern Hubei: Novel Blank Roasting–Acid Leaching Processes. *Appl. Sci.* **2025**, *15*, 9100.
8. Zhu, L.; Han, R.; Zhang, Y.; Fu, H.; Luo, J.; Luo, Y. Enhancing Prospecting Target Prediction Precision: A Multi-Source Data Mining Approach in Gansu's Beishan Area. *Appl. Sci.* **2025**, *15*, 5430.
9. Feng, B.; Zhang, Z.; Xu, M.; Mao, S. A Thermodynamic Model for the Solubility of SO2 in Multi-Ion Electrolyte Solutions and Its Applications. *Appl. Sci.* **2025**, *15*, 3927.

References

1. Huang, J.; Xu, S.; Liu, Y.; Yan, T.; Gong, Q. An Integrated Method for Assessing Pollution Risk of Heavy Metals Specified or Unspecified in China's National Standard for Agricultural Soils: Illustration on a Case Study. *Environ. Monit. Assess.* **2026**, *198*, 435. [CrossRef]
2. Miletić, A.; Lučić, M.; Onjia, A. Exposure Factors in Health Risk Assessment of Heavy Metal(loid)s in Soil and Sediment. *Metals* **2023**, *13*, 1266. [CrossRef]
3. Håkanson, L. An Ecological Risk Index for Aquatic Pollution Control. A Sedimentological Approach. *Water Res.* **1980**, *14*, 975–1001. [CrossRef]
4. Akpambang, V.O.E.; Ebuzeme, G.C.; Akinola, J.O. Heavy Metal Contamination of Topsoil Around a Cement Factory—A Case Study of Obajana Cement Plc. *Environ. Pollut. Bioavailab.* **2022**, *34*, 12–20. [CrossRef]
5. Kostecki, J.; Dronia, W.; Polomka, J.; Jedrczak, A. Effective Bio-Waste Recycling: Minimizing the Risk of Heavy Metal Emission. *Sustainability* **2026**, *18*, 3531. [CrossRef]
6. Zoller, W.H.; Gladney, E.S.; Duce, R.A. Atmospheric Concentrations and Sources of Trace Metals at the South Pole. *Science* **1974**, *183*, 198–200. [CrossRef]
7. Green-Ruiz, C.; Páez-Osuna, F. Heavy Metal Anomalies in Lagoon Sediments Related to Intensive Agriculture in Altata-Ensenada del Pabellón Coastal System (SE Gulf of California). *Environ. Int.* **2001**, *26*, 265–273. [CrossRef] [PubMed]
8. Islam, M.S.; Hossain, M.B.; Matin, A.; Islam Sarker, M.S. Assessment of Heavy Metal Pollution, Distribution and Source Apportionment in the Sediment from Feni River Estuary, Bangladesh. *Chemosphere* **2018**, *202*, 25–32. [CrossRef]

https://doi.org/10.3390/app16125739

9. Khatita, A.A. Risk Assessment of Potentially Toxic Elements in Street Dust from Mahd Ad Dhahab Gold Mine, Saudi Arabia. *J. Taibah Univ. Sci.* **2024**, *18*, 2281067. [CrossRef]
10. Muller, G. Index of Geoaccumulation in Sediments of the Rhine River. *Geo J.* **1969**, *2*, 108–118.
11. Roy, A.; Bhattacharya, T.; Kumari, M.; Kumar, A. Exploring Heavy Metal Dynamics and Risks from Dust and Soil in Urban Cities of Jharkhand, India. *Sci. Rep.* **2024**, *14*, 32101. [CrossRef] [PubMed]
12. Hou, Q.; Yang, Z.; Yu, T.; Xia, X.; Cheng, H.; Zhou, G. *Soil Geochemical Parameters in China*; Geological Publishing House: Beijing, China, 2020; pp. 1–3172. (In Chinese)
13. Taghavi, M.; Darvishiyan, M.; Momeni, M.; Eslami, H.; Fallahzadeh, R.A.; Zarei, A. Ecological Risk Assessment of Trace Elements (TEs) Pollution and Human Health Risk Exposure in Agricultural Soils Used for Saffron Cultivation. *Sci. Rep.* **2023**, *13*, 4556. [CrossRef]
14. Brice, C.; Montero-Serrano, J.-C.; Saint-Louis, R. Regional Geochemical Survey of Canadian Arctic Sediments: Insights into Provenance, Sediment Dynamics and Trace Metal Enrichment. *Appl. Geochem.* **2025**, *189*, 106432. [CrossRef]
15. Lupu, A.; Damian, G.; Maftei, A.E.; Buzatu, A.; Bereș, I. Geochemical Assessment of Potentially Toxic Elements in River Sediments from a Highly Contaminated Mining Area—Cavnic, Romania. *Mar. Pollut. Bull.* **2026**, *229*, 119678. [CrossRef] [PubMed]
16. Banat, K.M.; Howari, F.M.; Al-Hamad, A.A. Heavy Metals in Urban Soils of Central Jordan: Should We Worry About Their Environmental Risks? *Environ. Res.* **2005**, *97*, 258–273. [CrossRef]
17. Wang, Q.Z.; Zhang, Y.Q.; Wang, L.; Liang, Y.X. Pollution and Health Risk Evaluation at an Abandoned Industrial Site. *Toxics* **2026**, *14*, 49. [CrossRef]
18. Zhong, W.; Zhu, P.; Deng, J.; Wei, Y.; Fan, W.; Zhang, L. Distribution Characteristics and Health Risks of Heavy Metals in Roadside Dust along Typical Roads in the Guanzhong Region. *Environ. Res. Commun.* **2026**, *8*, 045024. [CrossRef]
19. Liu, H.; Probst, A.; Liao, B. Metal Contamination of Soils and Crops Affected by the Chenzhou Lead/Zinc Mine Spill (Hunan, China). *Sci. Total Environ.* **2005**, *339*, 153–166. [CrossRef]
20. Briffa, J.; Sinagra, E.; Blundell, R. Heavy Metal Pollution in the Environment and their Toxicological Effects on Humans. *Heliyon* **2020**, *6*, e04691. [CrossRef]
21. Hasimuna, O.J.; Gweon, H.S.; Yang, H. Heavy Metal Contamination and Ecological Risk in Kabompo River, Northwestern Zambia. *Environ. Pollut. Bioavailab.* **2025**, *37*, 2549084. [CrossRef]
22. *GB15618-2018*; Soil Environmental Quality, Risk Control Standard for Soil Contamination of Agricultural Land. SEPA, Ministry of Ecology and Environment of the People's Republic of China: Beijing, China, 2018.
23. Gong, Q.; Deng, J.; Xiang, Y.; Wang, Q.; Yang, L. Calculating Pollution Indices by Heavy Metals in Ecological Geochemistry Assessment and A Case Study in Parks of Beijing. *J. China Univ. Geosci.* **2008**, *19*, 230–241. [CrossRef]
24. Cheng, J.; Shi, Z.; Zhu, Y. Assessment and Mapping of Environmental Quality in Agricultural Soils of Zhejiang Province, China. *J. Environ. Sci.* **2007**, *19*, 50–54. [CrossRef] [PubMed]
25. Hu, B.; Jia, X.; Hu, J.; Xu, D.; Xia, F.; Li, Y. Assessment of Heavy Metal Pollution and Health Risks in the Soil-Plant-Human System in the Yangtze River Delta, China. *Int. J. Environ. Res. Public Health* **2017**, *14*, 1042. [CrossRef]
26. Rajasekar, A.; Murava, R.T.; Norgbey, E.; Zhu, X. Spatial Distribution, Risk Index, and Correlation of Heavy Metals in the Chuhe River (Yangtze Tributary): Preliminary Research Analysis of Surface Water and Sediment Contamination. *Appl. Sci.* **2024**, *14*, 904. [CrossRef]
27. Hao, X.; Gao, Y.; Hao, L.; Zheng, S.; Wang, S.; Chen, Y.; Sui, Y.; Zhang, Z.; Jiao, X. Accurate Determination and Comprehensive Evaluation of Heavy Metals in Farmland Soil of Heilongjiang Province. *Appl. Sci.* **2026**, *16*, 2257. [CrossRef]
28. Melo, H.d.N., Jr.; Paula Filho, F.J.d.; Coelho Menezes, J.M.; Melo Coutinho, H.D.; Pereira Teixeira, R.N.; Sobral Melo, H.V. Application of Sediment Quality Guidelines and Geochemical Indices to Evaluate the Impact of Trace Metals on Sediments from a Fish Farming Area in the Rosário Reservoir, Brazil. *Appl. Geochem.* **2026**, *199*, 106726. [CrossRef]
29. Gu, F.; Wu, G.J.; Zhang, C.L.; Yan, N.; Huang, J. Geochemistry of Surface Soil in the Eastern Pamirs and its Implications for Muztagata Ice Core Dust Provenance. *Appl. Geochem.* **2020**, *121*, 104724. [CrossRef]
30. Jean-Lavenir, N.M.; Cyrille, S.; Omar, E.D.; Yiika, L.P.; Junior, E.E.M. Geochemical and Gold-Ore Potential Assessment in Stream Sediments of Bindiba Gold District, Eastern Cameroon: Implications for Gold Exploration, Sediment Provenance, Paleoenvironment, and Tectonic Setting. *Min. Metall. Explor.* **2025**, *42*, 2415–2439. [CrossRef]
31. Chen, Y.; Wen, H.; Tao, N.; Xu, F.; Ye, Q. Mineralogical and Geochemical Investigations of the Li-rich Clay Strata from Central Yunnan, Southwest China. *Ore Geol. Rev.* **2025**, *181*, 106614. [CrossRef]
32. Cullers, R.L.; Podkovyrov, V.N. Geochemistry of the Mesoproterozoic Lakhanda Shales in Southeastern Yakutia, Russia: Implications for Mineralogical and Provenance Control, and Recycling. *Precamb. Res.* **2000**, *104*, 77–93. [CrossRef]
33. Grizelj, A.; Peh, Z.; Tibljaš, D.; Kovačić, M.; Kurečić, T. Mineralogical and Geochemical Characteristics of Miocene Pelitic Sedimentary Rocks from the South-Western Part of the Pannonian Basin System (Croatia): Implications for Provenance Studies. *Geosci. Front.* **2017**, *8*, 65–80. [CrossRef]

https://doi.org/10.3390/app16125739

34. Xu, H.; Shen, J.; Yang, F.; Hu, Q.; Yu, Y.; Xiong, L.; Zhang, T.; Deng, H.; He, J. Provenance, Sedimentary Paleoenvironment and Organic Matter Accumulation Mechanisms in Shales from the Lower Cambrian Qiongzhusi Formation, SW Yangtze Block, China. *Mar. Pet. Geol.* **2025**, *181*, 107520. [CrossRef]
35. Gong, Q.; Yan, T.; Wu, X.; Li, R.; Wang, X.; Liu, N.; Li, X.; Wu, Y.; Li, J. Geochemical Gene: A Promising Concept in Discrimination and Traceability of Geological Materials. *Appl. Geochem.* **2022**, *136*, 105133. [CrossRef]
36. Wu, Y.; Gong, Q.; Liu, N.; Wu, X.; Yan, T.; Xu, S.; Li, W. Classification of Geological Materials on Geochemical Lithogenes: Illustration on a Case Study in Gejiu Area of Yunnan Province, China. *Appl. Geochem.* **2022**, *146*, 105460. [CrossRef]
37. Li, W. *Crustal-Derived Rock Gene: A New Idea of Tracing the Origin of Rocks*; China University of Geosciences: Beijing: China, 2022; pp. 1–46.
38. Yan, T.; Wu, X.; Quan, Y.; Gong, Q.; Li, X.; Wang, P.; Li, R. Heredity, Inheritance and Similarity of Element Behaviors among Parent Rocks and their Weathered Products: A Geochemical Lithogene. *Geoscience* **2018**, *32*, 453.
39. Gong, Q.; Liu, N.; Wu, X.; Yan, T.; Fan, T.; Li, X.; Liu, M.; Li, R.; Albanese, S. Using Regional Geochemical Survey Data to Trace Anomalous Samples through Geochemical Genes: The Tieshanlong Tungsten Deposit Area (Southeastern China) Case Study. *J. Geochem. Explor.* **2020**, *219*, 106637. [CrossRef]
40. Wu, Y.; Li, X.; Gong, Q.; Wu, X.; Yao, N.; Peng, C.; Chao, Y.; Wang, X.; Pu, X. Test and Application of the Geochemical Lithogene on Weathering Profiles Developed over Granitic and Basaltic Rocks in China. *Appl. Geochem.* **2021**, *128*, 104958. [CrossRef]
41. An, Y.; Yin, X.; Gong, Q.; Li, X.; Liu, N. Classification and provenance on geochemical lithogenes: A case study on rock–soil–sediment system in Wanquan area of Zhangjiakou, North China. *Appl. Sci.* **2023**, *13*, 1008. [CrossRef]
42. Liu, R.; Chi, Y.; Xie, Y.; Kang, C.; Sun, L.; Wu, P.; Wei, Z. Characteristics and Influencing Factors of the Granite Weathering Profile: A Case Study of a High Latitude Area in Northeastern China. *Minerals* **2024**, *14*, 17. [CrossRef]
43. Palme, H.; O'Neill, H.S.C. Cosmochemical Estimates of Mantle Composition. In *Treatise on Geochemistry*, 2nd ed.; Elsevier: Amsterdam, The Netherlands, 2014; Volume 3, pp. 1–39.
44. Xie, X.; Cheng, H. Sixty Years of Exploration Geochemistry in China. *J. Geochem. Explor.* **2014**, *139*, 4–8. [CrossRef]
45. Lin, X.; Wang, X.; Zhou, J.; Chi, Q.; Nie, L.; Zhang, B.; Xu, S.; Zhao, S.; Liu, H.; Sun, B.; et al. Concentrations, Variations and Distribution of Molybdenum (Mo) in Catchment Outlet Sediments of China: Conclusions from the China Geochemical Baselines Project. *Appl. Geochem.* **2019**, *103*, 50–58. [CrossRef]
46. Yao, W.; Xie, X.; Wang, X. Comparison of Results Analyzed by Chinese and European Laboratories for FOREGS Geochemical Baselines Mapping Samples. *Geosci. Front.* **2011**, *2*, 247–259. [CrossRef]
47. Wang, X.; the CGB Sampling Team. China Geochemical Baselines: Sampling Methodology. *J. Geochem. Explor.* **2015**, *148*, 25–39. [CrossRef]
48. Liu, P.F.; Wen, M.L. The Application of Geo-Electrochemical Integrated Technology to the Prospecting in the Xiyi Pb-Zn Deposit, Yunnan Province. *Geophys. Geochem. Explor.* **2016**, *40*, 655–660.
49. Liu, P.; Luo, X.; Wen, M.; Zhang, J.; Zheng, C.; Gao, W.; Ouyang, F. Geo-Electrochemical Anomaly Prospecting for Uranium Deposits in Southeastern China. *Appl. Geochem.* **2018**, *97*, 226–237. [CrossRef]
50. Wen, M.L.; Ouyang, F. Anomaly Characteristics, Components of Soil Ionic Conductivity and Prospecting Prediction of Tin Deposits in Pinggui Area, Guangxi. *Geol. Bull. China* **2013**, *32*, 784–789.
51. Zheng, C.J.; Luo, X.R. Comparison of Conventional-Warming RHg Mercury Release of Soil with Ladder-Warming RHg Mercury Release, and Ore Prediction in Jielagepu Au Deposit, Southern Tibet. *Mineral. Petrol.* **2018**, *38*, 99–110.
52. Wen, H.; Luo, C.; Du, S.; Yu, W.; Gu, H.; Ling, K.; Cui, Y.; Li, Y.; Yang, J. Carbonate-Hosted Clay-Type Lithium Deposit and its Prospecting Significance. *Chin. Sci. Bull.* **2020**, *65*, 53–59. [CrossRef]
53. Zuo, R.; Wang, J.; Xiong, Y.; Wang, Z. The Processing Methods of Geochemical Exploration Data: Past, Present, and Future. *Appl. Geochem.* **2021**, *132*, 105072. [CrossRef]
54. Parsa, M.; Sadeghi, M.; Grunsky, E. Innovative Methods Applied to Processing and Interpreting Geochemical Data. *J. Geochem. Explor.* **2022**, *237*, 106983. [CrossRef]
55. Xu, Y.; Zuo, R. Geochemical Survey Data Cube: A Useful Tool for Lithological Classification and Geochemical Anomaly Identification. *Geochemistry* **2024**, *84*, 125959. [CrossRef]
56. Arne, D.; Garrett, R. Exploratory Data Analysis of Geochemical Exploration Data. *Geochem. Explor. Environ. Anal.* **2026**, *26*, geochem2025-037. [CrossRef]

https://doi.org/10.3390/app16125739

applied sciences

MDPI

Article

Preliminary XRF Analysis of Coal Ash from Jurassic and Carboniferous Coals at Five Kazakh Mines: Industrial and Environmental Comparisons

Medet Junussov [1,*] and Sezim Mustapayeva [2,*]

1 Department of Geosciences, School of Mining and Geosciences, Nazarbayev University, Astana 010000, Kazakhstan
2 Department of Geological Survey, Search and Exploration of Mineral Deposits, Faculty of Geology and Oil and Gas Business, Satbayev University, Almaty 050013, Kazakhstan
* Correspondence: medet.junussov@nu.edu.kz (M.J.); s.mustapayeva@satbayev.university (S.M.)

Abstract: This study analyzes 105 coal ash samples from Jurassic and Carboniferous coals from five mines in Kazakhstan, Lenin, Saradyr, Bogatyr, Maikuben, and Shubarkol, focusing on the inorganic elemental compositions, their occurrence, and industrial and environmental implications. Methods include coal ash yield and volatile matter analysis, mineralogical characterization via low-temperature ash X-ray diffraction (LTA-XRD), and scanning electron microscopy with energy-dispersive X-ray spectroscopy (SEM-EDS). High-temperature ash (HTA) was analyzed using energy-dispersive X-ray fluorescence (EDXRF), highlighting XRF's potential for rapid multi-elemental analysis. Nine major elements (Al, Si, P, S, Fe, K, Ca, and Ti) and eleven trace elements (As, Cu, Cr, Zn, Pb, V, Ga, Mn, Ni, Y, Yb, and Zr) were identified in HTA samples through EDXRF. SEM and dendrogram analysis confirm their co-occurrence with quartz, kaolinite, pyrite, and accessory minerals such as chalcopyrite, zircon, rutile, and REE-bearing apatite. The elemental content of samples enhances industrial suitability by reducing emissions. Only Yb shows slight enrichment for economic benefits, along with La, Ce, and Nd, while concentrations of potentially toxic elements indicate minimal environmental risk. EDXRF demonstrates its efficiency for large-scale investigations, with all samples analyzed in a few days using automated overnight measurements. This approach shows promise for future studies focusing on trace elements, including REE.

Keywords: Kazakhstan; major elements; trace elements; coal; XRF

1. Introduction

Coal is a complex natural material with a diverse chemical structure, containing both organic and inorganic components. It includes all naturally occurring elements and over 200 minerals, each hosting one or more of these elements [1,2]. Major inorganic elements in coal exceed 0.1% by weight (e.g., Al, Si, Ti, Fe, and Ca), whereas trace elements (e.g., As, V, Cu, Zn, Pb, rare earth elements, and U) are typically present in concentrations below 0.1%, ranging from 1 ppb to several hundred ppm [2]. Understanding the occurrence of these elements is crucial as it provides insights into peat formation, coal diagenesis, and basin development. Practically, this knowledge affects coal mining, preparation, combustion, and utilization by helping to avoid technological issues (abrasion, corrosion, fouling, and slagging) and is crucial for evaluating environmental pollution control, health impacts, and its applications as an industrial raw material in cement manufacturing, wastewater treatment, and valuable element recovery [2–6].

Various analytical techniques, such as ICP-MS/AES, AAS, NAA, and DPASV, are employed to characterize coal combustion ash, each offering distinct advantages like low detection limits, high sensitivity, and wide dynamic ranges. Among these, X-ray fluorescence (XRF) techniques are particularly effective for simultaneous multi-element analysis [4]. XRF techniques offer several benefits, including reduced sample preparation time, lower costs, and the avoidance of hazardous chemicals. In recent years, X-ray fluorescence has emerged as a powerful method for analyzing environmental samples, including coal ash. XRF instrumentation is advantageous due to its lower capital cost compared to Neutron Activation Analysis (NAA), faster analysis times relative to ICP-AES, and widespread availability [4]. In contrast, ICP-AES requires time-consuming dissolution procedures, which can lead to the loss of volatile elements such as As, Pb, Se, Sb, and Zn [7–9]. XRF is an effective method for rapid routine analysis of major and minor elements, offering a significant advantage due to its potentially wide application in industry settings for quick assessments [10].

The application of XRF for analyzing the chemical composition of coal and coal ash has been widely adopted in industrial, economic, and environmental studies across several countries, including Russia (28 studies), Australia (17), China (15), the USA (11), India (11), Mongolia (5), Portugal (5), the UK (4), Turkey (4), Spain (4), Austria (2), Canada (2), Kazakhstan (2), Poland (2), and others. Publications span various years: before 2000—6; 2001–2005—12; 2006–2010—11; 2011–2015—19; 2016–2020—31; and 2021–2023—13. More than half of the articles are concentrated in 8 journals, while the remainder are spread across approximately 30 other publications [11].

The major advantages of XRF analysis are as follows: (i) its versatility and wide availability as a common laboratory instrument; (ii) modern instruments often feature full automation for up to 50 samples; (iii) easy sample preparation, typically involving only pulverization and pelletization; and (iv) a broad concentration range from over 1 ppm to 100 wt%, with detection limits of 0.04 wt% for Al, 0.05 wt% for Si, 100 ppm for minor elements, and 20 ppm for trace elements [12,13]. XRF, in particular, is recognized as highly effective for analyzing high-temperature ash (HTA) to determine both major and trace elements [13–15]. HTA of coals is generally preferred for elemental analysis over fly ash or bottom ash derived from power plants, as HTA offers a more complete recovery of elements, particularly for coal and its ashes [16].

Recent studies on the inorganic geochemical characteristics of coal deposits in Kazakhstan, specifically in the Shubarkol coal deposit and Karagandy coal basin focusing on the k7 coal seam, highlight the potential for industrial extraction of rare metals from coal and coal waste, emphasizing coal's value as a source of critical elements [17–19].

However, comprehensive geochemical investigations of other coal deposits in central and northeast Kazakhstan remain limited. The coal deposits in this region are primarily found in Carboniferous coal basins, which include the largest in Kazakhstan: the Karagandy, Tenyz–Korzhinkol, and Ekibastuz basins, covering a combined area of 4155 km^2 and containing the region's largest coal reserves, totaling 54.4 billion tons [20,21]. The Carboniferous coals within these basins exhibit a wide range of total coal seams, varying from 6 to 80, with seam thicknesses ranging from 1.5 to 100 m. The Karagandy basin supports approximately 10 underground coal mines, including the active Lenin mine. In the southwest, the Ekibastuz basin produces coal from the Bogatyr mine, while the eastern Tenyz–Korzhinkol basin produces coal from the Saradyr mine [20–22]. Additionally, the Jurassic coals of this region, found in the Shubarkol and Maikuben coal basins (named after the mines), cover around 1470 km^2 and hold approximately 7.5 billion tons of coal reserves [21–23]. The Shubarkol mine contains two coal measures (Upper and Lower), with thicknesses ranging from 55 to 70 m. Meanwhile, the Maikuben mine within the Shoptykol formation includes three coal measures (I, II, and III), with a total thickness of approximately 260 to 290 m [21–23]. The Carboniferous coals are predominantly classified as hard coals, distinguished by high ash content, relatively low volatile matter, and variable sulfur levels, and they frequently exhibit coking properties [21–23]. In contrast, Jurassic coals are characterized as brown coals and exhibit a lower ash yield and higher volatile matter. Petrographically, the coals from both Carboniferous and Jurassic periods show distinct variations, primarily consisting of vitrinite, liptinite (mainly sporinite), and inertinite. The thermal maturity of these coals, as evidenced by vitrinite reflectance measurements, ranges from approximately 0.5% to 1.3% across the five coal mines, with potential variations observed across different zones, as discussed by various authors [21–23]. Coals from these mines are utilized for power and heat generation, as well as steel manufacturing.

These five coal mines are the primary focus of this study, aimed at a preliminary inorganic elemental assessment of coal, and will serve as the coal mines for more detailed investigations. This study presents the first report on the inorganic elemental composition of coal ash from five different coal deposits in Kazakhstan, dating to the Jurassic and Carboniferous periods. Although the results are preliminary and based on limited inorganic chemical analysis, they offer new insights into the distribution and modes of occurrence of major and trace elements. These insights could have potential industrial and environmental implications for the coals from these Jurassic and Carboniferous coal deposits.

The aims of the present study are as follows: (a) to determine the elemental content of coal ashes from Jurassic and Carboniferous coals sampled at five coal mines; (b) to compare the major and trace element concentrations in coal ash from the five coal mines with the Clarke values of the Earth's crust and global coal ash, aiming to determine the enrichment or depletion of elements and evaluate their potential as a secondary source for recovering valuable elements; and (c) to determine the mode of occurrence and origin of elements for their industrial and environmental implications.

2. Sampling and Methods

2.1. Mine Sampling Locations

The study area encompasses three major Carboniferous coal mines of Lenin, Saradyr, and Bogatyr in northeast Kazakhstan (see Figure 1A–C) and two Jurassic coal mines of Maikuben and Shubarkol (Figure 1D,E). Sampling at these mines was specifically chosen due to their status as the country's oldest and relatively young coal deposits, featuring extensive sedimentary deposition, various ash yield, substantial reserves, and significant mining activity, which makes them crucial for both industrial applications and environmental studies. A total 105 samples were collected and studied from working and exposed coal seams at the five active coal mines located in central and northeast Kazakhstan. Sampling was conducted across five operational coal mines of Lenin, Bogatyr, Saradyr, Maikuben, and Shubarkol (Figure 1A–E). A total of 105 grab samples were collected, with approximately 5 samples composited to form each of the 21 composite samples. Sampling was conducted at vertical intervals of 0.2–0.5 m for all mines, except at Bogatyr, where horizontal intervals were used due to mining area constraints. All samples were stored in plastic bags to minimize potential contamination and oxidation. The dataset includes 69 samples of Carboniferous hard coals obtained from three of these mines, and 36 samples of Jurassic brown coals sourced from the remaining two mines. Using a grab sampling method, samples were collected from active and mining coal seams at each mine and then combined through composite sampling to conduct a comprehensive analysis, which involved integrated inorganic geochemical investigations within general views of mineralogical content.

A total of 105 coal samples were selected: 26 samples from the workable coal seam № 6 within the Dolyn coal formation, collected from the underground Lenin mine (located in Shahtinsk town) at a depth of approximately 700 m, and 43 samples from the Saradyr (near Yerementau town) and Bogatyr (near Ekibastuz city) open-pit coal mines, at depths ranging from 140 to 260 m. Additionally, 36 Jurassic coal samples were collected from the Shubarkol (150 km northeast of Jezgazgan city) and Maikuben (160 km southwest of Pavlodar city, and 65 km southeast of the Bogatyr mine) open-pit mines, where coal production occurs at depths of around 150 m. The coal samples are labeled as DL and DU (from the Lower and Upper Dolyn seams) from the Lenin mine, B1, B2, and B3 (from three coal seams) from the Bogatyr mine, and S (from a single seam) from the Saradyr mine. For the Jurassic coals, samples are labeled SH and SP (from three coal seams in the western and central parts) from the Shubarkol mine, and M sample label (from the Shoptykol coal formation) is from the Maikuben mine.

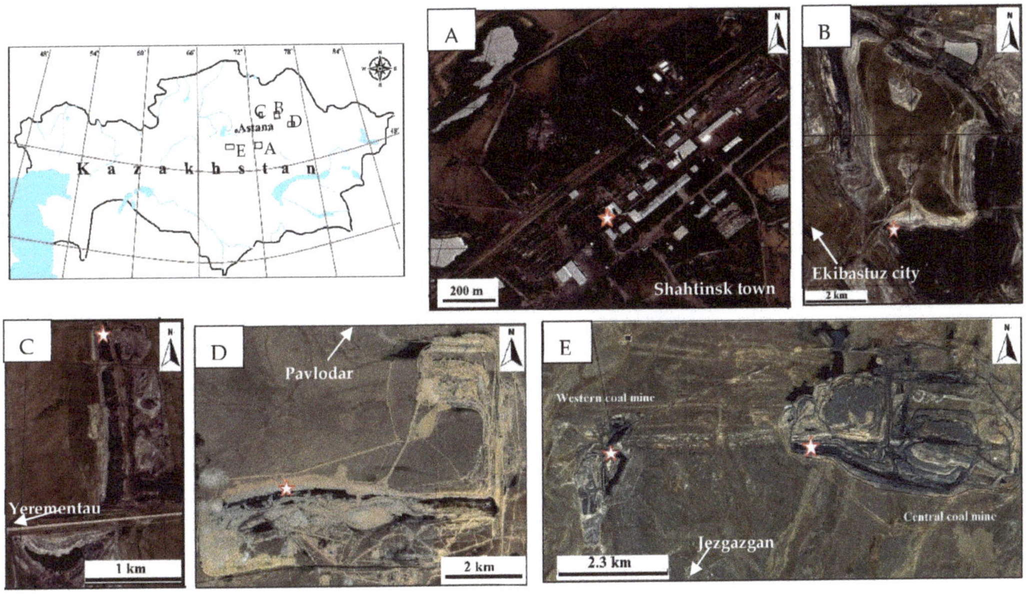

Figure 1. General overview map of three active coal mining areas: (**A**) Lenin underground mine (identified as Dolyn); (**B**) Bogatyr coal mine; (**C**) Saradyr coal mine; (**D**) Maikuben (identified as Shoptykol); (**E**) Shubarkol. The star symbol denotes the locations where coal sampling took place (a satellite map, Google Earth Pro 7.3).

2.2. *Analytical Methods*

2.2.1. Sample Preparation

Two sample types were prepared as powdered coals (for further coal combustion residuals) and polished blocks from selected samples.

An amount of 2 g of each coal sample was weighed and crushed to particle sizes ranging from ~50 to 200 microns using an agate mortar and pestle. Powdered coal samples were used to determine ash yield and volatile matter in the coals.

The powdered coal samples were prepared for geochemical analysis using the standard coal sample preparation procedures as described in [24]. For the preparation of finely powdered coal samples, for high-temperature ash (HTA) and low-temperature ash (LTA): 1g of powdered coal was precisely weighed in a porcelain crucible to obtain coal ash and heated in an open muffle furnace (Carbolite ELF 11, Carbolite Gero, Sheffield, UK). This process for HTA for elemental analysis involved two phases: first at 500 °C for 30 min and then at 815 °C for 60 min, followed by cooling after removal from the muffle furnace (i); meanwhile, the preparation of LTA for mineralogical composition analysis was conducted at 150 °C for 60 min in the muffle furnace based on protocols described by various authors (ii) [25,26].

To obtain polished blocks, each sample underwent preparation based on [27], facilitating subsequent analysis through scanning electron microscopy.

2.2.2. Ash Yield and Volatile Matter

A proximate analysis was performed to determine ash yields and volatile matter content for all 105 samples. Ash yields and volatile matter were measured using a Simultaneous Thermal Analyzer (STA 6000, PerkinElmer, Waltham, MA, USA), in accordance with protocols described by several authors [28–30]. Both parameters were determined by heating the sample at 900 °C for 7 min in a covered crucible under a nitrogen atmosphere.

2.2.3. EDXRF

Energy dispersive X-ray fluorescence (EDXRF) was employed to measure major and trace elements for all 105 samples using an Epsilon 4 spectrometer (Malvern PANalytical, Malvern, UK, max 50 kV/2 mA/10 W). High-temperature coal ash samples were used for this analysis. The procedures followed [31], and the equipment's reliability has been discussed by [32].

XRF was selected as a rapid and routine method to process a large number of samples. After obtaining full data, further investigations are planned using long-term digestion experiments and advanced techniques to analyze individual elements from promising sample units.

2.2.4. LTA-XRD and SEM-EDS

Minerals in the coal were determined using five selected LTA (low-temperature ash) samples on a Rigaku SmartLab X-ray diffractometer (XRD) machine (Rigaku, Tokyo, Japan) with a Ni-filtered Cu Kα X-ray source (energy 8.04 keV and X-ray wavelength 1.5406 Å) and the XRD pattern was recorded with a step size of 0.01°, over a 2θ interval of 2.6–70° [33]. The XRD evaluation of the LTA was conducted at room temperature. The X-ray diffractograms were subjected to quantitative mineralogical analysis using Crystal impact software (Match), based on methods established in the literature [34].

Nine selected polished blocks were coated with carbon, and these samples were investigated under a scanning electron microscope in conjunction with energy-dispersive X-ray spectroscopy (SEM-EDS, Jeol JSM-IT200 LA, Jeol, Tokyo, Japan) to observe minor mineral characteristics and to determine elemental distribution in the coal samples. The operating parameters were set to 20 kV acceleration voltage and high vacuum mode.

3. Results

3.1. Results of Ash Yield and Volatile Matter

Results from the ash yield and volatile matter of all coal samples are presented in Tables 1 and 2. The coal samples from the Saradyr and Bogatyr mines exhibit relatively higher ash yields, averaging up to 31.9%, followed by the Lenin mine samples, with an average of 14.8%. Volatile matter content is nearly consistent across all three mines, averaging up to 21.8%. In contrast, the coal samples from the Maikuben and Shubarkol mines exhibit comparatively low ash yields, averaging between 2.7% and 8.8%, and higher volatile matter content, ranging from 40.9% to 46.6% on average.

3.2. EDXRF Results

Tables 1 and 2 present the concentrations of both major and trace elements in coal ash samples from the five coal mines. Mean values for all samples were calculated and are provided in Table 3 for comparison. The chemical compositions are compared with the compositions of the Clarke value of the crust (CVC) [35–38] and worldwide hard coal combustion (WHC) [38].

Table 1. Ash yield, volatile matter, and elemental composition (major and trace elements) of Carboniferous coal samples from three mines.

Samples	Ash (d, %)	VM (daf, %)	Major Elements (Unit: %)									Trace Elements (Unit: ppm)										
			Al	Si	P	S	Fe	K	Ca	Ti	Mn	Cu	Cr	Zn	Pb	V	Ga	As	Ni	Y	Yb	Zr
Saradyr mine																						
S1	19.3	24.7	1.4	5.1	0.06	0.6	0.1	0.5	0.06	0.3	n.a.	24.1	11.0	45.0	4.4	67.5	5.3	0.7	n.a.	n.a.	5.0	n.a.
S2	40.6	22.1	6.5	12.5	0.1	0.5	3.4	1.7	0.2	1.1	0.10	146.0	7.8	32.3	40.7	253.0	77.5	8.1	5.0	n.a.	3.7	n.a.
S3	18.7	28.6	7.3	17.3	0.1	0.06	11.5	1.2	0.4	0.7	0.31	162.0	n.a.	111.0	15.6	253.0	53.8	8.8	2.0	79.8	58.2	287.8
S4	39.1	8.8	5.6	18.6	0.1	n.a.	0.6	2.3	0.3	1.3	0.001	26.2	31.1	34.0	33.0	735	48.1	95.8	47.8	132.4	n.a.	0.1
S5	17.0	23.7	6.4	16.3	0.1	0.2	12.0	0.6	0.6	0.9	0.08	264.0	65.3	99.3	53.0	n.a.	61.5	39.9	n.a.	122.6	101.0	n.a.
Mean	**26.9**	**21.5**	**5.4**	**13.9**	**0.1**	**0.2**	**5.5**	**1.2**	**0.3**	**0.9**	**0.1**	**124.4**	**23.0**	**64.3**	**29.3**	**261.7**	**49.2**	**30.6**	**10.9**	**66.9**	**40.2**	**57.5**
Lenin mine																						
DU0	12.2	22.3	1.2	2.0	0.5	0.01	0.3	0.9	2.6	0.5	n.a.	0.4	0.1	0.3	0.04	2.2	0.09	0.7	0.1	0.3	0.03	0.4
DU1	21.9	22.6	1.5	2.8	0.1	1.9	2.5	0.2	1.1	0.5	0.05	213.5	11.3	112.7	13.5	398.3	31.1	5.6	n.a.	n.a.	63.5	n.a.
DU2	21.6	24.6	1.1	1.7	0.1	1.9	2.5	0.4	0.3	0.4	0.001	185.6	n.a.	76.0	11.0	n.a.	23.4	8.6	42.2	36.7	39.6	117.5
DU3	3.8	23.3	1.9	14.9	0.1	0.03	1.2	0.3	0.2	0.3	0.05	110.6	19.2	29.1	4.2	168.6	38.0	6.3	n.a.	31.0	4.8	389.6
DU4	6.1	22.6	2.2	3.2	0.05	0.03	0.9	0.3	0.1	0.1	0.002	27.0	n.a.	12.4	n.a.	n.a.	3.5	1.3	5.1	5.4	3.6	13.5
DU5	11.3	24.4	2.5	3.5	n.a.	0.6	4.8	0.1	8.8	0.3	0.02	60.8	18.5	231.8	1.4	199.4	9.6	2.1	12.8	36.8	19.4	72.2
DU6	6.6	19.8	7.4	9.2	2.1	0.02	1.2	0.4	5.3	2.4	n.a.	326.9	7.4	61.7	11.8	121	39.5	4.9	n.a.	95.8	n.a.	180.8
DU7	12.2	21.0	5.2	6.4	0.5	0.5	5.5	0.4	4.1	1.7	0.05	32.3	23.6	79.4	24.5	242.8	65.2	9.0	32.3	64.1	21.2	296.4
DU8	6.3	22.3	1.9	2.3	0.5	0.04	0.9	0.09	1.1	0.6	0.003	19.0	4.5	8.4	n.a.	132.5	5.9	n.a.	7.5	3.0	3.7	21.5
DU9	6.7	22.5	0.5	0.5	0.2	0.01	0.06	0.02	0.3	0.09	n.a.	6.1	n.a.	1.5	n.a.	n.a.	n.a.	n.a.	n.a.	n.a.	n.a.	1.8
DU10	6.1	20.7	1.9	3.0	0.09	0.06	0.3	0.1	2.9	0.6	n.a.	80.5	8.7	20.1	4.5	346.9	19.3	2.4	30.8	38.2	n.a.	117.6
DU11	9.8	22.8	2.8	5.0	0.09	0.07	0.5	0.3	0.1	1.0	n.a.	201.5	78.2	64.9	43.5	100	45	5.5	39.7	49.4	n.a.	657.4
DU12	18.8	19.3	5.9	9.9	0.1	0.7	4.9	0.7	0.1	1.8	0.005	346	133.4	506	103.3	170	61.1	65.4	221.4	72.1	1.5	140
DU13	7.2	21.9	7.4	12.3	0.1	0.5	4.0	0.6	5.5	1.8	0.005	326.4	30.6	126.4	31.0	698	55.4	12.0	73.7	134.2	37.0	634
DU14	5.2	21.2	5.4	8.5	0.1	0.1	2.2	0.3	0.4	0.7	0.001	114.4	10.6	56.9	12.7	231.8	25.8	5.2	31.1	50.6	16.0	153.5
DU15	6.2	22.8	8.8	5.9	4.1	1.0	11.0	0.4	4.7	n.a.	0.04	212	n.a.	93.2	n.a.	n.a.	40.5	3.7	50.0	83.8	57.1	n.a.
DU16	8.9	20.4	6.9	8.8	1.0	0.05	2.8	0.4	2.4	1.0	0.005	181.1	n.a.	70.3	n.a.	366	34.3	n.a.	53.4	97.2	19.7	148.3

Table 1. *Cont.*

Samples	Ash (d, %)	VM (daf, %)	Major Elements (Unit: %)									Trace Elements (Unit: ppm)										
			Al	Si	P	S	Fe	K	Ca	Ti	Mn	Cu	Cr	Zn	Pb	V	Ga	As	Ni	Y	Yb	Zr
DU17	4.2	22.6	7.2	11.0	0.4	0.04	3.2	0.8	0.5	1.3	n.a.	212	20.8	194	18.4	768	35.9	9.3	50.3	50.6	16.2	175
DU18	6.0	23.2	8.7	10.9	1.0	0.6	6.0	0.7	5.4	1.3	0.02	180	29.5	84.6	17.9	407	57.8	4.1	73.1	50.9	17.8	307.8
DU19	4.6	22.4	0.6	1.0	0.01	0.04	0.3	0.07	0.1	0.08	n.a.	15.0	3.0	28.5	n.a.	50.1	1.4	n.a.	3.8	3.3	1.5	8.3
DL1	91.8 *	8.9	5.0	18.3	0.17	2.3	4.2	2.3	0.4	0.9	0.03	144.7	98.0	183.2	2.6	225.4	50.4	10.2	52.8	66.5	32.4	337.6
DL2	47.6	24.6	1.4	4.0	0.04	0.2	2.2	0.9	1.0	0.2	0.05	116.4	40.3	95.8	5.8	162.0	40.4	21.4	35.3	60.0	15.4	395.1
DL3	4.5	23.7	8.9	12.1	0.3	0.04	5.4	0.6	0.6	1.7	0.001	395.7	52.4	188.6	39.9	527.1	66.0	14.5	121.7	69.9	9.6	258.7
DL4	3.7	24.3	0.5	0.5	0.1	0.02	0.5	n.a.	0.1	0.04	0.007	11.2	n.a.	8.0	n.a.	n.a.	n.a.	n.a.	n.a.	n.a.	3.0	n.a.
DL5	43.0	20.4	9.2	12.2	0.1	0.1	16.2	0.5	1.0	3.1	n.a.	627.9	n.a.	216.0	22.6	487.4	109.6	25.9	6.6	80.5	101.5	312.6
DL6	9.7	22.5	9.3	10.9	1.1	0.1	18.7	0.3	4.4	0.7	n.a.	222.6	n.a.	250.5	n.a.	292.5	37.7	10.8	6.5	67.4	81.8	137.4
Mean	**14.8**	**21.8**	**4.1**	**6.5**	**0.5**	**0.4**	**3.2**	**0.5**	**1.9**	**0.9**	**0.01**	**159.5**	**22.7**	**98.1**	**14.2**	**223.3**	**33.0**	**8.4**	**36.3**	**45.4**	**18.6**	**182.3**
Bogatyr																						
B1/0	16.5	16.0	1.1	4.1	0.1	0.2	0.2	0.2	0.1	0.1	0.0003	21.9	n.a.	26.3	3.3	48.9	6.5	7.0	6.9	6.3	1.0	n.a.
B1/1	25.2	29.7	0.8	7.1	0.2	0.1	0.3	0.2	0.9	0.5	0.008	67.0	9.4	29.2	1.8	133.0	13.5	2.1	17.6	n.a.	0.6	n.a.
B1/2	24.9	21.0	4.2	20.2	0.1	0.03	1.5	1.0	0.2	0.6	0.02	209.0	45.3	51.4	13.9	430.0	49.8	11.8	47.2	54.0	12.5	n.a.
B1/3	29.7	19.3	4.1	20.2	0.1	0.06	3.8	1.1	0.2	0.6	0.06	144.0	20.4	100.2	15.3	161.5	39.5	6.4	21.8	43.8	26.1	223.8
B1/4	2.7	17.1	4.3	18.5	0.1	n.a.	0.38	0.9	0.1	0.6	0.007	107.5	11.5	27.6	9.3	177.0	29.0	2.6	6.5	16.4	6.6	72.5
B1/5	11.2	23.7	1.8	8.1	0.08	0.2	1.4	0.1	0.8	0.5	0.01	95.6	18.4	20.3	4.6	182.8	12.9	2.1	21.3	19.4	6.5	n.a.
B1/6	28.5	20.1	6.8	16.4	0.2	0.1	1.2	0.5	0.4	0.8	0.05	13.6	13.6	42.8	9.9	271.0	50.1	11.1	13.2	25.4	4.7	231.8
B1/7	24.3	23.1	5.4	16.0	0.2	0.01	0.72	0.5	0.2	1.6	0.01	410.0	51.3	43.0	n.a.	653	82.3	18.4	51.6	57.0	n.a.	n.a.
B1/8	16.6	23.1	5.9	20.1	0.2	0.06	6.1	0.5	0.2	0.8	0.01	263.0	50.7	64.2	9.3	469.0	52.1	16.1	44.0	42.3	25.4	245.7
B1/9	40.6	22.8	0.6	0.9	0.1	0.01	0.2	0.1	0.1	0.1	n.a.	16.9	n.a.	30.8	0.8	n.a.	n.a.	0.8	3.9	3.0	2.7	5.4
B2/1	56.8 *	13.7	17.6	0.1	0.1	n.a.	3.0	0.4	0.1	0.5	0.01	153.6	n.a.	23.6	45.7	213.2	40.3	32.4	21.6	40.2	27.1	323.0
B2/2	20.8	20.9	4.6	17.6	0.1	0.2	4.5	0.3	0.5	0.5	0.03	194.0	16.3	96.9	16.9	224.4	53.5	26.4	6.7	n.a.	21.4	116.1
B2/3	9.7	23.5	5.9	10.9	0.4	0.07	3.4	0.3	0.5	1.1	0.06	263.5	n.a.	212.2	n.a.	n.a.	60.7	39.1	74.2	34.8	n.a.	824
B2/4	12.0	22.3	5.5	17.5	0.2	0.04	9.8	0.4	0.2	0.9	0.04	292.6	27.2	122.7	19.2	360.8	64.8	37.1	19.8	90.5	41.4	476.8

Table 1. *Cont.*

Samples	Ash (d, %)	VM (daf, %)	Major Elements (Unit: %)									Trace Elements (Unit: ppm)										
			Al	Si	P	S	Fe	K	Ca	Ti	Mn	Cu	Cr	Zn	Pb	V	Ga	As	Ni	Y	Yb	Zr
B2/5	14.1	25.3	1.1	3.6	0.09	0.07	5.9	0.1	0.1	0.2	0.04	63.1	n.a.	77.9	1.5	109	12.3	15.7	11.6	16.7	21.2	129.3
B2/6	12.0	23.8	2.2	9.5	1.2	1.5	19.7	0.06	2.1	n.a.	0.06	133.4	n.a.	122.3	86	n.a.	10.8	n.a.	n.a.	93	69.6	n.a.
B2/7	34.5	18.9	3.2	10.7	0.1	0.08	0.6	0.5	0.2	0.4	0.008	84.9	3.6	289.3	9.0	123.0	26.3	5.3	2.3	18.4	7.0	168.9
B2/8	14.4	23.7	5.0	18.0	0.3	0.1	8.6	0.8	0.6	1.0	0.01	362.6	n.a.	79.6	n.a.	296.6	64.4	187	91.8	51.9	21.8	475.6
B2/9	15.4	20.2	3.4	15.4	0.3	0.1	0.9	0.4	0.2	0.7	0.01	208.9	n.a.	72.5	20.9	223.4	64.4	187	91.8	32.5	21.8	245.5
B2/10	30.8	20.2	3.1	16.2	0.1	0.8	3.0	0.8	0.2	0.7	0.05	197.7	n.a.	78.4	0.1	203.4	28.0	11.4	22.9	59.2	24.4	549.4
B2/11	26.0	23.8	4.7	15.5	0.1	0.1	4.8	1.0	0.3	0.8	0.03	360.1	18.9	124.5	20.5	314.3	43.1	17.8	31.8	75.3	21.9	389.4
B3/1	31.7	20.8	3.2	20.0	0.1	0.1	0.6	0.6	0.2	0.3	0.01	192.3	n.a.	44.2	9.4	100.9	22.8	6.6	8.0	48.7	6.6	145.1
B3/2	24.8	23.9	4.3	19.9	0.2	0.03	2.7	0.5	0.2	1.2	0.01	244.0	n.a.	57.9	15.9	307.7	43.9	36.4	38.6	53.9	11.4	416.9
B3/3	33.3	20.4	1.4	6.4	0.1	0.7	0.3	n.a.	3.2	0.4	0.01	66.9	10.4	52.5	8.9	129.6	16.0	12.0	n.a.	71.8	33.7	212.1
B3/4	53.5	16.1	6.6	17.4	0.1	0.06	0.2	0.5	0.3	1.5	0.008	88.4	8.2	11.2	16.9	253.4	69.1	14.3	n.a.	42.2	n.a.	266.6
B3/5	19.9	21.5	5.4	19.7	0.2	0.5	3.0	0.9	1.3	0.7	0.004	150.9	n.a.	38.3	15.7	343.6	52.8	46.5	33.2	120.8	28.2	285.0
B3/6	61.6	17.5	5.0	14.8	0.1	0.06	5.1	0.7	0.4	0.6	0.006	186.7	9.6	30.2	33.1	210.2	69.5	31.7	15.7	70.5	n.a.	1220
B3/7	39.1	15.9	5.7	18.3	0.2	0.05	4.9	0.7	0.4	0.67	0.006	179.7	13.3	63.9	27.7	193.6	58.0	27.2	21.3	52.4	34.4	908
B3/8	18.9	16.6	0.6	9.5	0.2	0.06	0.4	0.1	0.1	n.a.	0.01	85.3	n.a.	25.5	8.6	134	5.9	10.1	17.7	5.7	n.a.	68.4
B3/9	65.6	22.5	6.5	17.8	0.3	0.04	1.2	0.5	0.4	1.3	0.02	228.8	n.a.	99.5	40.8	238.8	73.9	52.6	13.7	57.2	3.5	457.5
B3/10	42.9	15.4	5.3	12.8	0.1	n.a.	0.6	0.4	0.1	1.1	0.009	92.0	n.a.	113.9	43.0	294.0	81.1	25.7	12.8	149.2	11.8	1160
B3/11	12.5	20.0	3.1	5.8	0.05	0.02	1.2	0.1	0.06	0.5	0.006	63.2	n.a.	58.1	12.5	n.a.	28.8	8.6	n.a.	22.8	13.6	n.a.
B3/12	37.4	17.0	6.5	19.1	0.2	0.05	3.1	0.4	0.2	0.8	0.01	135.9	n.a.	71.9	35.6	224.9	66.7	18.6	8.0	70.3	26.5	405.7
B3/13	52.5	20.8	8.3	15.8	0.2	0.1	2.1	0.4	0.2	1.0	0.01	84.4	n.a.	282.1	87.8	289.0	91.5	18.3	5.0	73.4	11.1	603.1
B3/14	62.5 *	14.9	6.5	17.3	0.1	0.1	26.3	0.3	0.3	0.5	0.05	226.6	n.a.	390.1	12.5	334.7	80.7	44.8	n.a.	102.8	116.8	586.3
B3/15	33.2	22.3	2.6	5.8	0.1	n.a.	0.4	0.1	0.07	0.6	0.003	20.9	2.9	128.3	114.2	138.7	62.7	n.a.	n.a.	90.8	n.a.	629.6
B3/16	65.1 *	14.4	5.2	11.2	0.1	0.02	0.5	0.1	0.08	0.7	0.007	25.4	5.1	19.2	7.4	111.3	50.4	5.9	n.a.	12.8	n.a.	144.3
B3/17	29.0	20.3	4.6	11.6	0.1	0.1	3.2	0.3	0.2	0.3	0.02	149.8	n.a.	341.4	24.2	128.4	36.7	36.3	7.9	99.8	29.9	345.3
Mean	**31.9**	**20.8**	**4.9**	**13.7**	**0.2**	**0.1**	**4.2**	**0.4**	**0.4**	**0.7**	**0.02**	**166.7**	**7.7**	**104.0**	**24.2**	**258.9**	**49.2**	**31.3**	**21.5**	**56.3**	**20.1**	**376.1**

Note: VM—volatile matter; d—dry basis; daf—dry and ash free. *—coal-bearing shale or rock samples.

Table 2. Ash yield, volatile matter, and elemental composition (major and trace elements) of Jurassic coal samples from two mines.

Samples	Ash (d,%)	VM (daf, %)	Major Elements (%)									Trace Elements (ppm)										
			Al	Si	P	S	Fe	K	Ca	Ti	Mn	Cu	Cr	Zn	Pb	V	Ga	As	Ni	Y	Yb	Zr
Maikuben																						
M1	4.1	44.7	2.5	3.9	n.a.	0.4	18.1	0.6	10.4	0.2	0.02	168.9	n.a.	192.5	n.a.	45.5	14.3	94.5	117.2	45.5	71.5	77.9
M2	1.9	52.4	3.1	5.3	n.a.	1.4	7.9	0.4	11.0	n.a.	0.03	146.5	88.8	187.1	1.1	n.a.	n.a.	59.3	107.5	23.5	n.a.	7.7
M3	2.6	45.1	4.2	1.4	n.a.	0.8	14.0	n.a.	18.5	0.3	0.01	319.2	n.a.	245.7	8.1	158.4	19.0	92.7	61.7	128.0	n.a.	66.3
M4	2.4	44.2	5.0	5.3	n.a.	1.1	14.6	0.4	18.8	0.1	0.01	243.8	107.9	157.4	n.a.	90.1	n.a.	36.9	64.1	61.1	n.a.	31.8
Mean	**2.7**	**46.6**	**3.7**	**4.0**	**n.a.**	**0.9**	**13.6**	**0.3**	**14.6**	**0.1**	**0.02**	**219.6**	**49.1**	**195.6**	**2.3**	**73.5**	**8.3**	**70.8**	**87.6**	**64.5**	**17.8**	**45.9**
Shubarkol																						
SH1	4.4	25.7	0.7	8.1	0.08	0.02	0.09	0.04	0.1	0.6	0.003	20.7	4.3	5.4	4.5	90.6	10.0	3.1	6.2	25.2	4.9	175.9
SH2	4.5	45.9	4.0	5.7	0.07	1.7	6.9	0.1	2.1	0.3	0.04	116.2	37.2	492.1	29.4	76.5	16.5	69.0	9.23	39.1	n.a.	87.8
SH3	12.8	33.8	3.2	5.2	0.05	0.1	0.4	0.2	0.4	0.3	0.006	126.6	50.1	109.2	14.8	270.7	18.5	11.8	30.1	52.3	2.3	78.0
SH6	5.5	41.4	1.4	2.7	0.1	0.04	3.4	0.3	0.5	0.3	0.01	144.1	150	336.4	54.3	109	68.6	191.3	134	39.6	n.a.	n.a.
SH8	1.9	43.8	0.6	0.6	0.02	0.7	0.1	0.08	0.6	0.05	0.007	13.2	n.a.	10.4	1.3	n.a.	n.a.	2.6	41.1	n.a.	n.a.	3.3
SH10	5.3	36.4	3.8	2.9	0.1	2.0	5.3	0.2	3.1	0.4	0.01	153.1	60.2	167.5	38.1	n.a.	13.8	115.1	51	44.8	n.a.	50.7
SH12	4.0	41.4	3.3	5.0	0.05	0.8	0.5	0.2	1.8	0.3	0.04	61.9	21.4	50.4	9.2	103.0	7.9	9.1	170.6	62.1	n.a.	40.0
SH13	1.9	41.8	3.9	3.1	0.1	1.5	0.6	0.1	2.9	0.3	0.01	105.8	175.6	33.8	15.3	617.8	12.0	17.1	382.8	n.a.	n.a.	33.8
SH15	11.2	37.7	4.7	8.3	0.1	1.3	3.2	0.7	2.9	0.8	0.06	169.5	114.8	184.3	107	373.9	33.2	44.3	121.6	n.a.	n.a.	218.8
SH16	10.6	44.1	2.3	5.5	0.1	1.0	5.6	0.4	0.9	0.3	0.09	91.3	43.5	222.8	38.4	n.a.	24.0	33.8	146.7	34.5	n.a.	133.0
SH18	12.6	40.6	1.0	2.8	0.06	1.7	2.1	0.5	1.3	0.07	0.09	59.0	n.a.	113.1	5.4	n.a.	n.a.	10.4	297.2	n.a.	n.a.	7.6
SH20	1.4	45.6	2.6	2.5	0.05	3.3	2.7	0.4	4.9	0.3	n.a.	81.4	n.a.	147.3	9.2	n.a.	n.a.	13.9	127	n.a.	n.a.	24.7
SH21	1.8	48.3	2.1	2.7	0.03	0.2	0.3	0.1	0.6	0.1	0.001	40.6	96.0	108.6	7.0	574.9	7.6	6.2	22.8	25.9	n.a.	28.0
SH23	1.1	42.8	1.3	0.8	0.01	1.6	1.8	0.08	1.6	0.08	0.03	38.3	n.a.	37.6	2.6	n.a.	n.a.	3.6	60.2	n.a.	2.1	8.0
SH24	3.4	44.3	2.5	2.6	0.1	3.1	15.4	0.3	2.8	0.2	0.01	119.1	n.a.	182.6	50.7	n.a.	6.7	60.1	207.1	n.a.	n.a.	59.8
SH26	2.7	40.7	6.2	4.0	n.a.	4.8	4.2	0.2	7.2	0.3	0.08	186.1	n.a.	121.7	23.5	n.a.	9.8	10.4	118	n.a.	n.a.	20.0
SH27	8.9	43.7	0.5	1.1	0.03	1.8	5.4	0.1	1.8	0.1	0.02	158.9	n.a.	93.7	n.a.	n.a.	n.a.	22.2	230.4	13.6	n.a.	26.9
SH28	1.3	41.7	1.6	0.9	0.04	0.7	0.4	0.07	2.1	0.08	0.007	27.8	17.7	22.1	4.9	n.a.	n.a.	6.2	24.3	20.0	3.4	10.8

Table 2. *Cont.*

Samples	Ash (d,%)	VM (daf, %)	Major Elements (%)								Trace Elements (ppm)											
			Al	Si	P	S	Fe	K	Ca	Ti	Mn	Cu	Cr	Zn	Pb	V	Ga	As	Ni	Y	Yb	Zr
SH30	1.3	42.1	1.6	2.8	0.08	3.2	0.9	0.1	3.7	0.2	0.01	69.5	18.4	137.1	7.1	223.5	n.a.	5.5	59.0	n.a.	12.3	27.9
SH31	2.5	38.3	8.2	11.9	0.1	0.3	1.3	0.9	2.0	0.5	0.008	314.6	376.3	102	48.7	n.a.	60.4	32.8	163.9	n.a.	36.4	n.a.
SP1	1.1	46.7	5.0	2.8	0.05	3.0	1.2	0.1	4.6	0.35	0.04	99.2	37.1	91.0	7.7	n.a.	11.9	9.3	40.2	n.a.	n.a.	13.9
SP2	2.1	44.8	2.3	2.1	0.04	3.1	2.8	0.3	3.4	0.4	0.03	172.2	60.4	n.a.	15.6	75.3	12.8	24.3	57.4	35.4	11.8	79.9
SP3	2.0	39.1	4.3	4.7	0.05	1.3	1.7	0.1	1.8	0.1	0.009	28.9	21.8	18.4	5.4	34.6	3.1	6.6	27.2	5.0	1.2	n.a.
SP4	1.7	44.6	3.0	1.6	0.03	3.0	0.8	0.2	3.4	0.2	0.02	74.1	18.5	38.3	1.5	n.a.	2.9	5.2	25.9	6.2	0.5	8.8
SP6	9.8	36.6	4.4	10.8	0.4	0.2	0.9	0.7	0.9	1.3	0.001	80.8	n.a.	30.1	9.9	n.a.	50.7	12.6	73.4	21.4	2.6	346.1
SP8	2.4	43.2	1.4	2.8	0.07	0.3	0.1	0.03	0.5	0.2	n.a.	27.0	10.5	21.4	5.5	n.a.	17.1	6.1	85.8	n.a.	n.a.	n.a.
SP9	1.0	43.6	2.2	1.8	0.08	2.3	0.9	0.09	2.3	0.2	n.a.	82.8	n.a.	55.6	8.7	n.a.	7.8	20.1	92.3	14.9	n.a.	31.5
SP14	5.0	35.8	1.5	4.5	0.04	0.8	0.4	0.1	1.1	0.6	0.004	56.9	n.a.	72.6	36.8	n.a.	22.0	10.0	88.1	n.a.	n.a.	n.a.
SP15	8.0	40.2	3.0	7.1	0.06	0.2	0.9	0.6	0.7	1.1	0.006	266.5	80.6	45.3	35.4	258.8	29.3	14.2	38.6	n.a.	n.a.	n.a.
SP16	12.3	37.4	3.5	5.6	0.2	0.2	0.4	0.4	0.6	0.3	n.a.	42.3	n.a.	19.0	18.7	n.a.	20.7	13.0	33.3	n.a.	3.3	249.8
SP17	5.2	40.2	6.2	9.2	0.4	0.5	1.1	0.6	1.7	0.5	0.003	145.0	107.2	51.7	57.5	n.a.	81.0	22.2	43.6	n.a.	n.a.	358.5
SP18	1.2	37.6	1.5	0.9	0.05	2.4	0.7	0.01	1.9	0.2	0.01	51.1	20.7	40.9	3.4	n.a.	n.a.	10.1	13.8	n.a.	n.a.	10.2
Mean	**8.8**	**40.9**	**2.9**	**3.9**	**0.1**	**1.5**	**2.3**	**0.2**	**2.0**	**0.3**	**0.02**	**100.3**	**47.4**	**98.6**	**21.0**	**84.9**	**16.8**	**25.5**	**94.4**	**12.9**	**2.3**	**61.4**

Note: VM—volatile matter; d—dry basis; daf—dry and ash free.

Table 3. Geochemical composition and comparison of mean major and trace element values in coal ash from the five coal mines in Kazakhstan, with enrichment factors normalized to CVC (Clarke value of the crust) and WHC (worldwide hard coal ash) values. Unit: ppm for trace elements.

№	Elements	Jurassic Kazakh Coal Ash (J)		Carboniferous Kazakh Coal Ash (C)		[a] CVC	CVC vs. J and C		[b] WHC	WHC vs. J and C	
		Range	Mean	Range	Mean	Mean	EF (J)	EF (C)	Mean	EF (J)	EF (C)
1	Al (%)	0.5–5	3.3	0.5–17.6	4.8	8.23	0.4	0.5	n.a.	n.a.	n.a.
2	Si (%)	0.9–11.9	3.9	0.1–20.2	11.3	28.1	0.1	0.4	n.a.	n.a.	n.a.
3	P (%)	0.01–0.08	0.05	0.01–2.1	0.2	0.10	0.5	2.6	0.15	0.3	1.7
4	S (%)	0.09–2.3	0.9	0.01–0.6	0.2	n.a.	n.a.	n.a.	n.a.	n.a.	n.a.
5	Fe (%)	0.09–18.1	7.9	0.06–26.3	4.2	5.63	1.4	0.7	n.a.	n.a.	n.a.
6	K (%)	0.03–0.7	0.2	0.02–1.7	0.7	2.09	0.1	0.3	n.a.	n.a.	n.a.
7	Ca (%)	0.1–18.8	8.3	0.06–8.8	0.8	4.15	2.0	0.2	n.a.	n.a.	n.a.
8	Ti (%)	0.3–0.6	0.2	0.04–3.1	0.8	0.57	0.3	1.4	0.53	0.3	1.5
9	Mn (%)	0.001–0.09	0.02	0.001–0.08	0.04	0.08	0.2	0.5	0.04	0.4	1.0
10	Cu	13.2–319	159.9	0.4–410.0	150.2	55	2.9	2.7	110	1.4	1.3
11	Cr	4.3–376.3	48.2	0.1–133.4	17.8	100	0.4	0.1	120	0.4	0.1
12	Zn	5.4–492.1	147.1	0.3–390.1	88.8	70	2.1	1.2	170	0.8	0.5
13	Pb	1.1–107.0	11.6	0.04–114.2	22.5	12.5	0.9	1.8	55	0.2	0.4
14	V	34.6–617.8	79.2	22–768.0	247.9	135	0.5	1.8	170	0.4	1.4
15	Ga	2.9–81.0	12.5	0.09–109.6	43.8	15	0.8	2.9	36	0.3	1.2
16	As	2.6–191.3	48.1	0.7–187	23.4	7.6	6.3	3.0	46	1.0	0.5
17	Ni	6.2–328.8	91.0	0.1–221.4	22.9	37	2.4	0.6	100	0.9	0.2
18	Y	5.0–128.0	38.7	0.3–134.2	56.2	29	1.3	1.9	57	0.6	0.9
19	Yb	1.2–71.5	10.0	1.0–116.8	26.3	2.0	5.0	13.1	6.9	1.4	3.8
20	Zr	3.3–358.5	53.6	0.1–657.4	205.3	165	0.3	1.2	230	0.2	0.8

Note: EF—enrichment factor (EF, J for Jurassic coal and EF, C for Carboniferous coal); [a] Clarke value of the crust (CVC) [35–38] and [b] worldwide hard coal ash (WHC) [38].

Figure 2 presents a comparison of the average concentrations of major and trace elements in Jurassic and Carboniferous coal ash with Clarke values of the Earth's crust (CVCs, [35–38]. The results indicate that the concentrations of major elements such as Al, Si, and K are generally lower than the CVC. In contrast, other major elements, including P, Ca, Ti, and Mn, as well as trace elements like Cu, Zn, Pb, V, Ga, As, Ni, Y, Yb, and Zr, are relatively higher than the CVC.

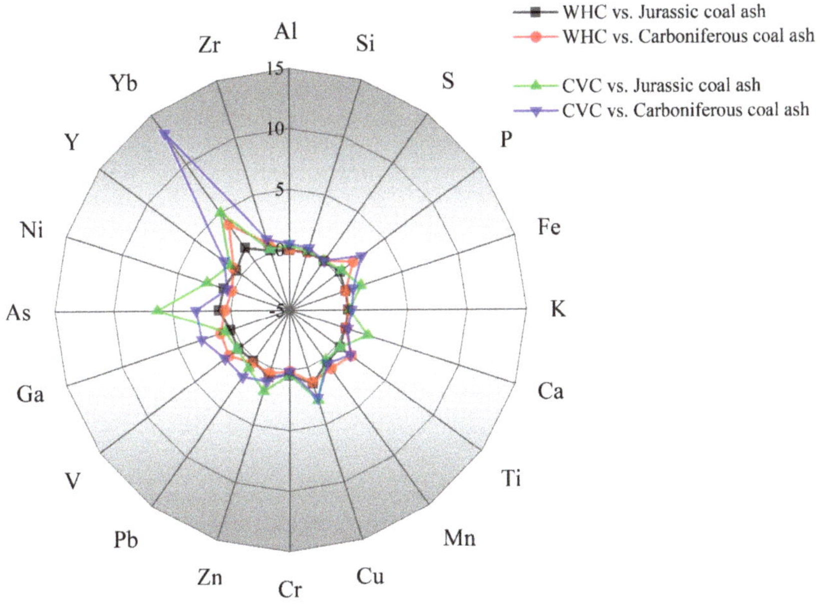

Figure 2. Spider plot of major and trace element concentrations in Jurassic and Carboniferous coal ash normalized to Clarke values of the Earth's crust (CVCs) [35–38] and worldwide hard coal ash (WHC) [38] values.

Figure 2 compares the average concentrations of major and trace elements in Jurassic and Carboniferous samples with the world hard coal (WHC) averages [38]. The results show that Yb averages 10 ppm in Jurassic samples and 26.3 ppm in Carboniferous samples, while Cu averages 159 ppm and 150 ppm, respectively. These values are higher than the WHC averages for these elements. In contrast, most other elements have lower averages compared to WHC, except for P (0.2%), Ti (0.8%), Ga (43.8 ppm), and V (247.9 ppm), which are higher than the WHC values and Jurassic samples. Notably, As (48.1 ppm) is higher in Jurassic samples compared to Carboniferous samples and WHC values.

3.3. XRD and SEM-EDS Results

Based on the results of XRD analysis (Figure 3 and Table 4) and SEM-EDS, the samples primarily consist of kaolinite, quartz, and pyrite. Additionally, SEM-EDS data also revealed the presence of accessory minerals such as apatite, chalcopyrite, and rutile. The mineralogical contents of the samples from the Carboniferous coal mines show slight variations compared to each other, but are generally comparable to those from the Jurassic coal mines. The differences in the Jurassic samples include the presence of apatite containing various rare-earth elements (REE with La, Nd, and Ce), pyrite with arsenic content filling the edges of sporinite macerals, and tiny rutile particles within woody fragments (Figures 4 and 5).

The dominant silicate mineral phases varied among the studied samples in the five coal mines, with quartz and kaolinite being particularly prominent. Quartz and kaolinite were identified as the most dominant minerals in all samples. Kaolinite is present as pore and cell fillings within the coal, whereas quartz is observed as angular grains and occasionally as particles exhibiting good abrasion, appearing as subangular or angular quartz grains, as shown in Figures 4 and 5. Additionally, quartz and kaolinite were the only major minerals detected in a few samples by XRD. Accessory minerals, including chalcopyrite, rutile, and REE-bearing phosphate minerals, were detected in all samples, as identified during SEM-EDX studies. The accessory minerals, consistently found in association with kaolinite crystal grains, were identified through SEM-EDS analysis as cleat/pore-infilling minerals within the coal.

SEM-EDS analysis shows results from samples collected from Carboniferous coal mines: The samples from the Bogatyr mine the samples of Bogatyr (with B1/0 and B1/2) are mainly composed of quartz grains ranging in size from 5 to 10 μm. Additionally, kaolinite is distributed along the bedding planes and in the cleats of the coals within these samples. Minor amounts of pyrite are also present, occurring in two forms: euhedral (7 μm) and framboidal (10 μm). All minerals were identified as fracture-infilling minerals within the coals (Figure 4a,c). The samples of the Lenin mine (with DL4 and DU2) are enriched in quartz and kaolinite, with minor amounts of pyrite. Accessory minerals, such as apatite, were rarely observed in these samples (Figure 4c,d). Quartz occurs as grains of variable roundness (around 5–10 μm), kaolinite is mainly distributed along the bedding planes and in the cleats of the coals, while pyrite occurs as isolated euhedral pyrite particles with sizes ranging from 0.5 to 2 μm. The samples of Saradyr (S1 and S2) primarily consist of quartz grains and kaolinite distributed with a size of around 25 μm, along the bedding planes and in the cleats of the coals. Additionally, minor amounts of euhedral pyrite, with sizes of up to 10 μm, are observed in the samples. Within the kaolinite grains, accessory minerals such as chalcopyrite (10 μm) and rutile (4–5 μm) were identified within the coals (Figure 4e,f).

Jurassic coal mine samples show that the samples from Maikuben (M1) are enriched with quartz and kaolinite within coal fractures, with chalcopyrite (approximately 10 μm) present as an accessory mineral (Figure 5a,b). Shubarkol coal samples (SH1 and SH26) are primarily composed of quartz and kaolinite, along with apatites containing rare-earth elements (La, Ce, and Nd). Additionally, sporinite macerals are observed within vein-like pyrite structures that contain arsenic (Figure 5c,d).

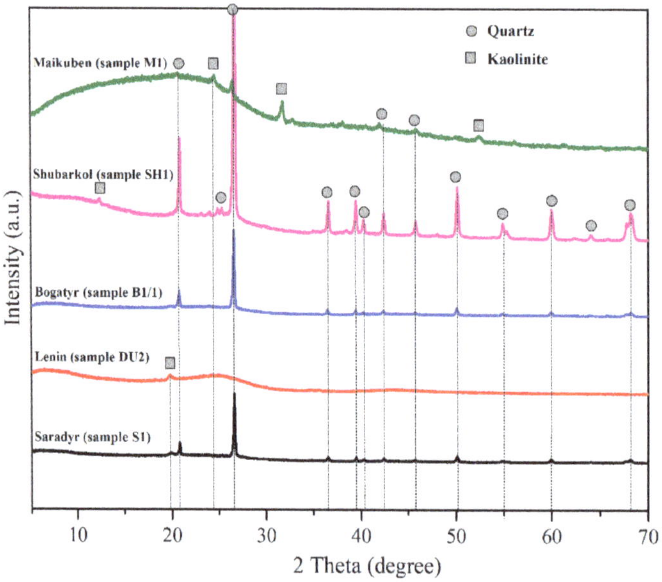

Figure 3. X-Ray diffraction (XRD) patterns of five coal samples from five active mines.

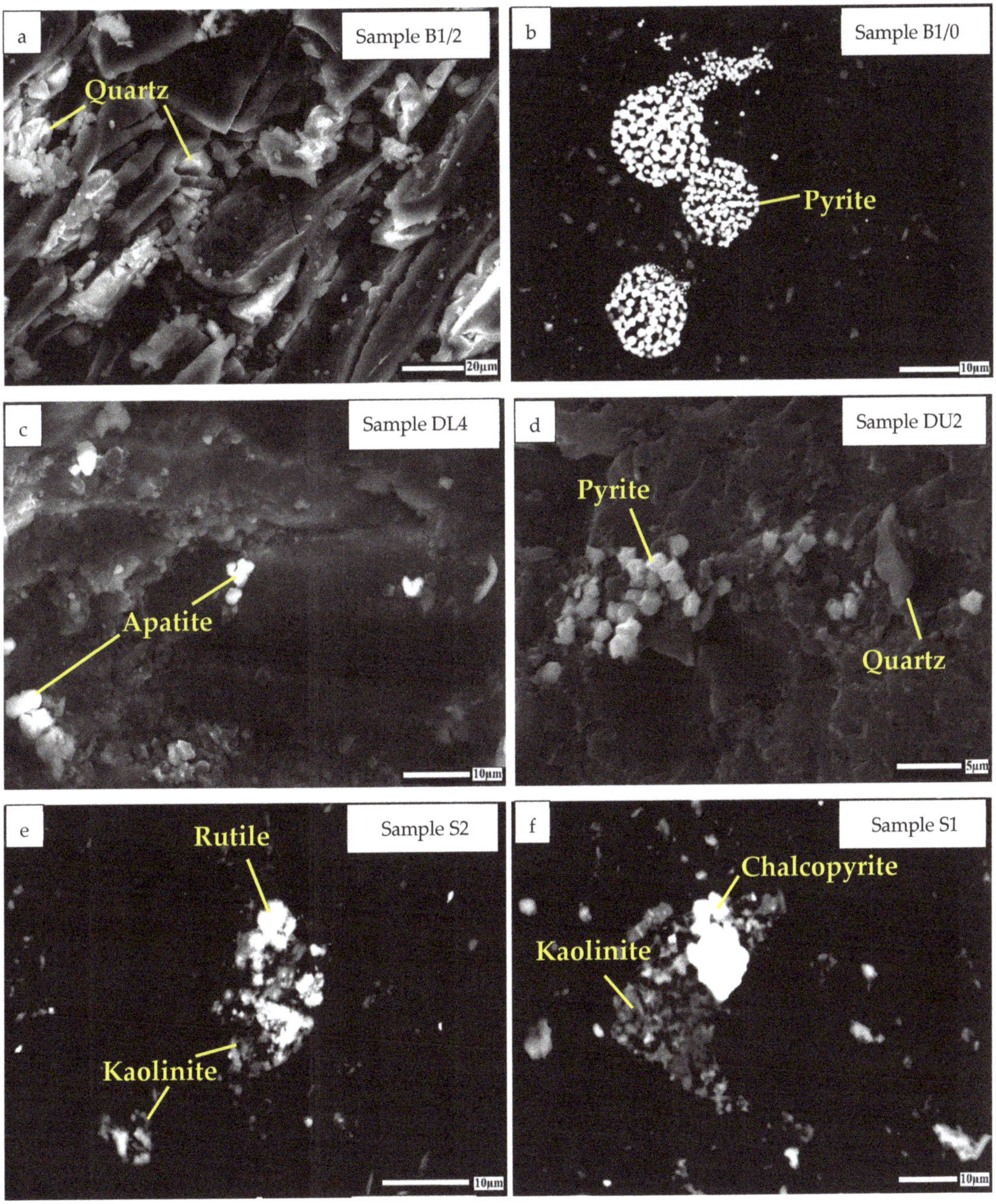

Figure 4. Selected SEM-BSE images of crystalline phases and EDS spectra of Carboniferous coal samples: (**a**) quartz growth occurring within coal fractures; (**b**) framboidal pyrite growth within coal matrix; (**c**) crystal growth of apatite among coal matrix; (**d**) euhedral pyrites associated with surrounding quartz; (**e**) individual crystals of rutile with kaolinite; and (**f**) chalcopyrite grains with kaolinite.

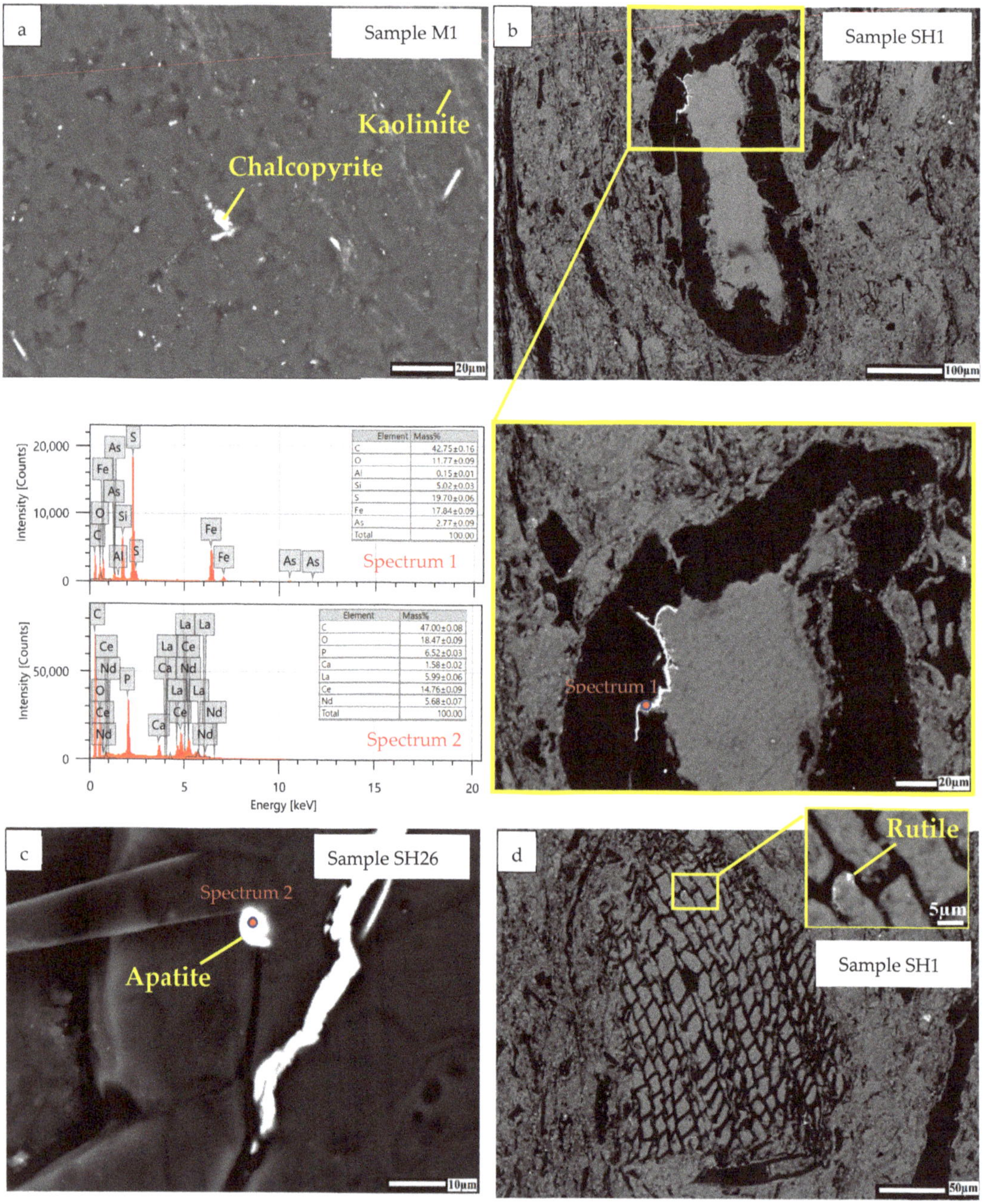

Figure 5. Selected SEM-BSE images of crystalline phases and EDS spectra of Jurassic coal samples: (**a**) chalcopyrite grains with kaolinite; (**b**) sporinite macerals observed within vein-like pyrite structures containing arsenic, with the EDS spectrum (1) showing elemental concentrations; (**c**) individual apatite crystals within coal fractures, with the EDS spectrum (2) showing rare-earth element concentrations (La, Ce, and Nd); and (**d**) rutile-filled primary organic matter pores in cellular woody material.

Table 4. Mineral contents in the five selected coal samples from the five coal mines measured by low-temperature ashing X-ray diffraction (LTA-XRD) and SEM (+++—dominant phase >20 wt%; ++— abundant phase 20–5 wt%; +— minor phase <5 wt%; and a—accessory mineral detected by SEM-EDS).

Minerals	Coal Mines				
	Lenin	Bogatyr	Saradyr	Maikuben	Shubarkol
Quartz	+++	+++	+++	+++	+++
Kaolinite	++	++	++	++	++
Pyrite	+	+	+	+	+
Chalcopyrite			a	a	
Rutile			a		a
Apatite	a				a

4. Discussion

4.1. Mode of Occurrence of Major and Trace Elements in Coals

Figure 6 presents the dendrogram analysis, which provides valuable insights into the clustering of elements based on their similarities, highlighting their modes of occurrence and affinities. The close association of Al and Si signifies their predominant presence in silicate minerals like quartz and clay minerals (kaolinite), illustrating their essential roles in aluminosilicate phases and emphasizing their strong inorganic affinity within mineral structures. Other authors [1,39,40] noted that Si and, to a lesser extent, Al are the most abundant elements in coal and primary components of its combustion products, identifying 63 silicate and aluminosilicate minerals, including quartz and clay minerals, among more than 200 total minerals.

Ti, Ga, and Cu are associated with accessory minerals, with titanium found in rutile and gallium substituting for Al in kaolinite minerals, and Cu prevalent in sulfide minerals such as chalcopyrite. Their clustering highlights shared associations with specific mineral phases, emphasizing their inorganic nature. As other authors have indicated, titanium is found in rutile, anatase, and other heavy minerals [41], and can also exist as Ti-gel-like phases within kaolinite particles [42]. Gallium is primarily associated with diaspore and kaolinite in coals [2], while copper exhibits a complex distribution in coals, occurring in carbonates, sulfides, silicates, and organic matter, with no single mode of occurrence predominating [43].

The grouping of V, Y, Zr, and K suggests their common occurrence in accessory minerals. This relationship indicates their contributions to the mineralogical complexity of environmental formations, supported by their inorganic affinities. Various authors indicate that Zr is almost exclusively found in zircons, while potassium is present in K-feldspars and clay minerals [41]. Additionally, V is primarily associated with clay minerals [44], and Y in coals is mainly linked to clay minerals and organic compounds [45].

S's central role in forming sulfide minerals demonstrates its critical interactions with metals like Pb and Fe, confirming its strong inorganic affinity. Pb and Fe's connection highlights their occurrences in sulfide minerals, with lead typically found in galena and iron in minerals like pyrite. This association is significant for understanding heavy mineral concentrations and sulfide mineralization, emphasizing their strong inorganic affinities. Many authors have noted that Fe in coal exhibits various modes of occurrence, including in sulfides, carbonates, oxides, chlorite, and Fe-bearing silicates [2]. Additionally, Pb has multiple inorganic associations, including carbonates, oxides, and sulfides, which can coexist in coal [46].

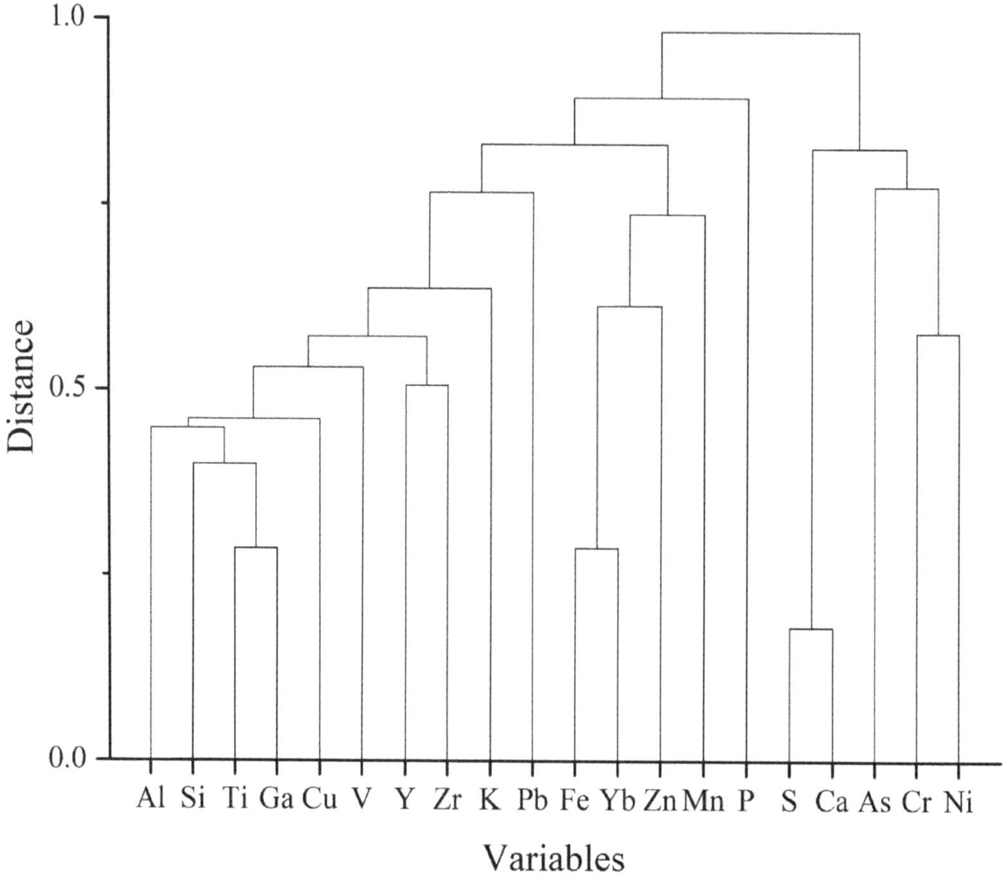

Figure 6. Dendrogram developed from cluster analysis on the elemental data of the complex Jurassic and Carboniferous coals from five coal mines.

Yb's link to rare earth elements emphasizes its presence in accessory minerals like apatite, with a distinct inorganic affinity that underscores its role in trace mineral phases. Zn and Mn are associated with sulfide and oxide minerals, reflecting their roles in less abundant mineral assemblages while maintaining strong inorganic affinities. Many authors emphasize that apatite, a common accessory mineral, is particularly important for hosting REE, especially Yb [47,48]. Mn in coal primarily occurs in carbonates and silicates [2], while Zn is mainly found as an accessory sulfide mineral [2]. Zn is associated with various mineral phases, including silicate-bound, sulfide-bound, and carbonate-bound fractions [49].

P is primarily found in phosphate minerals, and its occurrence in sediments highlights its importance in biogeochemical processes. Although it primarily shows an inorganic affinity, it may interact with organic matter in specific environments. Ward et al. (1996) [50] mentioned that P in coal occurs primarily as crystalline mineral particles, mainly apatite and aluminophosphates.

Ca is commonly found in carbonate minerals, phosphate, and silicates, emphasizing its widespread occurrence in geological settings and its strong inorganic affinity. As noted by Sánchez-Navas et al. (2013) [51], calcium (Ca) in coals is commonly found in carbonate, phosphate, and silicate minerals.

The grouping of As, Cr, and Ni indicates their associations with sulfide and oxide minerals, reflecting a shared inorganic affinity. While As primarily exhibits inorganic behavior, it may show organic affinity in reducing environments. The organic affinity of As, Cr, and Ni in coal varies by element and coal type, with studies suggesting moderate to strong organic associations for Cr and Ni, while As exhibits variable affinities, ranging from strong organic bonds [52,53] to inorganic associations [54], influenced by factors such as element abundance, properties, and coal-forming processes [53].

The analysis indicates that the elements in the dendrogram are predominantly associated with inorganic mineral phases, specifically silicates, oxides, sulfides, and phosphates. Although some elements may exhibit organic affinity under specific conditions, their primary associations are within inorganic mineral frameworks. The use of SEM-EDS further supports these findings, confirming the presence of various minerals, including aluminosilicates (clay minerals), oxides (e.g., quartz and rutile), sulfides (e.g., pyrite and chalcopyrite), and phosphates (e.g., apatite).

4.2. Technological Implications

Silicon (Si) and sulfur (S) can significantly impact coal utilization, primarily through their effects on combustion and ash behavior. Si in coals, as a major component of coal mineral matter, contributes to ash formation and slagging issues. S in coals, often present as pyrite, can react with silica during combustion to form clinkers, potentially inhibiting coal combustion [2,55].

While silicon (Si) in the form of quartz (hardness 7) causes significant abrasion to mining and grinding equipment, it also occurs in softer minerals such as clays and other silicates, which contribute less to abrasion [2]. In the studied samples, silicon content ranges from 0.9% to 11.9% (average 3.9%) in Jurassic coal, and from 0.1% to 20.2% (average 11.3%) in Carboniferous coal. The high silicon values of all samples may suggest a technological impact, but the effects may be mitigated by the presence of kaolinite, in addition to quartz, as observed through SEM analysis.

In the coal ash samples from the five coal mines, the average sulfur concentration is 0.9% in the Jurassic samples and 0.2% in the Carboniferous samples. According to [56], coal with sulfur levels ranging from 1.42% to 5.3% (average 3.36%) by weight is considered high-sulfur and low-quality. This indicates that the sulfur content in these samples is relatively low, which positively affects their technological suitability.

Research on high-ash coal combustion highlights complex interactions between coal properties and combustion efficiency. A higher volatile matter content reduces unburned carbon, facilitating easier ignition and more complete combustion. However, it can also increase emissions of elements like Hg, Pb, Zn, and Cd, which behave differently depending on furnace temperature [57,58]. Additionally, high-ash fusion temperatures and the presence of refractory minerals such as alumino-silicates, iron oxides, and titanium oxides impact combustion efficiency, as well as the potential for slagging and fouling [59]. Optimal combustion typically occurs with $25 \pm 1.75\%$ moderate volatile matter and 20–35% moderate ash content [60] is ideal for combustion because it provides efficient burning with fewer problems like emissions and slagging. The coal samples from the five mines show varying ash yields, with lower values in Jurassic samples and higher values in Carboniferous samples, along with higher volatile matter (Table 5). These differences may result in deviations from the optimal combustion range (20–35% ash yield) and the ideal volatile matter content (around 25%). This suggests that different combustion technologies may be needed for the coal from the five mines to handle these variations effectively.

Table 5. Comparison of ash yield, volatile matter, and inorganic sulfur content in coal samples from five mines with optimal combustion values [60] and impact sulfur level threshold [56].

Coal Mines	Coal Ash Yield (Dry Basis, Unit %)		Volatile Matter (Dry and Ash-Free Basis, Unit %)		Inorganic Sulfur (Dry Basis, Unit %)	
	Range	Average	Range	Average	Range	Average
			Jurassic coals			
Maikuben	1.9–4.1	2.7	44.2–52.4	46.6	0.4–1.4	0.9
Shubarkol	1.1–12.8	8.8	33.8–45.6	40.9	0.02–3.1	1.5
			Carboniferous coals			
Saradyr	17.0–40.6	26.9	8.8–28.6	21.5	0.06–0.6	0.2
Lenin	3.7–47.6	14.8	8.9–24.6	21.8	0.01–2.3	0.4
Bogatyr	2.7–42.9	31.9	13.7–29.0	20.8	0.01–0.8	0.1
	OCV: range 20–35		OCV: 25 ± 1.75		IS range 1.42–5.3	

Note: optimal combustion values (OCVs); IS (impact sulfur).

4.3. Economical Implications

According to the method proposed by [61] for assessing the relative enrichment of trace elements, concentrations of trace elements in coal can be categorized into six groups: highly enriched ($CC \geq 100$, where CC represents the concentration coefficient, indicating the ratio of trace element concentrations in the samples being studied compared to the averages for WHC), significantly enriched ($10 \leq CC < 100$), enriched ($5 \leq CC < 10$), slightly enriched ($2 \leq CC < 5$), normal ($0.5 \leq CC < 2$), and depleted ($CC < 0.5$).

The concentration coefficient (CC) of P and Ti major elements and trace elements in our samples was determined by referencing the values established for world hard coal (WHC) by [38]. This CC helps ascertain whether elements are enriched or depleted by comparing the concentration of trace elements in our coal ash samples to the corresponding WHC values (Figure 3). While we aimed to calculate CCs for all major elements, most lacked corresponding WHC values for normalization. Consequently, we specifically measured the CCs for the major elements P, Ti, and Mn. The element P exhibits a CC ranging from 0.3 to 1.7 in Jurassic and Carboniferous coal ash samples, placing it within the normal concentration range. Similarly, Ti has a CC between 0.3 and 1.5, also falling within the normal concentration category. Mn has a CC of 0.4 in Jurassic samples, placing it in the depleted category, while in Carboniferous samples, its CC is 1.0, indicating a normal concentration.

Trace elements in the five mine samples, including Yb, have a CC ranging from 1.4 to 3.8 in Jurassic and Carboniferous coal ash samples, indicating they fall into the slightly enriched category. Other trace elements, such as Cu, As, and Y, with CC values ranging from 0.5 to 1.4, fall into the normal concentration category across the samples from the five coal mines. Elements such as Cr, Pb, V, Ga, and Zr in Jurassic samples have CC values below 0.5, placing them in the depleted category. Similarly, in Carboniferous samples, Cr, Zn, Pb, and Ni have CC values below 0.5 and are also categorized as depleted. However, in the same Carboniferous samples, V, Ga, and Zr have CC values greater than 0.5, placing them in the normal concentration category (see Figure 7).

The concentration coefficients (CCs) of certain elements in the studied coal ash samples provide insights into their potential economic value. Elements like Yb, slightly enriched across Jurassic and Carboniferous coal ashes, may attract economic interest due to their applications in high-tech industries. Although elements such as Cu, As, and Y fall within normal concentration ranges, they could still hold value depending on the feasibility of extraction. Conversely, the depleted levels of elements like Cr, Pb, V, Ga, and Zr suggest limited economic recovery potential from these samples. These findings highlight the selective recovery potential of specific trace elements, with implications for resource utilization and potential economic benefits in targeting enriched or slightly enriched elements for extraction. Additional rare-earth elements, including La, Ce, and Nd, were identified through EDS analysis, indicating a promising economic presence of valuable elements as well. Yb was detected in coal ash samples using XRF, while La, Ce, and Nd were identified in unchanged coal matrices using SEM-EDS analysis, which allows for detailed elemental analysis of the coal structure without alteration. This distinction highlights the different analytical capabilities of XRF and SEM-EDS in characterizing rare-earth elements in coal and its by-products.

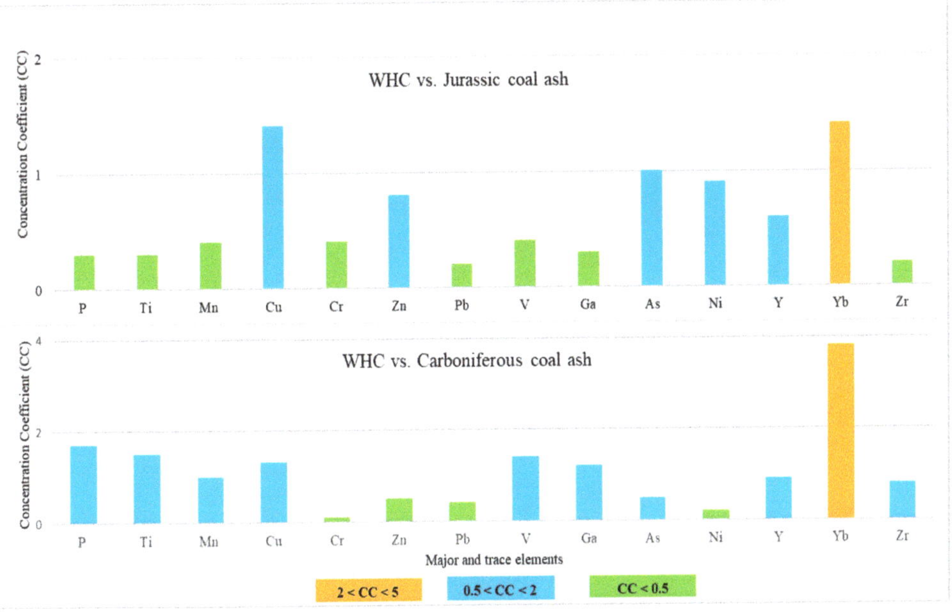

Figure 7. Concentration coefficient (CC) of major elements and trace elements in Carboniferous and Jurassic coal ash from five coal mines, normalized by the average concentrations of major/trace elements in world hard coal (WHC, [38]).

4.4. Environmental Implications

As [62–64] noted, the environmental impact elements associated with coals include As (100 ppm), Sb (45 ppm), Hg (1500 ppm), Pb (500 ppm), Cu (1–1.5%) Cd (0.1%), Bi (0.8%), Ag (1%), Zn (0.3–0.5%), Cr (0.5%), Mn (0.5%), and Ba (0.4%); shown in parentheses are the concentrations considered harmful to the environment. These chemical elements originate from both associated minerals and transformation processes during combustion [62]. It is well established that environmental impact elements in coal tend to be associated with mineral matter, particularly in high-ash coal containing significant amounts of heavy-metal-bearing minerals like pyrite, barite, and rutile/ilmenite. During coal combustion, ashes become enriched with trace elements, often reaching concentrations 3–10 times higher than those in the original coal. Remarkably, these elements are not released during combustion, as evidenced by the consistent concentration ratios found in both ash and original coal [64]. The comparison in Figure 8 illustrates the concentrations of trace elements in samples from three mines against standard harmful thresholds, based on protocols described in the literature [62]. The comparison of potentially toxic elements in ash samples from the five coal mines with harmful threshold standards suggests that the trace element-bearing coal ash poses minimal environmental impact. Potentially toxic elements such as As, Pb, Cu, Zn, and Cr, including the highest concentration of Mn across all samples, exhibit concentrations below 433 ppm, which are ten times lower than harmful levels.

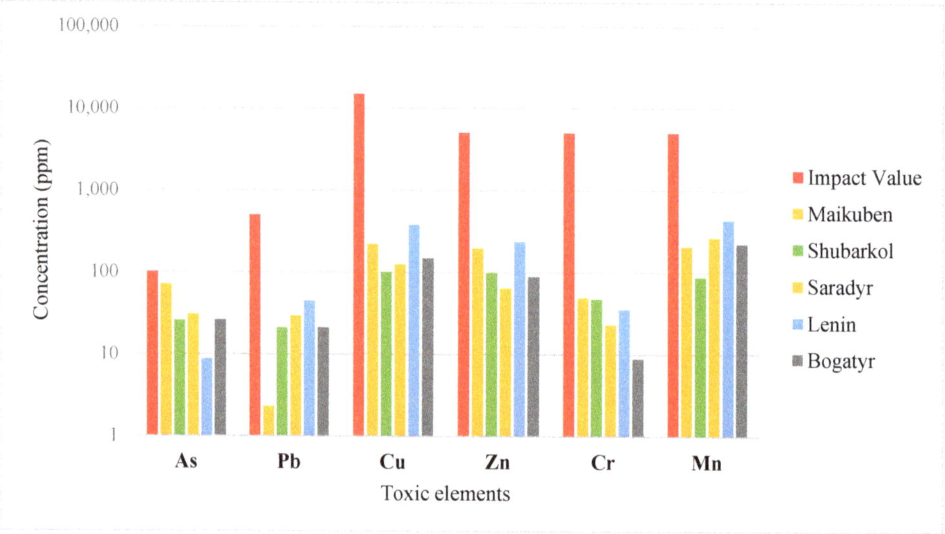

Figure 8. Comparative environmental impact assessment of selected potentially toxic elements in three coal mines against standard harmful thresholds (referred to as impact value) (impact value for comparison as per [62–64]).

5. Conclusions

XRF was selected as a fast and routine tool for analyzing 105 coal ash samples from Jurassic and Carboniferous coals sourced from five mines in Kazakhstan: Lenin, Saradyr, Bogatyr, Maikuben, and Shubarkol. A total of 20 elements were identified in the coal samples, including 9 major elements (Al, Si, P, S, Fe, K, Ca, and Ti) and 11 trace elements (As, Cu, Cr, Zn, Pb, V, Ga, Mn, Ni, Y, Yb, and Zr). Dendrogram analysis shows strong inorganic affinities, with Al and Si linked to aluminosilicates such as quartz and kaolinite, while Fe, Pb, Cu, Zn, and Mn are associated with sulfide minerals. Elements like Ti, Ga, Cu, Yb, V, Zr, and Y are found in accessory minerals, including rutile, zircon, and apatite, contributing to the mineralogical complexity alongside kaolinite. P and As may interact with organic matter. Quartz and kaolinite are the dominant minerals, with minor pyrite present in both euhedral and framboidal forms, accompanied by accessory minerals such as chalcopyrite, rutile, and REE-bearing apatite, as observed through SEM-BSE with EDS analysis.

Technological impacts arise from the significant variations in silicon and sulfur content in the coal samples from the five mines, affecting combustion and suitability for industrial use. High Si levels, particularly in Carboniferous coals, may cause abrasion and ash-related issues, though kaolinite could help mitigate these effects. The low S content improves the coals' suitability by reducing emissions. Variations in ash yield and volatile matter highlight the need for customized combustion technologies to optimize performance, minimize slagging, and improve efficiency for both Carboniferous and Jurassic coals.

The concentration coefficient (CC) analysis shows that most major elements, including P and Ti, fall within normal concentrations in both Jurassic and Carboniferous coal ash samples. Mn is depleted in Jurassic samples but normal in Carboniferous ones. Trace elements like Yb are slightly enriched, while Cu, As, and Y show normal levels. Elements such as Cr, Pb, and Zn are depleted in both coal types, highlighting differences in elemental distribution with potential environmental and industrial implications. Moreover, EDS analysis revealed additional rare-earth elements (La, Ce, and Nd), as identified in unchanged coal matrices using SEM-EDS, while Yb was detected in coal ash samples via XRF.

The comparison of trace element concentrations in coal ash samples from the five mines against harmful environmental thresholds suggests minimal environmental impact. Potentially toxic elements such as As, Pb, Cu, Zn, and Cr, including the highest levels of Mn, are found in concentrations significantly below harmful thresholds, often ten times lower. These findings indicate that the coal ash from the studied samples poses a low risk of environmental contamination.

XRF analysis may prove to be highly effective for large-scale studies involving multiple coal samples, as demonstrated by the successful analysis of 105 samples in 1.5 days using automated overnight measurements. This efficiency greatly exceeds that of traditional digestion-derived techniques, which require longer processing times. Additionally, the XRF results highlight the potential for further detailed investigations, particularly focused on trace elements, including REE.

Author Contributions: Methodology, M.J.; Investigation, M.J.; Resources, M.J.; Writing—original draft, M.J.; Writing—review & editing, M.J.; Visualization, M.J. and S.M.; Funding acquisition, S.M. All authors have read and agreed to the published version of the manuscript.

Funding: This study was supported by Satbayev University (Almaty, Kazakhstan), project no. AP14871618 "Major boundaries of the Lower Carboniferous in the sections of Central Kazakhstan: development of a zonal standard".

Institutional Review Board Statement: Not applicable.

Informed Consent Statement: Not applicable.

Data Availability Statement: Data is contained within the article.

Conflicts of Interest: The authors declare no conflict of interest.

References

1. Finkelman, R.B.; Dai, S.; French, D. The importance of minerals in coal as the hosts of chemical elements: A review. *Int. J. Coal Geol.* **2019**, *212*, 103251. [CrossRef]
2. Dai, S.; Finkelman, B.R.; French, D.; Hower, J.C.; Graham, I.T.; Zhao, F. Modes of occurrence of elements in coal: A critical evaluation. *Earth-Sci. Rev.* **2021**, *222*, 103815. [CrossRef]
3. Dogan, O.; Simsek, O.; Ertugrul, M.; Kobya, M. X-ray fluorescence spectrometry analysis of trace elements in fly ash samples of Yenikoy thermal power plants. *Instrum. Sci. Technol.* **2001**, *29*, 433–439. [CrossRef]
4. Tiwari, M.; Sahu, S.K.; Bhangare, R.C.; Ajmal, P.Y.; Pandit, G.G. Elemental characterization of coal, fly ash, and bottom ash using an energy dispersive X-ray fluorescence technique. *Appl. Radiat. Isot.* **2014**, *90*, 53–57. [CrossRef] [PubMed]
5. Junussov, M. Characteristics, Distribution and Morphogenesis of Gold-Bearing Sulfide Minerals in the Gold Black Shale Deposit of Bakyrchik. In *International Multidisciplinary Scientific GeoConference Surveying Geology and Mining Ecology Management*; SGEM: Vienna, Austria, 2018; Volume 18, pp. 643–650.
6. Junussov, M.; Madai, F.; Kristály, F.; Tóth, T.; Fintor, K.; Muránszky, G.; Prekob, Á.; Hámor-Vidó, M. Preliminary analysis on roles of metal–organic compounds in the formation of invisible gold. *Acta Geochim.* **2021**, *40*, 1050–1072. [CrossRef]
7. Junussov, M.; Mohammad, A.; Longinos, S. Geochemical analysis of organic matter associated with gold in ore deposits: A study of Kazakhstan and Hungary. *Acta Geochim.* **2024**, 3–13. [CrossRef]
8. Junussov, M.; Mádai, F.; Földessy, J.; Hámor-Vidó, M. The role of organic matter in gold occurrence: Insights from Western Mecsek uranium ore deposit. *Econ. Environ. Geol.* **2024**, *57*, 371–386. [CrossRef]
9. Bettinelli, M.; Taina, P. Rapid analysis of coal fly ash by X-ray Fluorescence spectrometry. *X-Ray Spectrom.* **1990**, *19*, 227–232. [CrossRef]
10. Suarez-Fernandez, G.P.; Vega, J.M.G.; Fuertes, A.B.; Garcia, A.B.; Martínez-Tarazona, M.R. Analysis of major, minor and trace elements in coal by radioisotope X-ray fluorescence spectrometry. *Fuel* **2001**, *80*, 255–261. [CrossRef]
11. Revenko, A.G.; Pashkova, G.V. Study of the Chemical Composition of Coal and Coal Ash by X-ray Fluorescence Method: A Review. *X-Ray Spectrom.* **2024**. [CrossRef]
12. Pearce, B.C.; Hill, J.W.F.; Kerry, I. Use of X-ray fluorescence spectrometry for the direct multi-element analysis of coal powders. *Analyst* **1990**, *115*, 1397–1403. [CrossRef]
13. Huggins, F.E. Overview of analytical methods for inorganic constituents in coal. *Int. J. Coal Geol.* **2002**, *50*, 169–214. [CrossRef]
14. Yan, Z.; XinLei, Z.; WenBao, J.; Qing, S.; YongSheng, L.; DaQian, H.; Da, C. Online X-ray Fluorescence (XRF) Analysis of Heavy Metals in Pulverized Coal on a Conveyor Belt. *Appl. Spectrosc.* **2016**, *70*, 272–278. [CrossRef] [PubMed]

15. Junussov, M.; Tarikhov, F.; Abildakhanov, A.; Zhanaidar, D.; Mekenbek, G.; Assambayeva, A. Mineralogical and elemental analysis of Kazakh coals from three mines: Preliminary insights from mode of occurrence to environmental impacts. *Open Geosci.* **2024**, *16*, 1–21.
16. Low, F.; Zhang, L. Microwave digestion for the quantification of inorganic elements in coal and coal ash using ICP-OES. *Talanta* **2012**, *101*, 346–352. [CrossRef] [PubMed]
17. Kopobayeva, A.; Portnov, V.S.; Kim, S.P.; Amangeldykyzy, A.; Askarova, N. Naukovyi Visnyk Natsionalnoho Hirnychoho Universytetu. *Acta Geochim.* **2024**, *43*, 1–13. [CrossRef]
18. Kopobayeva, A.N.; Amangeldikyzy, A.; Blyalova, G.; Askarova, N.S. Mineralogical and Geochemical Features of Coals and Clay Layers of the Karaganda Coal Basin. *Minerals* **2024**, *14*, 349. [CrossRef]
19. Kopobayeva, A.; Baidauletova, I.; Amangeldikyzy, A.; Askarova, N. Trace Elements Distribution in the k7 Seam of the Karaganda Coal Basin, Kazakhstan. *Geosciences* **2024**, *14*, 143. [CrossRef]
20. Ammosov, I.I.; Burcev, D.N.; Gorsky, I.I.; Ivanov, G.A.; Kotlukov, V.A.; Kuznecov, I.A.; Lugovoy, G.I.; Mironov, K.V.; Molchanov, I.I.; Nekipelov, B.E.; et al. *Geology of Coal and Black Deposits*; Issue 5; Part 1; Academy of Sciences of the USSR: Moscow, Russia, 1973; pp. 13–718. (In Russian)
21. Azizov, T.M.; Vlasov, Y.I. *Coal and Oil Shale Basins and Deposits of Kazakhstan*; Institute of Geological Sciences of the RK: Almaty, Kazakhstan, 2013; pp. 6–106.
22. Matveev, A.K.; Vlasov, V.M.; Golycin, M.V.; Ignatshenko, N.A.; Mazor, Y.R.; Molshanov, I.I.; Nagorny, V.N.; Nagorny, Y.N.; Otten, F.F.; Penzin, Y.P. *Geology of Coal Deposits in SSSR*; Academy of Sciences of the USSR: Moscow, Russia, 1990; p. 352. (In Russian)
23. Ammosov, I.I.; Burcev, D.N.; Gorsky, I.I.; Ivanov, G.A.; Kotlukov, V.A.; Kuznecov, I.A.; Lugovoy, G.I.; Mironov, K.V.; Molchanov, I.I.; Nekipelov, B.E.; et al. *Geology of Coal and Black Deposits*; Issue 5; Part 2; Academy of Sciences of the USSR: Moscow, Russia, 1973; p. 219. (In Russian)
24. *ASTM STM D2013/D2013M-12*; Standard Practice for Preparing Coal Samples for Analysis. ASTM International: West Conshohocken, PA, USA, 2012.
25. Qi, J.; Liu, Y.; Li, S.; Hu, X. Transformation characteristics of minerals during low-temperature ashing of coal and the effect on element migration. *Int. J. Coal Geol.* **2021**, *238*, 103735.
26. Han, F.; Chen, J.; Zhang, L.; Wang, H. Mineralogical characteristics and chemical structure of low-temperature ash of bituminous coal under controlled atmosphere. *Fuel* **2022**, *309*, 122981.
27. *ISO 7404-2*; Methods for the Petrographic Analysis of Bituminous Coal and Anthracite—Part 2: Methods for Preparing Coal Samples. International Organization for Standardization: Geneva, Switzerland, 2009; 12p.
28. Warne, S.T. Thermal Analysis of Coal. In *Thermal Analysis*; Elsevier: Amsterdam, The Netherlands, 1991; Volume 1, pp. 235–258.
29. Beamish, B.B. Coal Thermal Analysis. In *Advanced Reservoir and Production Engineering for Coal Bed Methane*; Elsevier: Amsterdam, The Netherlands, 1994; pp. 81–94.
30. Donahue, R.; Rais, J. *Coal: Its Properties, Analysis, Classification, Geology, Extraction, Uses, and Distribution*, 2nd ed.; Society for Mining, Metallurgy, and Exploration; Wiley: New York, NY, USA, 2009.
31. Tian, X.; Zhao, L. Determination of concentrations of Sr and Ba in coal and coal combustion by-products: A comparison between results by ICP-MS and XRF techniques. *Talanta* **2024**, *266*, 124919. [CrossRef]
32. Matsunami, H.; Matsuda, K.; Yamasaki, S.I.; Kimura, K.; Ogawa, Y.; Miura, Y.; Tsuchiya, N. Rapid simultaneous multi-element determination of soils and environmental samples with polarizing energy dispersive X-ray fluorescence (EDXRF) spectrometry using pressed powder pellets. *Soil Sci. Plant Nutr.* **2010**, *56*, 530–540. [CrossRef]
33. Yang, Y.; Wei, X.; Wang, H.; Ma, D. Characterization of coal ash from thermal power plants in China and its environmental impact assessment. *Fuel* **2015**, *143*, 39–45.
34. Rietveld, H.M. A Profile Refinement Method for Nuclear and Magnetic Structures. *J. Appl. Cryst. Lography* **1969**, *2*, 65–71. [CrossRef]
35. Ronov, A.B.; Yaroshevsky, A.A.; Migdisov, A.A. *Chemical Composition of the Earth's Crust and Geochemical Balance of Main Elements, Science Pub*; Nauka: Moscow, Russia, 1990; p. 192.
36. Zhao, Z.G. *Research on Rare Earth Element Geochemistry of Coal bearing Strata*; Coal Industry Publishing House: Beijing, China, 2002; 37p. (In Chinese)
37. Nader, T.; Zahra, S.M. Geochemistry and origin of elements of Upper Triassic Olang coal deposits in northeastern Iran. *Geopersia* **2015**, *5*, 189–204.
38. Ketris, M.P.; Yudovich, Y.E. Estimations of Clarkes for carbonaceous biolithes: World averages for trace element contents in black shales and coals. *Int. J. Coal Geol.* **2009**, *78*, 135–148. [CrossRef]
39. Ward, C.R. Analysis and significance of mineral matter in coal seams. *Int. J. Coal Geol.* **2002**, *50*, 135–168. [CrossRef]
40. Ward, C.R. Analysis, origin and significance of mineral matter in coal: An updated review. *Int. J. Coal Geol.* **2016**, *165*, 1–27. [CrossRef]
41. Craigie, N.W. Geochemistry and Mineralogy. *Princ. Elem. Chemostratigr.* **2018**, *1*, 39–83.
42. Malengreau, N.; Muller, J.-P.; Calas, G. Spectroscopic Approach for Investigating the Status and Mobility of Ti in Kaolinitic Materials. *Clays Clay Miner.* **1995**, *43*, 615–621. [CrossRef]
43. Riley, K.W.; French, D.H.; Farrell, O.P.; Wood, R.A.; Huggins, F.E. Modes of occurrence of trace and minor elements in some Australian coals. *Int. J. Coal Geol.* **2012**, *94*, 214–224. [CrossRef]

44. Rimmer, S.M. Distributions and associations of selected trace elements in the lower Kittanning seam, western Pennsylvania, USA. *Int. J. Coal Geol.* **1991**, *17*, 189–212. [CrossRef]
45. Kang, J.; Zhao, L.; Wang, X.; Song, W.; Wang, P.; Wang, R.; Li, T.; Sun, J.; Jia, S.; Zhu, Q. Abundance and Geological Implication of Rare Earth Elements and Yttrium in Coals from the Suhaitu Mine, Wuda Coalfield, Northern China. *Energy Explor. Exploit.* **2014**, *32*, 873–889. [CrossRef]
46. Mihaljevič, M.; Ettler, V.; Strnad, L.; Šebek, O.; Vonásek, F.; Drahota, P.; Rohovec, J. Isotopic composition of lead in Czech coals. *Int. J. Coal Geol.* **2009**, *78*, 38–46. [CrossRef]
47. Aide, M.T.; Aide, C. Rare Earth Elements: Their Importance in Understanding Soil Genesis. *Int. Sch. Res. Not.* **2012**, *2012*, 783876. [CrossRef]
48. Mao, M.; Rukhlov, A.S.; Rowins, S.; Spence, J.; Coogan, L.A. Apatite Trace Element Compositions: A Robust New Tool for Mineral Exploration. *Econ. Geol.* **2016**, *111*, 1187–1222. [CrossRef]
49. Zhou, C.J.; Liu, G.; Dun, W.; Ting, F.; Ruwei, W.; Xiang, F. Mobility behavior and environmental implications of trace elements associated with coal gangue: A case study at the Huainan Coalfield in China. *Chemosphere* **2014**, *95*, 193–199.
50. Ward, C.R.; Corcoran, J.F.; Saxby, J.D.; Read, H.W. Occurrence of phosphorus minerals in Australian coal seams. *Int. J. Coal Geol.* **1996**, *30*, 185–210. [CrossRef]
51. Sánchez-Navas, A.; Martín-Algarra, A.; Román, M.S.; Jiménez-López, C.; Nieto, F.; Ruiz-Bustos, A. Crystal Growth of Inorganic and Biomediated Carbonates and Phosphates. In *Advanced Topics on Crystal Growth*; IntechOpen: London, UK, 2013.
52. Kojima, T.; Furusawa, T. Behaviour of elements in coal ash with sink-float separation of coal and organic affinity of the elements. *J. Fuel Soc. Jpn.* **1986**, *65*, 143–149. [CrossRef]
53. Zhao, F. Quantitative Study of Organic Affinity of Elements in Low Rank Coals. *J. China Univ. Min. Technol.* **2003**, *32*, 18–22.
54. Benson, S.A.; Pavlish, J.H.; Zygarlicke, C.J. *Trace Elements in Low-Rank Coals*; USDOE: Washington, DC, USA, 1998.
55. Spears, D.A.; Zheng, Y. Geochemistry and origin of elements in some UK coals. *Int. J. Coal Geol.* **1999**, *38*, 161–179. [CrossRef]
56. Hashan, M.; Howladar, M.F.; Jahan, L. Sulphur Concentration in Imported Coal Around the Tamabil Area, Sylhet, Bangladesh. *Int. J. Environ. Monit. Anal.* **2016**, *4*, 27. [CrossRef]
57. Huang, X.; Gordon, T.; Rom, W.N.; Finkelman, R.B. Interaction of iron and calcium minerals in coals and their roles in coal dust-induced health and environmental problems. *Rev. Mineral. Geochem.* **2006**, *64*, 153–178. [CrossRef]
58. Yılmaz, S.O. The Relationship Between Unburned Carbon Levels in Coal Combustion Ash and Volatile Matter Content in Coal. *Combust. Sci. Technol.* **2019**, *193*, 716–725. [CrossRef]
59. Sahoo, M.; Dey, S. A comparative study on the characterisation and combustion behaviour of high ash coals from two different geographical origins. *Fuel* **2021**, *286*, 119397. [CrossRef]
60. Aich, S.; Behera, D.U.; Nandi, B.K.; Bhattacharya, S. Relationship between proximate analysis parameters and combustion behaviour of high ash Indian coal. *Int. J. Coal Sci. Technol.* **2020**, *7*, 766–777. [CrossRef]
61. Dai, S.; Jiang, Y.; Ward, C.R.; Li, T.; Hower, J.C. Chemical and mineralogical compositions of coals from the Guanbanwusu Mine, Inner Mongolia, China: Insights into the effects of stratigraphic variation and coal rank on coal quality. *Int. J. Coal Geol.* **2015**, *146*, 1–15.
62. Bugaru, V.; Rujoiu, V. Geochemistry and environmental impact of trace elements in coal and coal combustion residues from the Oltenia region, Romania. *Environ. Earth Sci.* **2014**, *71*, 3833–3844.
63. Zhou, D.X. Investigation of chronic arsenic poisoning caused by high arsenic coal pollution. *Zhonghua Yu Fang Yi Xue Za Zhi* **1993**, *3*, 147–150. (In Chinese) [PubMed]
64. Flues, M.; Mierzwa-Hersztek, M.; Chauvel, J.J.; Maurin, M. Concentrations and spatial distribution of heavy metals in soils near three coal-fired power plants in Poland. *Environ. Earth Sci.* **2013**, *68*, 2099–2108.

applied sciences

MDPI

Article

Risk Assessment of Soil Heavy Metals in the Jiahe River Basin of Yantai City, China

Xizhuo Chen [1,2,3], Pengfei Zhao [1,3,*], Jiaxin Huang [2], Jun Liu [1,3], Xiaoli Cao [1,3], Jing Che [1,3], Hui Liao [1,3], Xiaolong Zhu [1,3] and Qingjie Gong [2,*]

1 Yantai Center of Coastal Zone Geological Survey, China Geological Survey, Yantai 264000, China; chenxizhuo@mail.cgs.gov.cn (X.C.); ljun_qds@mail.cgs.gov.cn (J.L.); caoxiaoli@mail.cgs.gov.cn (X.C.); chejing@mail.cgs.gov.cn (J.C.); liaohui@mail.cgs.gov.cn (H.L.); zhuxiaolong@mail.cgs.gov.cn (X.Z.)

2 School of Earth Sciences and Resources, China University of Geosciences, Beijing 100083, China; 3001220024@email.cugb.edu.cn

3 Ministry of Natural Resources Observation and Research Station of Land-Sea Interaction Field in the Yellow River Estuary, Yantai 264000, China

* Correspondence: zhaopengfei@mail.cgs.gov.cn (P.Z.); qjiegong@cugb.edu.cn (Q.G.)

Abstract: The issues related to soil environmental contamination caused by heavy metals have garnered increasing attention. In particular, the soil pollution risk in the eastern coastal regions of China has attracted widespread concern. This study surveyed heavy metals in the soils near the Jiahe River Basin of Yantai City in Shandong Province, China. A total of 213 soils were sampled and analyzed for 12 items: Cr, Hg, As, Pb, Cd, Cu, Ni, Zn, Co, V, Mn, and pH. The 11 heavy metals were evaluated using the national standard GB15618-2018, with three risk levels of background, screening, and intervention, and using pollution indices, including the contamination factor (C_f), ecological risk factor (Er), enrichment factor (EF), and index of geo-accumulation (I_{geo}), with different respective risk levels. The results indicate a strong consistency between the evaluations both for the index I_{geo} and for GB15618-2018 on five metals (i.e., Cr, Hg, As, Pb, and Cd). Therefore, the index I_{geo} may serve as a supplementary indicator for assessing the pollution risks of heavy metals in agricultural soils regarding samples of Cu, Ni, and Zn that exceed the screening values in GB15618-2018, as well as for Co, V, and Mn, which have not yet been established in GB15618-2018. According to the three-level classification of risk in GB15618-2018, the seven commonly used levels of the index I_{geo} are also incorporated into the three levels of background, screening, and intervention. The overall pollution risk of 11 heavy metals in the soils of the Jiahe River Basin of Yantai City belongs to the background level. Specifically, Hg and Pb in the total area are classified at the background level. Manganese, V, Co, Zn, Ni, and Cr are recognized at the screening level sporadically, while Cu, As, and Cd are found at the screening level in small areas. No areas within the region are classified at the intervention level.

Keywords: heavy metals; soil; geo-accumulation index; pollution indices; GB15618-2018

1. Introduction

With the rapid development of industrialization and urbanization, the issue of soil heavy metals has become increasingly prominent, prompting intensified research and advancements in this field [1–3]. Heavy metal pollution not only degrades the quality of soil, crops, the atmosphere, and water bodies but also accumulates in the food chain, posing a threat to human health [4–6]. Since the 1970s, global environmental legislation and policies have facilitated systematic investigations into soil heavy metal pollution. The heavy metals of common concern are Cr, Hg, As, Pb, Cd, Cu, Ni, Zn, Co, V, Mn, etc. [7–9].

Methods for assessing heavy metal pollution are generally categorized into two types. The first type involves evaluations based on established standards, such as the national standard GB15618-2018 for agricultural soils in China [10–12]. The second type employs pollution indices, including the contamination factor (C_f), ecological risk factor (Er), enrichment factor (EF), and index of geo-accumulation (I_{geo}) [13–15].

In recent years, a series of soil environmental quality assessments have been conducted by the China Geological Survey in Shandong Province, China. These efforts aim to enhance the role of geological work in the integrated management of natural resources. This study was based on agricultural soils sampled from the Jiahe River Basin of Yantai City in Shandong Province, China, during a comprehensive geological survey. In this paper, risk assessments of heavy metals were conducted using both national standards and pollution indices. Then, the results of these two evaluation methods were compared to identify an appropriate pollution assessment method for heavy metals that lack national standards. Finally, a comprehensive risk evaluation on 11 heavy metals was performed.

2. Materials and Methods

2.1. Geological Settings

The study area (37°20′ to 37°28′ N, 121°05′ to 121°18′ E) is in the Jiahe River Basin of Yantai city in Shandong Province, China (Figure 1), which is situated in a warm temperate continental monsoon climate zone with the characteristics of moderate precipitation, humid air, and a mild climate, compared to inland areas at the same latitude. The mean annual temperature is about 11.8 °C, and the annual rainfall is about 651.9 mm. The topography of the study area is low mountains and hills that are high in the southwest and low in the northeast [16]. The main types of soil are brown soil (with the largest distribution area), fluvo-aquic soil, cinnamon soil, and so on. The dominant land use type is dry land, followed by forest land, and a small amount is grassland [17]. In the study area, the main developed strata include the Quaternary, Neogene, Sinian, Paleoproterozoic, and Archean (Figure 1b) [18–20]. Faults are mainly distributed in the directions of NW-SE, NE-SW, and E-W [21–23]. The developed intrusive rocks mainly including Cretaceous quartz diorite porphyrite and granite porphyry, Mesoproterozoic medium-fine grained monzogranite, and Paleoproterozoic amphibolite (metagabbro) [24,25].

2.2. Materials

The sampling scale for soil samples was 1:50,000, which means a 500 m × 500 m sampling grid. In each grid cell, five soil subsamples weighing ca. 300 g were collected at a depth of 15–20 cm, and roots, stones, straw, and other debris were removed. Then, a composite sample was formed in each grid cell, with the five subsamples equally mixed.

The composite samples were naturally air-dried and crushed. Subsequently, they were sieved through a nylon sieve of 10 mesh. Before the chemical analysis, the samples were ground to a particle size of ≤0.074 mm with a pollution-free planetary ball mill. A total of 213 composite samples were collected in this basin.

2.3. Analytical Methods

All chemical analyses were performed at the analysis and testing laboratory of the Yantai Center of Coastal Zone Geological Survey, China Geological Survey. In total, 12 items were analyzed: Cr, Hg, As, Pb, Cd, Cu, Ni, Zn, Co, V, Mn, and pH. The preparation and analysis followed the Specification of Land Quality Geochemical Assessment (DZ/T 0295-2016) [26]. As and Hg were determined using the AFS (atomic fluorescence spectrometry) method, with detection limits of 0.2 μg/g and 5 ng/g, respectively. The pH value was assessed using the ISE (ion-selective electrode) method, with a detection limit of 0.1. Cd, Cr, Pb, Ni, Cu, Zn, Co, V, and Mn were analyzed using the ICP-MS (inductively coupled plasma mass spectrometry) method, with detection limits of 21 ng/g, 2 μg/g, 0.5 μg/g, 0.6 μg/g, 0.6 μg/g, 1.1 μg/g, 0.02 μg/g, 3 μg/g, and 0.2 μg/g, respectively. During the sample analysis, first-grade standard samples (GSS-3 and GSS-13) were utilized to control the analytical quality. The quality of the analytical data, including the analytical methods, detection limits, accuracy, and precision, were all within the requirements outlined in the Specification of Land Quality Geochemical Assessment (DZ/T 0295-2016) [26]. The detection rate was 100%, with accuracy greater than 99.85% and precision exceeding 99.83%.

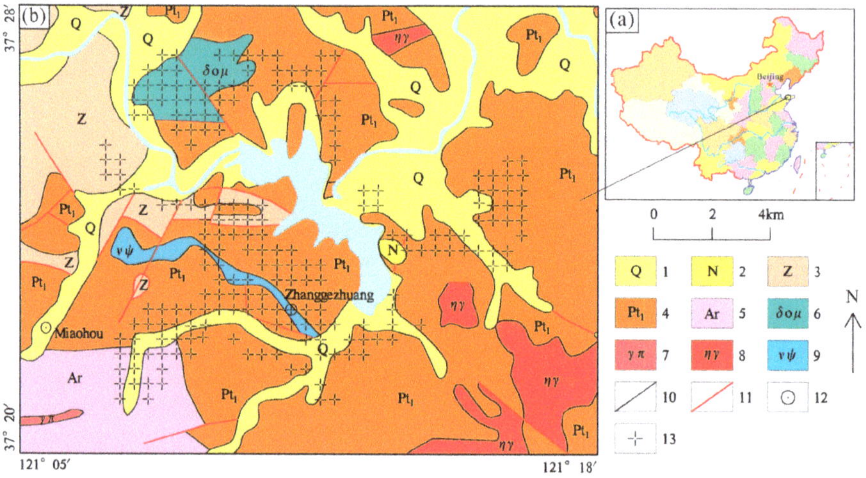

Figure 1. Location of the study area in China (**a**) and its geological sketch map (modified from the J-51–20 geological map from China Geological Survey) with the locations of samples (**b**). Notes in (**b**): 1—Quaternary clay, sand, gravel; 2—Neogene gabbro; 3—Sinian slate, marble, marl, quartzite, phyllite, metasandstone; 4—Lower Proterozoic schist, leptite, granulite, marble, tremolite, diopsidite; 5—Archean gneiss, amphibolite, leptite, granulite, schist; 6—Mesozoic quartz diorite porphyry; 7—Mesozoic granite porphyry; 8—Mesoproterozoic fine-grained monzogranite; 9—Paleoproterozoic plagioamphibolite (metagabbro); 10—Lithological boundary; 11—Fault; 12—Main residential place; 13—Location of soil sample.

3. Results

3.1. Concentrations of Heavy Metals

Histograms with box plots of the concentrations of 11 heavy metals and the pH in 213 soil samples are illustrated in Figure 2. Basically, the concentrations of the 11 heavy metals follow a log-normal distribution. However, significant outliers, particularly larger ones, are evident in the box plots. Therefore, geometric means without outliers were used herein to represent the average values of the heavy metal concentrations; these are listed in Table 1, along with other statistical parameters.

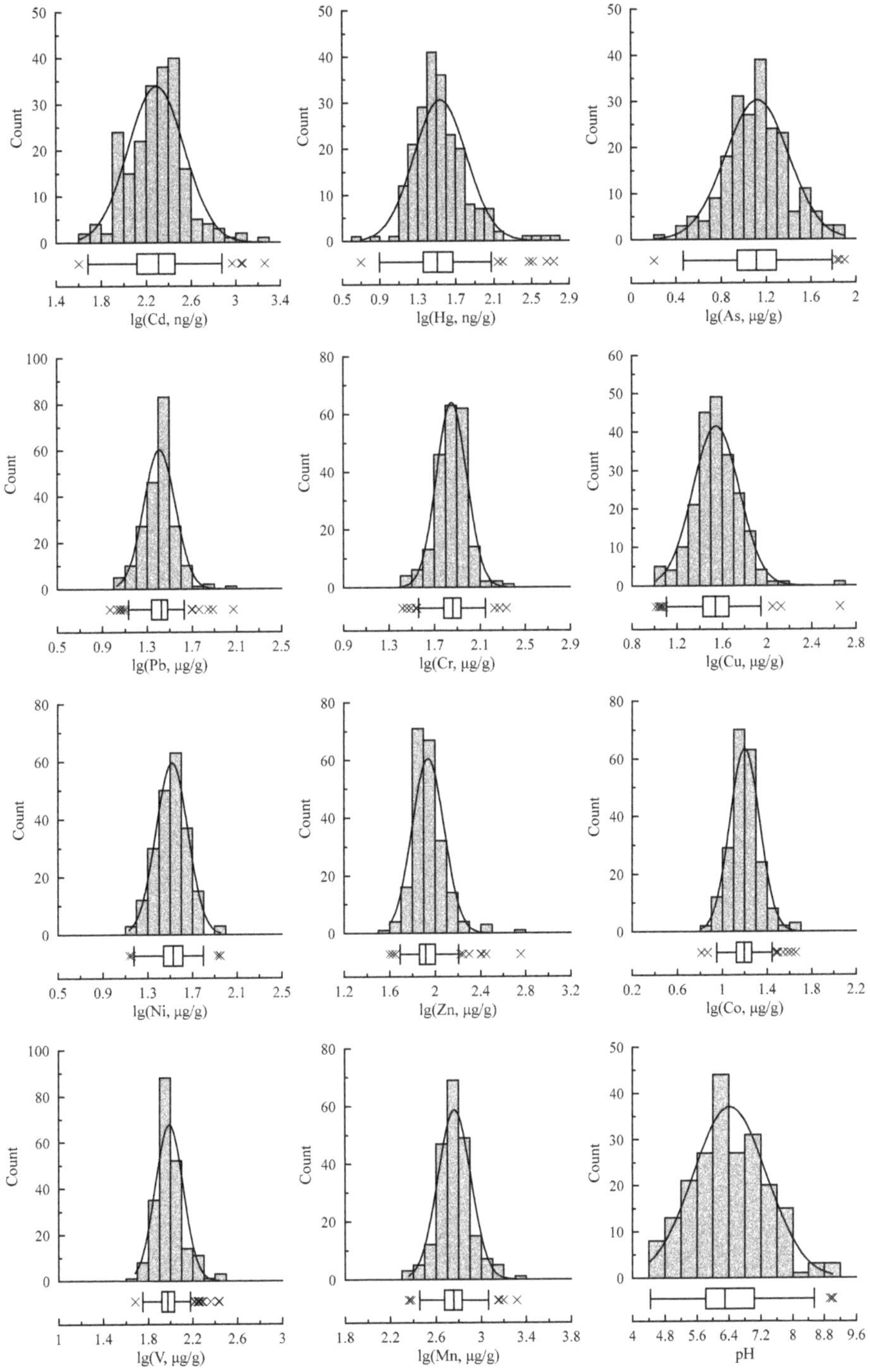

Figure 2. Histograms with box plots of concentrations of heavy metals and pH.

Table 1. Statistical parameters of heavy metal concentrations and pH.

No.	Item	Min.	Q_1	Q_2	Q_3	Max.	Avg.	Avg._cn
1	Cd	40	130	204	285	1820	202	150
2	Hg	5	22.45	32	46.85	542	33	50
3	As	1.6	8.67	12.9	19.75	79.1	12.9	9.1
4	Pb	9.26	21.7	26.59	30.15	117.7	26	25
5	Cr	26.2	60.57	72.72	85.74	214.5	73	63
6	Cu	10.3	26.81	34.4	46	448.4	37	23
7	Ni	13.7	27.27	33.45	41.05	89.17	34	26
8	Zn	39.6	72.44	82.21	102.1	569.5	85	67
9	Co	6.47	13.34	15.71	18.33	45.1	15.8	11.7
10	V	48	82.87	93.81	108.5	274.8	94	79
11	Mn	230	470.5	571	675.4	2071	575	552
12	pH	4.44	5.8	6.3	7.04	9.01	6.42	—

Note: The number of samples was 213. The units of concentrations are μg/g, except for Cd and Hg, which are in ng/g. Q_1, Q_2, and Q_3 represent the first, second, and third quartiles of the concentrations and pH. Avg. is the geometric mean calculated without outliers, and Avg._cn refers to the element abundance of surface soil in China [27].

The pH values of the 213 soil samples from the Jiahe River Basin range from 4.44 to 9.01, with an average of 6.42. Regarding the concentrations of heavy metals in the 213 samples compared with those in greater China, the level of Hg in the Jiahe River Basin is lower, that of Pb is comparable, and the others are slightly greater (Table 1).

3.2. Risk Assessment Using GB15618-2018

According to GB15618-2018, agricultural land is categorized into three types: paddy fields, orchards, and other agricultural land. In the study area, land use includes two types, which are orchards and other agricultural land. After the land use types are determined, GB15618-2018 provides criteria for risk assessments of heavy metals based on different pH levels. That is to say, three risk levels are classified for land use types and pH levels in GB15618-2018: background level, screening level, and intervention level [28]. According to Huang et al. [28], the PI (or the single pollution index) for heavy metals in soils of agricultural land, based on GB15618-2018, is presented as follows:

$$\mathrm{PI} = \begin{cases} 0 & \mathrm{C} < \mathrm{C_{RS}}, \quad \text{the background level} \\ 1 & \mathrm{C_{RS}} \le \mathrm{C} < \mathrm{C_{RI}}, \text{the screening level} \\ 10 & \mathrm{C_{RI}} \le \mathrm{C}, \text{the intervention level} \end{cases} \tag{1}$$

where C represents the concentration of a heavy metal in soil samples, and C_{RS} and C_{RI} are its risk screening value and risk intervention value specified in GB15618-2018 for different land use types and pH levels (Table A1).

Based on Equation (1), the PI values of five heavy metals (Cd, Hg, As, Pb, and Cr) for 213 soil samples from the study area were calculated, and their counts on each level are listed in Table 2. It is worth noting that no intervention-level values for Cu, Ni, and Zn are presented in GB15618-2018, except only screening-level values. Furthermore, Co, V, and Mn are not mentioned in GB15618-2018.

Table 2. Statistical counts of samples on each risk level in the 213 soil samples.

Element	Lv.	Cd	Hg	As	Pb	Cr	Cu	Ni	Zn	Co	V	Mn
Evaluation method	Background reference value (C_{ri})	150	50	9.1	25	63	23	26	67	11.7	79	552
PI	0 (background level)	176	213	194	213	211	195	211	211	-	-	-
	1 (screening level)	37	0	19	0	2	-	-	-	-	-	-
	10 (intervention level)	0	0	0	0	0	-	-	-	-	-	-
C_f	$C_f^i < 1$ (low risk)	66	165	58	87	69	28	47	38	33	44	92
	$1 \leq C_f^i < 3$ (moderate risk)	135	43	131	124	143	169	163	171	176	166	120
	$3 \leq C_f^i < 6$ (considerable risk)	8	2	19	2	1	15	3	3	4	3	1
	$C_f^i \geq 6$ (very high risk)	4	3	5	0	0	1	0	1	0	0	0
Er	toxic-response factor (Tr^i)	30	40	10	5	2	5	-	1	-	-	-
	$Er^i < 40$ (low risk)	103	165	197	213	213	212	-	213	-	-	-
	$40 \leq Er^i < 80$ (moderate risk)	94	35	15	0	0	0	-	0	-	-	-
	$80 \leq Er^i < 160$ (considerable risk)	12	9	1	0	0	1	-	0	-	-	-
	$160 \leq Er^i < 320$ (high risk)	3	2	0	0	0	0	-	0	-	-	-
	$Er^i \geq 320$ (very high risk)	1	2	0	0	0	0	-	0	-	-	-
EF	EF < 2 (depletion to normal)	171	203	162	204	207	157	199	195	194	205	-
	2 ≤ EF < 5 (moderate enrichment)	37	6	41	8	6	54	14	17	19	8	-
	5 ≤ EF < 20 (significant enrichment)	5	4	10	1	0	1	0	1	0	0	-
	20 ≤ EF < 40 (very high enrichment)	0	0	0	0	0	1	0	0	0	0	-
	EF > 40 (extremely high enrichment)	0	0	0	0	0	0	0	0	0	0	-
I_{geo}	$I_{geo} \leq 0$ (unpolluted)	124	191	110	195	187	109	152	160	146	174	192
	$0 < I_{geo} \leq 1$ (from unpolluted to moderately polluted)	77	17	79	16	25	89	58	49	63	36	20
	$1 < I_{geo} \leq 2$ (moderately polluted)	8	2	19	2	1	14	3	3	4	3	1
	$2 < I_{geo} \leq 3$ (from moderately to strongly polluted)	3	3	5	0	0	0	0	1	0	0	0
	$3 < I_{geo} \leq 4$ (strongly polluted)	1	0	0	0	0	1	0	0	0	0	0
	$4 < I_{geo} \leq 5$ (from strongly to extremely polluted)	0	0	0	0	0	0	0	0	0	0	0
	$I_{geo} > 5$ (extremely polluted)	0	0	0	0	0	0	0	0	0	0	0

Note: The background reference values (C_{ri}) for pollution indices were derived from the element abundance of soil in China without outliers [27]. The units of elements are μg/g, except for Cd and Hg, which are in ng/g. -, not applicable.

For the five heavy metals—Cd, Hg, As, Pb, and Cr—with screening and intervention values in GB15618-2018, no samples with PI values of 10 were found in this area. The PI values for Hg and Pb were 0 in all of the 213 samples. For Cr, As, and Cd, there were 2, 19, and 37 samples, respectively, with a PI value of 1. These results indicate that Hg and Pb in this area are at the background level; Cr, As, and Cd basically belong to the background level except for a few samples. Regarding the three heavy metals—Cu, Ni, and Zn—with only screening-level values in GB15618-2018, it was found that 211 samples of Ni and Zn, as well as 195 samples of Cu, had PI values of 0. That is to say, only a few samples of Cu, Ni, and Zn among the 213 samples were not at the background level. Because the three heavy metals Co, V, and Mn are not mentioned in GB15618-2018, their risk assessment is not discussed here using the criteria in GB15618-2018.

3.3. Assessment Using Pollution Indices

Four single pollution indices [29], C_f, Er, EF, and I_{geo}, were used to assess the pollution risk of heavy metals in the study area.

3.3.1. Contamination Factor (C_f)

The contamination factor (C_f^i) was originally proposed by Håkanson [30] and subsequently defined by Cabrera et al. [31] as follows:

$$C_f^i = \frac{C_i}{C_{ri}} \tag{2}$$

where C_i is the concentration of heavy metal i, while C_{ri} denotes the reference value for heavy metal i, which was taken herein as the element abundance of soils in China (Table 1). $C_f^i < 1$, $1 \leq C_f^i < 3$, $3 \leq C_f^i < 6$, and $C_f^i \geq 6$ correspond to low, moderate, considerable, and very high contamination risk, respectively (Table 2). The C_f values of heavy metals in the 213 soil samples were calculated, and their counts on each risk level are listed in Table 2.

For Hg, the majority of the soils were at a low risk level, with a small number of samples at a moderate level and only a few at considerable or very high risk levels. Regarding Cd, As, Pb, Cr, Cu, Ni, Zn, Co, V, and Mn, most samples were at a moderate risk level, with a small number at a low level and only a few at a considerable risk level. Additionally, few samples of Cd, As, Cu, and Zn were identified at the very high risk level.

3.3.2. Ecological Risk Factor (Er)

The ecological risk factor (Er^i) is an index used to quantitatively describe the potential ecological risk associated with a specific pollutant [30]; it is defined as follows:

$$Er^i = Tr^i \cdot C_f^i \tag{3}$$

where C_f^i represents the contamination factor in Equation (2), while Tr^i is the toxic-response factor for a specific substance. The Tr^i values for the heavy metals Cd, Hg, As, Pb, Cr, Cu, and Zn are 30, 40, 10, 5, 2, 5, and 1, respectively. $Er^i < 40$, $40 \leq Er^i < 80$, $80 \leq Er^i < 160$, $160 \leq Er^i < 320$, and $Er^i \geq 320$ correspond to low, moderate, considerable, high, and very high ecological risk levels, respectively (Table 2). The Er values of seven heavy metals (i.e., Cd, Hg, As, Pb, Cr, Cu, and Zn) in the 213 soil samples were calculated, and their counts on each risk level are listed in Table 2.

All samples were at the low potential ecological risk level for Pb, Cr, and Zn. For Cu, except for one sample at the considerable risk level, all samples were at the low risk level. For As, most samples were at the low potential ecological risk level, but a few were moderate and a few were at the considerable risk level. Regarding Cd and Hg, most samples were at the low risk level, with some at the moderate or considerable risk level and few at the high or very high levels of potential ecological risk.

The risk levels of Ni, Co, V, and Mn are not discussed here due to the absence of toxic-response factors.

3.3.3. Enrichment Factor (EF)

The enrichment factor (EF) [32–34] is presented here to evaluate the degree of enrichment of a heavy metal relative to a selected reference element through the ratio of their concentrations in the soil. The calculation equation is as follows:

$$EF = (\frac{C_i}{C_{ie}})_S / (\frac{C_i}{C_{ie}})_{RS} \tag{4}$$

where C_i represents the concentration of element i, and C_{ie} denotes the concentration of an immobile element. S indicates the analyzed sample, and RS refers to the selected reference sample. Here, the reference sample was the virtual Chinese soil represented by the elemental abundances (Table 1). The immobile elements that are commonly used are Al [35,36], Zr [37], Ti [38], Fe [39], and Mn [40]. Here, Mn was used as the immobile element. EF < 2, 2 $\leq$ EF < 5, 5 $\leq$ EF < 20, 20 $\leq$ EF < 40, and EF > 40 correspond to the levels of depletion to normal, moderate enrichment, significant enrichment, very high enrichment, and extremely high enrichment, respectively (Table 2). The EF values of heavy metals except for Mn in the 213 soil samples were calculated, and their counts on each level are listed in Table 2.

For Cr, Ni, Co, and V, the majority of the samples were at the level of depletion to normal, with a few at moderate enrichment level. Regarding Pb, Cu, Zn, Cd, Hg, and As, most samples were at the level of depletion to normal, with some at the moderate enrichment level and few at the significant enrichment level.

The enrichment level of Mn is not discussed here because it was used as the immobile element.

3.3.4. Index of Geo-Accumulation (I_{geo})

The index of geo-accumulation (I_{geo}) [41,42] uses the ratio of the total concentration of a heavy metal to its background value to determine the pollution level. The calculation formula is as follows:

$$I_{geo} = \log_2[C_i/(1.5C_{ri})] \quad (5)$$

where C_i represents the concentration of element i, while C_{ri} denotes the geochemical background value or reference value for element i. Here, the reference value is the elemental abundance in soil in China (Table 1). I_{geo} is categorized as follows: $I_{geo} \leq 0$ indicates class 0 (unpolluted), $0 < I_{geo} \leq 1$ corresponds to class 1 (from unpolluted to moderately polluted), $1 < I_{geo} \leq 2$ represents class 2 (moderately polluted), $2 < I_{geo} \leq 3$ signifies class 3 (from moderately to strongly polluted), $3 < I_{geo} \leq 4$ corresponds to class 4 (strongly polluted), $4 < I_{geo} \leq 5$ indicates class 5 (from strongly to extremely polluted), and $I_{geo} > 5$ represents class 6 (extremely polluted). The I_{geo} values of heavy metals in the 213 soil samples were calculated, and their counts in each class are listed in Table 2.

For Pb, Cr, Ni, Co, V, and Mn, the majority of the samples were classified as class 0, unpolluted; some were classified as class 1, indicating a range from unpolluted to moderately polluted; and a few were recognized as class 2, moderately polluted. Regarding Hg, As, and Zn, most samples were also classified as class 0, some were classified as class 1, and a few were classified as class 2 or class 3. For Cu and Cd, the majority were classified as class 0, some were classified as class 1, and a few were class 2 to class 4.

Despite the differing qualitative classifications of the four single pollution indices C_f, Er, EF, and I_{geo}, which are categorized into four, five, five, and seven levels, respectively, the assessment results of these indices indicate that the pollution risk of the heavy metals in this area is low to moderate, except for a few samples at a high (or significant, or strong) level.

4. Discussion

4.1. Comparison Between GB15618-2018 and Pollution Indices

In GB15618-2018, the screening and intervention risk values are given for five elements (i.e., Cr, Hg, As, Pb, and Cd), and only screening values are listed for three elements (i.e., Cu, Ni, and Zn). However, other heavy metals such as Co, V, and Mn are not mentioned in GB15618-2018. Therefore, a comparison was made between GB15618-2018 and the four pollution indices C_f, Er, EF, and I_{geo}, mainly for the five elements Cr, Hg, As, Pb, and Cd. This comparison aims to identify a suitable assessment method for other heavy metals in this area.

In comparisons between GB15618-2018 and C_f, Pb and Hg were at the background level (the lowest level) in GB15618-2018, while Pb was at a considerable risk level and Hg was at a very high risk level (the highest level) in C_f. Therefore, the assessment results are inconsistent for Pb and Hg. The method of C_f is not suitable in this area.

In comparisons between GB15618-2018 and Er, Hg was at the background level (the lowest level) in GB15618-2018 but at a very high risk level (the highest level) in Er. Cd was at the screening level (the medium level) in GB15618-2018 but at a very high risk level (the highest level) in Er. Therefore, the assessment results are inconsistent for Hg and Cd. Furthermore, Er cannot be used for Ni, Co, V, and Mn due to the lack of toxic-response factor values. These results indicate that the Er method is also unsuitable in this area.

In comparisons between the three levels in GB15618-2018 and the five levels of EF, five metals (i.e., Cr, Hg, As, Pb, and Cd) were not at the highest risk level. Therefore, no clear inconsistency was found in the results of these two methods. However, Mn could not be assessed because it was used as the immobile element. The assessment results may be different if another element is used as the immobile element. These results indicate that the EF method does not meet our aims in this area.

In comparisons between the three levels of GB15618-2018 and the seven levels of I_{geo}, the five metals Cr, Hg, As, Pb, and Cd were not at the highest risk level. Therefore, no clear inconsistency was found in the results of GB15618-2018 and I_{geo}. When the first two levels in I_{geo} were viewed as one level ($I_{geo} \leq 1$) corresponding to the background level in GB15618-2018, Pb was at the background level in GB15618-2018 and also at the background level in I_{geo}, except for two samples. Hg was at the background level in GB15618-2018 and at the background level in I_{geo}, except for five samples. Cr was at the background level except for two samples in GB15618-2018 and one in I_{geo}. Cd and As were mainly at the background level, with some outliers in the results for both GB15618-2018 and I_{geo}. When the last two levels in I_{geo} were viewed as one level ($4 < I_{geo}$) corresponding to the intervention level in GB15618-2018, the five metals (i.e., Cr, Hg, As, Pb, and Cd) were not at the intervention level. When the middle three levels in I_{geo} were viewed as one level ($1 < I_{geo} \leq 4$) corresponding to the screening level in GB15618-2018, the assessment results were basically consistent between GB15618-2018 and I_{geo} for the five metals.

Cu, Ni, and Zn were mainly at the background level in GB15618-2018, except for 18, 2, and 2 samples, respectively. With the three merged levels of I_{geo}, Cu, Ni, and Zn were also mainly at the background level, except for 14, 3, and 3 samples, respectively.

In summary, the I_{geo} method can be selected as a suitable method in this area to evaluate the risk of heavy metals, whether or not they are mentioned in GB15618-2018. Here, we suggest an idea that the risk of heavy metals can be evaluated using the criteria in GB15618-2018 with I_{geo} as a supplement for metals without criteria. If the seven levels in I_{geo} are merged into three levels corresponding to those in GB15618-2018, the merging method and description is as follows:

$$I_{geo} = \begin{cases} 0 & I_{geo} \leq 1, \text{the background level} \\ 1 & 1 < I_{geo} \leq 4, \text{the screening level} \\ 10 & I_{geo} > 4, \text{the intervention level} \end{cases} \quad (6)$$

which can be viewed as a supplement to GB15618-2018 to evaluate the pollution risk of heavy metals.

4.2. Assessment Results in the Jiahe River Basin

The pollution risk of heavy metal in the Jiahe River Basin of Yantai City was assessed using the criteria of GB15618-2018 with I_{geo} as a supplement for metals without criteria. Specifically, the following assessment approaches were employed: (1) Cd, Hg, As, Pb, and Cr were assessed using GB15618-2018, which is presented in Equation (1); (2) samples with concentrations of Cu, Ni, and Zn lower than the risk screening values were also evaluated using GB15618-2018, which is detailed in Equation (1); (3) samples with contents of Cu, Ni, and Zn exceeding the risk screening values were assessed using the newly established I_{geo}, which is described in Equation (6); and (4) Co, V, and Mn were assessed using Equation (6).

The assessment results of the pollution risk of heavy metals in agricultural soils in this area are presented in Table 3. Hg and Pb were at the background level in all samples. Cr, Ni, Zn, Co, V, and Mn were at the background level, except for a few samples at the screening level. Cu, As, and Cd were basically at the background level, except for some samples at the screening level. In other words, the overall pollution risk of heavy metals in the agricultural soils in the Jiahe River Basin was at the background level. Specifically, Hg and Pb in the whole area were at the background level. There were sporadic areas at the screening level for Cr, Ni, Zn, Co, V, and Mn, while some regions were at the screening level for Cu, As, and Cd. No heavy metals were at the intervention level in this area.

Table 3. Assessment results of 11 heavy metals in agricultural soils of the Jiahe River Basin of Yantai City.

Evaluation Method	Risk Level	Cd	Hg	As	Pb	Cr	Cu	Ni	Zn	Co	V	Mn
PI	0 (background level)	176	213	194	213	211	195	211	211	-	-	-
	1 (screening level)	37	0	19	0	2	-	-	-	-	-	-
	10 (intervention level)	0	0	0	0	0	-	-	-	-	-	-
I_{geo}	0 (background level)	-	-	-	-	-	-	-	-	209	210	212
	1 (screening level)	-	-	-	-	-	18	2	2	4	3	1
	10 (intervention level)	-	-	-	-	-	0	0	0	0	0	0

The aforementioned assessment results are only statistical results and do not indicate the spatial distribution. To spatially illustrate the pollution risk of heavy metals in the study area, the assessment results from Equation (6) were mapped as shown in Figure 3. The results in Figure 3 are consistent with the statistical results that only Cu, As, and Cd were clearly at the screening level. Specifically, the regions with Cu at the screening level were mainly in the north part of the area; the screening regions of As were mainly distributed in the north and south parts; and the screening regions of Cd were found in many parts of the study area.

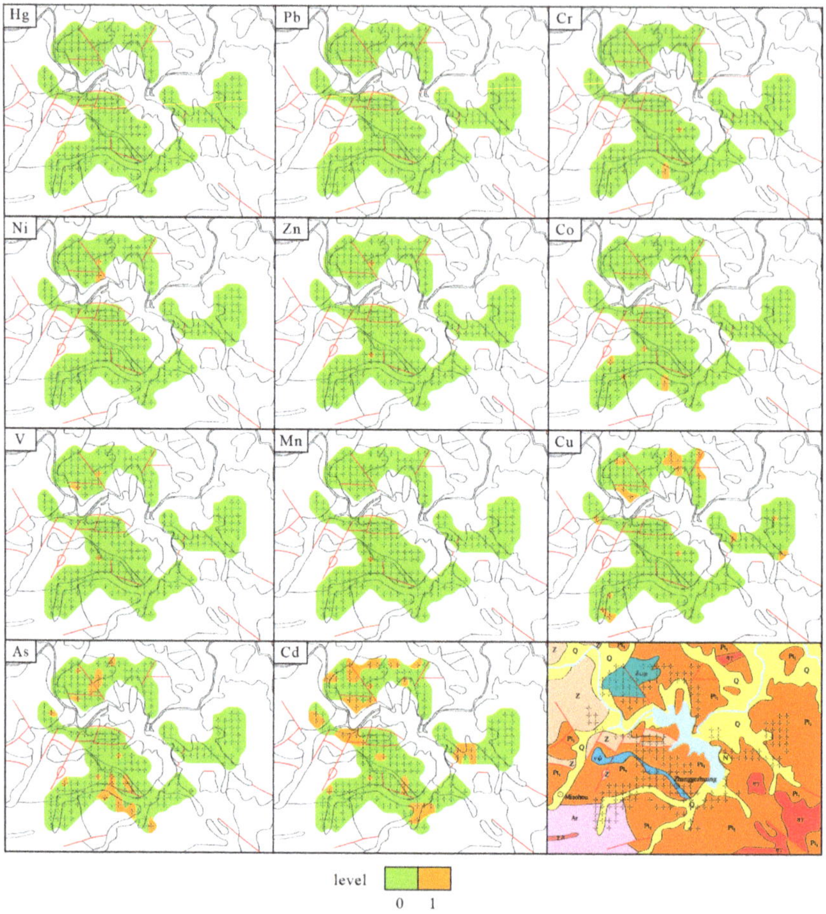

Figure 3. Pollution risk of 11 heavy metals in agricultural soils in the Jiahe River Basin of Yantai City.

5. Conclusions

(1) A comparison of the risk assessment results for five heavy metals, Cr, Hg, As, Pb, and Cd, in soils of the Jiahe River Basin using GB15618-2018 and four pollution indices indicated that the assessment results from GB15618-2018 and the index of geo-accumulation (or I_{geo}) were well consistent relative to those from the contamination factor (or C_f), ecological risk factor (or Er), and enrichment factor (or EF).

(2) In accordance with the three-level classification scheme established in GB15618-2018 for the pollution risks of heavy metals, the seven-level classification scheme of I_{geo} was also merged into three levels. Specifically, $I_{geo} \leq 1$ corresponded to the background level, $1 < I_{geo} \leq 4$ corresponded to the screening level, and $I_{geo} > 4$ aligned with the intervention level. This can be viewed as a supplement to GB15618-2018 to evaluate the pollution risk of heavy metals lacking criteria.

(3) The pollution risk of heavy metals in the Jiahe River Basin of Yantai City was assessed using the method in GB15618-2018 with the newly established I_{geo} as a supplement, and the results indicated that the pollution risks of eight heavy metals (i.e., Hg, Pb, Cr, Ni, Zn, Co, V, and Mn) in soils of agricultural land were at the background level, while Cu, As, and Cd were at the screening level.

Author Contributions: X.C. (Xizhuo Chen): conceptualization, data curation, writing—original draft. P.Z.: data curation, writing—review and editing. J.H.: conceptualization, data curation, formal analysis. J.L.: investigation, resources. X.C. (Xiaoli Cao): project administration, funding acquisition. J.C.: data curation. H.L.: data curation. X.Z.: formal analysis. Q.G.: conceptualization, methodology, writing—review and editing. All authors have read and agreed to the published version of the manuscript.

Funding: This work was supported by geological survey projects of the China Geological Survey (DD20242837, DD20220990).

Institutional Review Board Statement: Not applicable.

Informed Consent Statement: Not applicable.

Data Availability Statement: The raw data supporting the conclusions of this article will be made available by the authors on request.

Acknowledgments: We greatly appreciate the comments from the anonymous reviewers and editors to improve the quality of this manuscript.

Conflicts of Interest: The authors declare that they have no known competing financial interests or personal relationships that could have appeared to influence the work reported in this paper.

Appendix A

Table A1. Risk control standards of heavy metals for soil contamination of agricultural land, extracted from GB15618-2018 of China.

Element	Risk Value	Land Use Type	pH $\leq$ 5.5	5.5 < pH $\leq$ 6.5	6.5 < pH $\leq$ 7.5	pH > 7.5
Cd	RS	Others	300	300	300	600
	RI	All	1500	2000	3000	4000
Hg	RS	Others	1300	1800	2400	3400
	RI	All	2000	2500	4000	6000
As	RS	Others	40	40	30	25
	RI	All	200	150	120	100
Pb	RS	Others	70	90	120	170
	RI	All	400	500	700	1000
Cr	RS	Others	150	150	200	250
	RI	All	800	850	1000	1300
Cu	RS	Orchard	150	150	200	200
		Others	50	50	100	100
	RI	-	-	-	-	-
Ni	RS	All	60	70	100	190
	RI	-	-	-	-	-
Zn	RS	All	200	200	250	300
	RI	-	-	-	-	-

Notes: The units for As, Pb, Cr, Cu, Ni, and Zn are µg/g; those for Cd and Hg are ng/g. RS and RI are the Risk Screening value and the Risk Intervention value, respectively.

References

1. Chai, L.; Wang, Y.; Wang, X.; Ma, L.; Cheng, Z.; Su, L. Pollution characteristics, spatial distributions, and source apportionment of heavy metals in cultivated soil in Lanzhou, China. *Ecol. Indic.* **2021**, *125*, 107507. [CrossRef]
2. Adnan, M.; Xiao, B.; Ali, M.U.; Xiao, P.; Zhao, P.; Wang, H.; Bibi, S. Heavy metals pollution from smelting activities: A threat to soil and groundwater. *Ecotox. Environ. Safe.* **2024**, *274*, 116189. [CrossRef]
3. Zhang, Z.; Huang, S.; Chen, H.; Wang, J. Deciphering the pollution risks, sources and their links of heavy metals in soils. *Sci. Total Environ.* **2024**, *950*, 175331. [CrossRef]

4. Zhang, J.; Liu, Z.; Tian, B.; Li, J.; Luo, J.; Wang, X.; Ai, S.; Wang, X. Assessment of soil heavy metal pollution in provinces of China based on different soil types: From normalization to soil quality criteria and ecological risk assessment. *J. Hazard. Mater.* **2023**, *441*, 129891. [CrossRef] [PubMed]
5. Zhao, X.; Liu, X.; Xing, Y.; Wang, L.; Wang, Y. Evaluation of water quality using a Takagi-Sugeno fuzzy neural network and determination of heavy metal pollution index in a typical site upstream of the Yellow River. *Environ. Res.* **2022**, *211*, 113058. [CrossRef] [PubMed]
6. Angon, P.B.; Islam, M.S.; KC, S.; Das, A.; Anjum, N.; Poudel, A.; Suchi, S.A. Sources, effects and present perspectives of heavy metals contamination: Soil, plants and human food chain. *Heliyon* **2024**, *10*, e28357. [CrossRef]
7. Lin, H.; Wang, Z.; Liu, C.; Dong, Y. Technologies for removing heavy metal from contaminated soils on farmland: A review. *Chemosphere* **2022**, *305*, 135457. [CrossRef]
8. Tomczyk, P.; Wdowczyk, A.; Wiatkowska, B.; Szymańska-Pulikowska, A. Assessment of heavy metal contamination of agricultural soils in Poland using contamination indicators. *Ecol. Indic.* **2023**, *156*, 111161. [CrossRef]
9. Proshad, R.; Asharaful Abedin Asha, S.M.; Abedin, M.A.; Chen, G.; Li, Z.; Zhang, S.T.; Tan, R.; Lu, Y.N.; Zhang, X.F.; Zhao, Z. Pollution area identification, receptor model-oriented sources and probabilistic health hazards to prioritize control measures for heavy metal management in soil. *J. Environ. Manag.* **2024**, *369*, 122322. [CrossRef] [PubMed]
10. An, Y.; Yan, T.; Gong, Q.; Wang, X.; Huang, Y.; Zhang, B.; Yin, Z.; Zhao, X.; Liu, N. Chromium (Cr) geochemical mapping based on fixed-values' method: Case studies in China. *Appl. Geochem.* **2022**, *136*, 105168. [CrossRef]
11. Xu, S.; Huang, Z.; Huang, J.; Wu, S.; Dao, Y.; Chen, Z.; Yang, B.; Xu, Y.; Liu, N.; Gong, Q. Environmental Pollution Assessment of Heavy Metals in Soils and Crops in Xinping Area of Yunnan Province, China. *Appl. Sci.* **2023**, *13*, 10810. [CrossRef]
12. Zhang, Z.; Xu, D.; Huang, T.; Zhang, Q.; Li, Y.; Zhou, J.; Zou, R.; Li, X.; Chen, J. High levels of cadmium altered soil archaeal activity, assembly, and co-occurrence network in volcanic areas. *Sci. Total Environ.* **2024**, *924*, 171529. [CrossRef] [PubMed]
13. Mazurek, R.; Kowalska, J.; Gąsiorek, M.; Zadrożny, P.; Józefowska, A.; Zaleski, T.; Kępka, W.; Tymczuk, M.; Orłowska, K. Assessment of heavy metals contamination in surface layers of Roztocze National Park Forest soils (SE Poland) by indices of pollution. *Chemosphere* **2017**, *168*, 839–850. [CrossRef] [PubMed]
14. Ni, M.; Mao, R.; Jia, Z.; Dong, R.; Li, S. Heavy metals in soils of Hechuan County in the upper Yangtze (SW China): Comparative pollution assessment using multiple indices with high-spatial-resolution sampling. *Ecotox. Environ. Safe.* **2018**, *148*, 644–651. [CrossRef]
15. Liu, W.; Yuan, C.; Farooq, T.H.; Chen, P.; Yang, M.; Ouyang, Z.; Fu, Y.; Yuan, Y.; Wang, G.; Yan, W.; et al. Pollution index and distribution characteristics of soil heavy metals among four distinct land use patterns of Taojia River Basin in China. *Gondwana Res.* **2024**, *135*, 198–207. [CrossRef]
16. Lv, W.J. Genetic Mineralogy with Emphasis on Gold Prospecting of DuJiaya Gold Deposit in Yantai Fushan Area, Jiaodong Region. Ph.D. Thesis, China University of Geosciences (Beijing), Beijing, China, May 2010. (In Chinese with English Abstract)
17. Zheng, L.; Wu, M.; Zhou, M.; Zhao, L.; Wang, Q.; Luan, S. Research on land resource development under county land use change: A case study of Fushan District, Yantai City. *Ecol. Sci.* **2023**, *42*, 114–123. (In Chinese with English Abstract)
18. Guo, L.-N.; Deng, J.; Yang, L.-Q.; Wang, Z.-L.; Wang, S.-R.; Wei, Y.-J.; Chen, B.-H. Gold deposition and resource potential of the Linglong gold deposit, Jiaodong Peninsula: Geochemical comparison of ore fluids. *Ore Geol. Rev.* **2020**, *120*, 103434. [CrossRef]
19. Wang, H.; Lan, T.; Chen, Y.; Hu, H.; Shu, L. Mobilization of Au and Ag during Supergene Processes in the Linglong Gold Deposit: Evidence from SEM and LA-ICP-MS Analyses of Sulfides. *Minerals* **2022**, *12*, 367. [CrossRef]
20. Wang, H.; Lan, T.-G.; Fan, H.-R.; Zhao, L.-Y.; Hu, H.-L.; Peng, H.-W.; Xu, Y.; Shu, L.; Dai, Z.-H. Anomalous cobalt enrichment in pyrite from an altered mafic dike in the Linglong gold deposit, Jiaodong Peninsula, China: Mechanisms and implications. *Ore Geol. Rev.* **2024**, *170*, 106139. [CrossRef]
21. Li, Q.-L.; Chen, F.; Yang, J.-H.; Fan, H.-R. Single grain pyrite Rb-Sr dating of the Linglong gold deposit, eastern China. *Ore Geol. Rev.* **2008**, *34*, 263–270. [CrossRef]
22. Guo, L.-N.; Goldfarb, R.J.; Wang, Z.-L.; Li, R.-H.; Chen, B.-H.; Li, J.-L. A comparison of Jiaojia-and Linglong-type gold deposit ore-forming fluids: Do they differ? *Ore Geol. Rev.* **2017**, *88*, 511–533. [CrossRef]
23. Zhao, S.-R.; Li, Z.-K.; Lin, Z.-W.; Gao, J.-F.; Sun, H.-S.; Li, M.Y.H.; Zhao, X.-F. Magmatic fluids responsible for lode gold mineralization in the giant Linglong deposit at Jiaodong, North China Craton: Constraints from Li-O isotopes. *Chem. Geol.* **2023**, *638*, 121696. [CrossRef]
24. Yang, Q.; Shen, J.; Li, S.; Santosh, M.; Luo, Z.; Liu, Y. Oxygen, boron, chromium and niobium enrichment in native Au and Ag grains: A case study from the Linglong gold deposit, Jiaodong, eastern China. *J. Asian Earth Sci.* **2013**, *62*, 537–546. [CrossRef]
25. Liang, Y.; Shu, L.; Ma, P.; Zhang, C.; Ma, Y.; Khan, M.; Shen, C. Gold source and ore-forming process of the Linglong gold deposit, Jiaodong gold province, China: Evidence from textures, mineral chemical compositions and sulfur isotopes of pyrite. *Ore Geol. Rev.* **2023**, *159*, 105523. [CrossRef]
26. *DZ/T 0295-2016*; Specification of Land Quality Geochemical Assessment. Ministry of Land and Resources of the People's Republic of China: Beijing, China, 2016.

27. Hou, Q.; Yang, Z.; Yu, T.; Xia, X.; Cheng, H.; Zhou, G. *Soil Geochemical Parameters in China*; Geological Publishing House: Beijing, China, 2020; pp. 1–3172. (In Chinese)
28. Huang, J.; Xu, S.; Gong, Q.; Chen, Z.; Huang, Z. Environmental Assessment of Heavy Metal Contamination in the Northeast of Huize County, Yunnan Province. *Acta Geosci. Sin.* **2022**, *43*, 93–100. (In Chinese with English Abstract)
29. Gong, Q.; Deng, J.; Xiang, Y.; Wang, Q.; Yang, L. Calculating Pollution Indices by Heavy Metals in Ecological Geochemistry Assessment and a Case Study in Parks of Beijing. *J. China Univ. Geosci.* **2008**, *19*, 230–241.
30. Håkanson, L. An ecological risk index for aquatic pollution control: A sedimentological approach. *Water Res.* **1980**, *14*, 975–1001. [CrossRef]
31. Cabrera, F.; Clemente, L.; Barrientos, D.E.; L´opez, R.; Murillo, J.M. Heavy Metal Pollution of Soils Affected by the Guadiamar Toxic Flood. *Sci. Total Environ.* **1999**, *242*, 117–129. [CrossRef] [PubMed]
32. Zoller, W.H.; Gladney, E.S.; Duce, R.A. Atmospheric Concentrations and Sources of Trace Metals at the South Pole. *Science* **1974**, *183*, 198–200. [CrossRef]
33. Duce, R.A.; Hoffmann, G.L.; Zoller, W.H. Atmospheric Trace Metals at Remote Northern and Southern Hemisphere Sites: Pollution or Natural? *Science* **1975**, *187*, 59–61. [CrossRef] [PubMed]
34. Reimann, C.; de Caritat, P. Distinguishing between Natural and Anthropogenic Sources for Elements in the Environment: Regional Geochemical Surveys versus Enrichment Factors. *Sci. Total Environ.* **2005**, *337*, 91–107. [CrossRef] [PubMed]
35. Chatterjee, M.; Silva, F.E.V.; Sarkar, S.K.; Sella, S.M.; Bhattacharya, A.; Satpathy, K.K.; Prasad, M.V.R.; Chakraborty, S.; Bhattacharya, B.D. Distribution and Possible Source of Trace Elements in the Sediment Cores of a Tropical Macrotidal Estuary and Their Ecotoxicological Significance. *Environ. Int.* **2007**, *33*, 346–356. [CrossRef] [PubMed]
36. Sutherland, R.A. Bed Sediment-Associated Trace Metals in an Urban Stream, Oahu, Hawaii. *Environ. Geol.* **2000**, *39*, 611–627. [CrossRef]
37. Blaser, P.; Zimmermann, S.; Luster, J.; Shotykb, W. Critical Examination of Trace Element Enrichments and Depletions in Soils: As, Cr, Cu, Ni, Pb, and Zn in Swiss Forest Soils. *Sci. Total Environ.* **2000**, *249*, 257–280. [CrossRef] [PubMed]
38. Yan, T.; Gong, Q.; Li, J.; Chen, J.; Zhou, W.; Li, R. Al_2O_3/Ti: Protoliths tracer in the process of weathering and alteration. *Acta Petrol. Sin.* **2016**, *32*, 2425–2432. (In Chinese with English Abstract)
39. Zhang, L.; Ye, X.; Feng, H.; Jing, Y.; Ouyang, T.; Yu, X.; Liang, R.; Gao, C.; Chen, W. Heavy Metal Contamination in Western Xiamen Bay Sediments and Its Vicinity, China. *Mar. Pollut. Bull.* **2007**, *54*, 974–982. [CrossRef] [PubMed]
40. Liu, W.-H.; Zhao, J.-Z.; Ouyang, Z.-Y.; Soderlund, L.; Liu, G.-H. Impacts of Sewage Irrigation on Heavy Metal Distribution and Contamination in Beijing, China. *Environ. Int.* **2005**, *31*, 805–812. [CrossRef] [PubMed]
41. Banat, K.M.; Howari, F.M.; Al-Hamad, A.A. Heavy metals in urban soils of central Jordan: Should we worry about their environmental risks? *Environ. Res.* **2005**, *97*, 258–273. [CrossRef]
42. Buccolieri, A.; Buccolieri, G.; Cardellicchio, N.; Dell'Atti, A.; Di Leo, A.; Maci, A. Heavy metals in marine sediments of Taranto Gulf (Ionian Sea, Southern Italy). *Mar. Chem.* **2006**, *99*, 227–235. [CrossRef]

applied sciences

MDPI

Article

Paleo-Asian Ocean Ridge Subduction: Evidence from Volcanic Rocks in the Fuyun–Qinghe Area, Southern Margin of the Chinese Altay

Jixu Liu [1], Cui Liu [1,*], Qing Liu [1], Zhaohua Luo [1], Yong Liu [2], Chenghao Zhou [3], Xu Guo [4], Xianghui Yu [1] and Miao Wang [1]

1 School of Earth Sciences and Resources, China University of Geosciences, Beijing 100083, China
2 Institute of Geology, Chinese Academy of Geological Sciences, Beijing 100037, China
3 Geologic Party No. 216, China National Nuclear Corporation, Urumqi 830011, China
4 217 Geological Team of Shanxi Provincial Geological Prospecting Bureau, Datong 037008, China
* Correspondence: liucui@cugb.edu.cn

Abstract: The Chinese Altay is located in the western segment of the Central Asian Orogenic Belt (CAOB) and preserves critical records of the Paleo-Asian Ocean (PAO) Plate evolution during the Paleozoic era. This region also hosts significant mineral deposits, making it a focal point for geological research. In this paper, field investigation, petrology, mineralogy, and petrography studies were conducted on volcanic rocks in the Fuyun–Qinghe area, southern margin of the Chinese Altay, and the paper provided new zircon LA-ICP-MS dating data, Lu-Hf isotope data, and whole-rock geochemical data of the basaltic to andesitic volcanic rocks. Thus, the formation age, petrogenesis, and tectonic setting of these rocks were discussed, which was of great significance to reveal the nature of the PAO Plate. The findings showed that the basaltic andesitic volcanic breccia was formed at 382.9 ± 3.4 Ma, the basalt was 401.7 ± 4.7 Ma, and the andesites were 405.1 ± 5.6 Ma and 404.8 ± 6.7 Ma, which indicated that the above rocks were formed in the Early–Middle Devonian. The volcanic rock assemblages were hawaiite, mugearite, potassic trachybasalt, basaltic andesite, andesite, benmoreite, etc., which contained labeled magmatic rocks such as adakite, sub-boninite, niobium-enriched arc basalt (NEAB), picrite, high-magnesium andesite (HMA), and magnesium andesite (MA). Comprehensive analysis indicated that magma probably mainly originated from three sources: (1) partial melting of the PAO slab, (2) partial melting of the overlying garnet–spinel lherzolite mantle peridotite metasomatized by subducting-related fluids (melts), and (3) a possible input of the asthenosphere. Comparative analysis with modern analogs (e.g., Chile Triple Junction) indicates that ridge subduction of the PAO had existed in the Fuyun–Qinghe area during the Early–Middle Devonian. Based on available evidence, we tentatively named the oceanic plates in this region the central Fuyun–Qinghe Ridge and the Junggar Ocean Plates, separated by the ridge on both sides. Although the ocean had a certain scale, it had entered the climax period of transition from ocean to continent.

Keywords: Chinese Altay; Early–Middle Devonian; magmatic arc; ridge subduction; PAO

1. Introduction

The CAOB is the largest Phanerozoic accretionary orogenic belt in the world, documenting the subduction and extinction history of the PAO Plate [1–3]. Over its extensive geological history, the expansion, subduction, closure, and extinction of the PAO have profoundly influenced the CAOB. The Chinese Altay, located in northern Xinjiang Province, is a crucial component of the western section of the CAOB. Its complex multi-stage tectonic evolution and extensive magmatism–tectonism–mineralization have been a hot topic for global researchers. It is generally accepted that the PAO began subducting during the Neoproterozoic to Early Paleozoic [4–7], with the most intense subduction occurring in the early Late Paleozoic. The closure time of the western segment of the PAO remains contentious, with proposed dates ranging from the Late Devonian [8], Early Carboniferous [9], Late Carboniferous or Permian [10–13], Late Permian to Triassic [14], to the Middle to Late Triassic [15,16]. Thus, at the latest, the western part of the ocean closed during the Middle to Late Triassic.

The Chinese Altay was undoubtedly influenced by the evolution of the PAO during the Late Paleozoic. Most scholars agree that the PAO was in a subduction stage during this period [17–19]. However, debates persist regarding the specific subduction setting and the nature of the PAO slab. For instance, Zhang et al. [20] proposed that the Devonian volcanic rocks formed in an island arc setting with varying degrees of maturity, transitioning to an intraplate setting in the Permian. Liang et al. [21] suggested that the eastern Junggar and southern margin of the Chinese Altay were in an island arc setting during the Early and Middle Devonian, followed by a collision setting in the Late Devonian, post-collision in the Early Carboniferous, and intraplate extension from the late Early Carboniferous to the Late Carboniferous. Xu et al. [22] studied the ophiolites in the Kurt area and concluded that they were the products of a Late Paleozoic back-arc basin system. Zheng et al. [19] proposed that the Early Devonian Xiaodonggou granites in the Kelan basin formed in an active continental margin arc setting. Niu et al. [17] found volcanic rocks with diverse source characteristics in the limited Devonian strata on the southern margin of the Chinese Altay, attributed to the complex accretion and tectonic processes during the Devonian to Carboniferous. Recently, scholars have proposed that ridge subduction could be a new model to explain the Late Paleozoic subduction processes in the Chinese Altay, based on studies of mafic complexes and veins [2,23], basalts and dacites [24], gabbros [3], and granites [25]. They posited that ridge subduction played a significant tectonic role in the evolution of the Altay Orogenic Belt.

The ridge subduction model has gained attention in recent years, exemplified by the Chile Trench (the Nazca–Atlantic–South American ridge–trench–trench triple junction [26,27]), southern Alaska, western California [28] and so on (Figure 1). Mantle convection causes upwelling at the ridge axis, driving mid-ocean ridge expansion and crust formation. The basaltic oceanic crust formed at the ridge cools and subducts, with magma from the ridge intruding near the trench, marking the onset of ridge subduction [29]. In these regions, ridge subduction generates new magmatic arcs and distinctive rock assemblages with unique geochemical characteristics [26,27]. Shen et al. [30] proposed ridge subduction in the Chinese Altay and summarized the petrological evidence, though they did not address the spatial distribution of plates, the migration of triple junctions, or the pattern of ridge subduction.

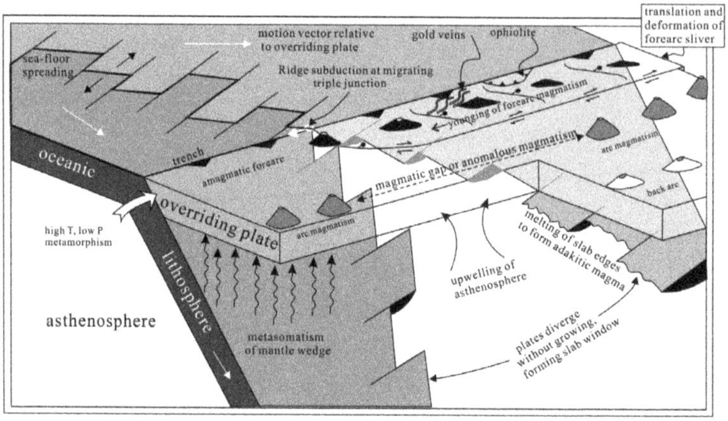

Figure 1. Spreading ridge subduction model [31]. Reprinted with permission from ref. [31].

The ridge subduction model challenges traditional views on magmatic arcs. Despite its significance, it remains understudied, particularly regarding ancient mid-ocean ridge subduction. Nonetheless, it offers a crucial avenue for exploring extinct ancient oceanic slabs.

This study, based on the concept of ridge subduction and magmatic arc examples, focuses on volcanic rocks in the Fuyun–Qinghe region, located at the junction of the southern margin of the Altay Orogenic Belt and northeast Junggar in China. The research aims to reveal the formation age, rock assemblages, and characteristics of these rocks, discuss their petrogenesis, and compare them with magmatic arcs formed by mid-ocean ridge subduction in Chile. This study seeks to infer the types of Fuyun–Qinghe magmatic arcs and the properties of the subducting oceanic slabs related to the PAO subduction. This research holds significant theoretical and practical importance for understanding the evolution of the PAO and regional magmatism and mineralization.

2. Geological Background

The Chinese Altay is located in the northern part of the Xinjiang Uygur Autonomous Region, between the Sayan Block to the north and the Junggar Block to the south (Figure 2a). According to Windley et al. [32] and Wang et al. [33], the Altay orogenic belt comprises six blocks (or units) (Figure 2b): the Northeast Altay Block, Northwest Altay Block, Central Altay Block, Qiongkuer–Abagong Block, Irtysh Block, and Buerjin–Ertai Block. The study area is situated in the Buerjin–Ertai Block.

Strata in this area were relatively fully exposed, primarily consisting of Proterozoic and Paleozoic strata. The oldest stratum was Paleoproterozoic Kumuqi Group, while the youngest was Quaternary stratum, Cretaceous strata were generally absent. In the Fuyun-Qinghe area, the major exposed Paleozoic strata included Late Ordovician Baihaba Formation, late Devonian Kangbutiebao Formation, Tuoranggekuduke Formation, Middle Devonian Altay Formation, Beitashan Formation, Ashele Formation, Yundukala Formation, and so on [34]. The Tuoranggekuduke Formation was a series of marine clastic rocks and volcanic sedimentary clastic rocks with lavas, containing brachiopods, corals, trilobites, and some other marine fossils, and this formation can be roughly divided into two lithological sections. The first section developed clastic and pyroclastic rocks, while the second section had more lavas; it was in unconformity contact with the overlying Beitashan Formation, and the lower part had no visible bottom. Previous studies obtained its formation age as the late Early Devonian. The Beitashan Formation was generally a set of pyroclastic rocks, volcanic lavas, and terrigenous clastic rocks, which can be subdivided into two segments. The lower segment was composed of pyroclastic rocks and terrigenous clastic rocks, and the upper segment mainly contained volcanic lavas and pyroclastic rocks with a small amount of terrigenous clastic rocks. It was in conformity contact with the overlying Yundukala Formation, and in angular unconformity contact with the lower Tuoranggekuduke Formation. Previous scholars believe its formation age is the early Middle Devonian.

The regional fault structures were well developed, exhibiting multi-stage characteristics. The primary faults included the Irtysh fault and the Kalaxiangeer fault (Figure 2c). The Irtysh fault strikes NW-SE, with a total length of 1500 km. The fault is narrow in the Fuyun–Qinghe area and is dextral strike–slip faulted by the Kalaxinger fault to the east. The eastern section is also called the Mayinebo fault, and the western section is called the Irtysh fault. The Kalaxinger fault, also known as the Ertai fault, is about 200 km in the NNW direction. In the Qinghe area, the fault transforms from the NNW strike–slip fault to the SE-SEE cleavage zone, presenting a large-scale broom structure [34]. Samples in this paper were mainly collected and compiled from the west of the Kalaxianger fault and south of the Irtysh fault. The main fold structures within the Fuyun–Qinghe area encompassed the Mengku composite syncline, the Suput anticline, and the Altay composite syncline, among others.

The study area experienced volcanic activities from the Proterozoic to the Cenozoic. Volcanic activities were weak before the Late Ordovician, with the most intense and frequent activities occurring during the Late Paleozoic, particularly in the Devonian. Volcanic activities declined during the Carboniferous–Permian, transitioning from marine to terrestrial facies and becoming more spatially restricted. Sporadic volcanic activities occurred during the Mesozoic to Cenozoic. Extensive magmatic emplacement took place during the Early and Middle Paleozoic (500 Ma to 370 Ma) [32,34,35]. Intrusive rocks are widely distributed in the region, predominantly granitoids with minor basic and intermediate rocks. Their formation ages primarily span the Early and Late Paleozoic, with minor occurrences in the Triassic and Jurassic. In the early Paleozoic, intrusive rocks in the Fuyun–Qinghe area mainly consisted of granitoids [36–39]. Intrusive rocks were most developed during the Devonian, primarily comprising granitoids, gabbros, and diorites [40–45]. During the Carboniferous to Permian, intrusive rocks mainly consisted of granitoids and diorite (porphyrite) rocks [46–48]. After the Triassic, intrusive rocks were rarely developed in the Fuyun–Qinghe area.

The study area lies at the junction of the southern edge of the Altay Metallogenic Province and the northern edge of the Junggar Metallogenic Province, Sawuer–Ertai Metallogenic Belt, and developed minerals such as iron, copper, gold, lead, zinc, etc. The main deposits include the Qiaoxiahala Fe-Cu-Au deposit, the Kalatongke Cu-Ni deposit, the Laoshankou Fe-Cu deposit, the Yulekenhalasu porphyry copper deposit, and so on.

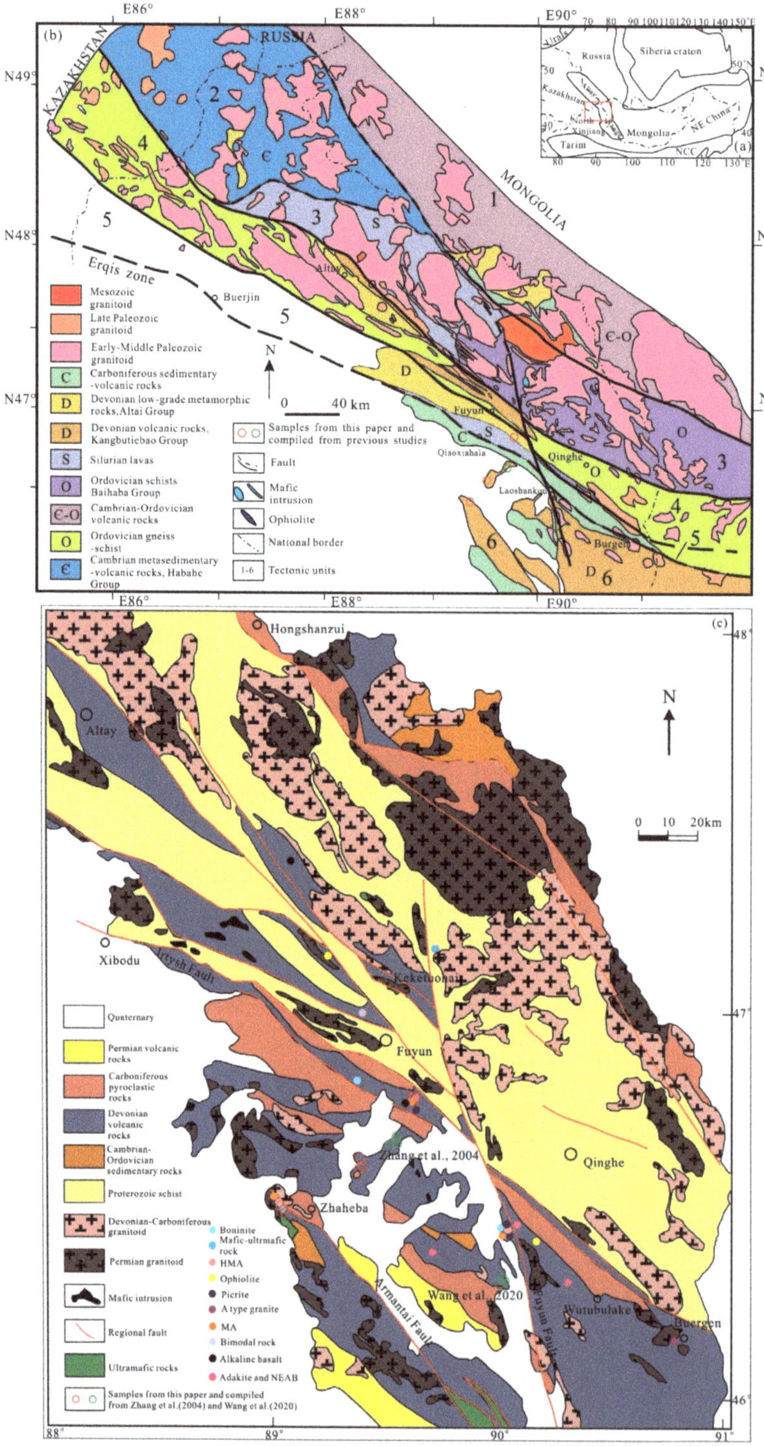

Figure 2. (**a**) Simplified tectonic divisions of the CAOB. (**b**) Geological sketch map of Altay (modified from [33]; 1, Northeast Altay Block; 2, Northwest Altay Block; 3, Central Altay Block; 4, Qiongkuer–

Abagong Block; 5, Irtysh Block; 6, Buerjin–Ertai Block). (**c**) Geological sketch map of the southern margin of the Chinese Altay, which contains special Devonian rock types related to ridge subduction (modified from [21], the compiled data are from [45] and [49]).

3. Volcanic Rock Geology and Sample Characteristics

Volcanic rocks and pyroclastic rocks are widely distributed in the Fuyun–Qinghe area and its vicinity. Volcanic rocks are found in thick layers in the field, primarily belonging to the Beitashan Formation as defined by previous researchers. Some parts of the rocks have undergone significant alteration, while the majority remain relatively fresh. Alteration is characterized by epidotization and chloritization (Figure 3a), a common phenomenon in the study region, particularly developed in mining areas such as the Qiaoxiahala and Laoshankou deposits. Volcanic rocks near the Ulungu River and the Qiaoxiahala mining area exhibit obvious orientations and developed cleavages. In the Qiaoxiahala mining area, strong cleavage and pronounced orientation are evident in the wall rocks where they contact the volcanic rocks. Although the overall deformation of the volcanic rocks is not pronounced, they exhibit a clear orientation. The overall degree of weathering is relatively weak, while epidotization alteration is pronounced. Closer to the ore body, the alteration becomes more severe, with heavier epidotization on rock fracture or joint surfaces. Away from the ore body or mineralization sites, epidotization is still present but much weaker. This suggests that the formation of the deposit was accompanied by fluids or hydrothermal solutions rich in epidotization. Additionally, some volcanic rocks contain relatively fresh diorite porphyry or diabase porphyry veins, which also exhibit epidotization. There is no distinct condensation or baking edge where the vein rocks contact the volcanic rocks, indicating that the vein rocks formed concurrently or slightly earlier than the orebody but later than the volcanic rocks. Local volcanic rocks include xenoliths, where epidotes are notably abundant, even forming individual rocks. Therefore, it is speculated that the volcanic rocks in this area originated from marine eruptions.

Figure 3. Field characteristics of volcanic rocks in the Fuyun–Qinghe area. (**a**) Epidotized volcanic rocks in Qiaoxiahala, (**b**) hornblende phenocrysts are oriented, (**c**) breccia in volcanic rocks, (**d**) quartz veins in cleavage zones.

On the north bank of the Ulungu River, near Kurt town, numerous directionally aligned hornblende grains appear in the phenocrysts of volcanic rocks. Some hornblende grains exhibit elongation (Figure 3b), with phenocryst particles becoming progressively larger towards the south. Breccia is present in certain volcanic rocks (Figure 3c), and quartz veins are visible in the cleavage zones (Figure 3d).

The collected samples primarily consist of basaltic to andesitic volcanic rocks and pyroclastic rocks (Figure 4). Some sample characteristics are described as follows:

Sample FY-80611, basalt (named from field and petrography, hereinafter the same), was collected from the hillside behind the eastern mining pit in the Qiaoxiahala mining area, Fuyun County (N46°48′60″, E89°38′39″). The basalt is blue-gray in color and exhibits a porphyritic texture and a massive structure (Figure 4a). The phenocrysts are mainly hornblende (~8%, 0.5~1.5 mm), with long plate-like plagioclase (~10%, 0.3~1.5 mm), and a small amount of pyroxene (~5%, 0.5 mm). The hornblende grains are euhedral and long columnar in shape, and show chloritization, while the plagioclase grains are euhedral to subhedral and have undergone sericitization and kaolinization. The groundmass shows tholeiitic texture and is composed of plagioclase microcrystals, hornblende microcrystals, opaque minerals, microcryptocrystalline grains, and glassy materials (Figure 4b).

Sample FY-80615, andesite, was collected from the same site as sample FY-80611. The andesite is dark gray in color, with calcite veins visible in the hand specimen. The sample exhibits a porphyritic texture and a massive structure (Figure 4c). Phenocrysts are mainly long columnar hornblende (~15%, 0.2~0.5 mm), with lath-shaped plagioclase (~10%, 0.5~1 mm), some of which show polysynthetic twins. Additionally, there are aggregates of calcite in this sample. The groundmass is composed of plagioclase microcrystals, opaque minerals, microcryptocrystalline grains, and glassy materials (Figure 4d).

Sample FY-80618, basalt, was also collected from the same site as sample FY-80611. The basalt is dark green in color and exhibits a porphyritic texture and a massive structure (Figure 4e). The phenocrysts are mainly short columnar pyroxene (~10%, 0.5~1.0 mm), long columnar hornblende (~20%, 0.2~0.5 mm), and long plate-like plagioclase (~20%, 0.3~0.7 mm). Some plagioclase grains show Carlsbad–albite twins and polysynthetic twins. The groundmass shows an intersertal texture and is composed of opaque minerals, microcryptocrystalline grains, and glassy materials (Figure 4f).

Sample FY-80703, andesite, was collected from the north bank of the Ulungur River, near Kurt town, with coordinates N46°28′38″, E89°07′06″. A schistosity belt developed in this region, with a cleavage attitude of 192°∠59°. The rock mass shows a dark brown weathered surface and a grayish-green fresh surface, containing a large number of hornblende phenocrysts (0.2~0.8 mm), which are elongated and oriented. Moreover, there are several quartz veinlets. The sample is greenish-gray in color, with a porphyritic texture and a massive structure (Figure 4g). The phenocrysts are mainly elongated hornblende grains and a small amount of plagioclase grains, with a content of about 20%. Hornblende phenocrysts are mostly elongated grains, with sizes of about 0.5~3 mm, and have undergone heavy chloritization. Plagioclase phenocrysts are about 0.2~1 mm and have also undergone sericitization. The groundmass shows an andesitic texture and is composed of microcrystalline plagioclases, microcrystalline hornblendes, and glassy materials (Figure 4h).

Sample FY-80704, andesite, was collected 30 m south of sample FY-80703. The hornblende phenocrysts here are more numerous and larger than those in sample FY-80703, also with orientation. Therefore, we speculated on a SW-NE direction compressive stress. The sample is greenish-gray in color and exhibits a porphyritic texture and a massive structure (Figure 4i). Phenocrysts are mainly hornblende and plagioclase, with a content of about 25%. Hornblende phenocrysts are deformed, and elongated grains have undergone chloritization and are about 2~5 mm in size. Plagioclase phenocrysts are about 0.2~1 mm and have undergone sericitization, and most grains are broken into particles. The groundmass shows an andesitic texture (Figure 4j).

Figure 4. Hand specimen photos and photomicrographs under cross-polarized light for some volcanic samples from the Fuyun–Qinghe area. (**a**,**b**) Basalt, (**c**,**d**) andesite, (**e**,**f**) basalt, (**g**,**h**) andesite, (**i**,**j**) andesite. Hb, hornblende; Px, pyroxene; Pl, plagioclase; Cal, calcite.

4. Analytical Methods

4.1. Zircon U–Pb Dating

Zircons were extracted from four samples (FY-80418, FY-80618, FY-80703, FY80704) by the Langfang Regional Geological Survey in Hebei Province, China. Standard density and magnetic separation methods were employed, after which zircon grains were manually selected under a binocular microscope. These grains were then analyzed using both transmitted and reflected light microscopy. To examine their internal structures and identify suitable targets for U-Pb dating, cathodoluminescence (CL) images were captured using a JEOL JSM6510 scanning electron microscope equipped with a GATAN Chroma CL detector at Gaonianlinghang Company in Beijing, China. U-Pb geochronological analysis was conducted using an LA-ICP-MS system located at the Inner Mongolia Geological Survey Institute in Hohhot, China. Zircon 91500 [50,51] was used as an external standard for age calibration Zircon standards GJ-1 [52] were analyzed as unknown samples that were inserted between 91500 and the samples, and NIST SRM 610 silicate glass was applied for instrument optimization. Detailed information on the instrument configurations and analytical protocols can be found in prior publications [53]. Corrections for common Pb were applied based on established methods [54], and Andersen [54] elaborated on a novel common lead correction method applicable to U-Pb analyses lacking ^{204}Pb data. This approach determines the radiogenic lead content, total lead content, common lead proportion, initial crystallization age, lead loss age, and lead loss magnitude in zircon or other U/Th-rich minerals by numerically solving a set of equations. In this study, we implemented common lead corrections using the Excel spreadsheet tool provided by the authors in their original publication. Data reduction was performed using ICP–MS–DataCal (Ver. 6.7 [55]), and concordia diagrams were generated using Isoplot (Ver. 3.0; [56]).

4.2. Major and Trace Element Determinations

For whole-rock major and trace element analysis, the samples were initially crushed using a jaw crusher and subsequently pulverized to 200 mesh with an agate ball mill at the Langfang Regional Geological Survey in Hebei Province, China. The analyses of major and trace elements were conducted at ALS Chemex (Guangzhou, China) Co., Ltd. Major elements were measured using a Rigaku ZXS100e X-ray fluorescence spectrophotometer (XRF) (Tokyo, Japan), achieving an analytical precision of better than 2%. Trace elements were quantified using an Agilent 7500a (Agilent, Santa Clara, CA, USA) inductively coupled plasma–mass spectrometer (ICP-MS). To ensure accuracy, two standard reference materials (AGV-2 and GSR-1) were employed, with the analytical precision maintained within 5–10%.

4.3. Hf Isotope Analyses

In situ zircon Hf isotope measurements were conducted using a Neptune multi-collector–inductively coupled plasma–mass spectrometer (MC-ICP-MS) coupled with a GeoLas HD 193 nm laser ablation system at the Analysis and Testing Center of the Inner Mongolia Geological Survey Institute in Hohhot, China. The standard zircon 91500 [50,51,57] was utilized during the analyses. Comprehensive details regarding the analytical methodology and instrument parameters are provided in the work by Wu et al. [57].

5. Analytical Result

5.1. Zircon U–Pb Dating

This study presents the LA–ICP–MS zircon U–Pb dating results for four samples (FY-80418, FY-80618, FY-80703, FY80704) collected from the Fuyun–Qinghe area (see Table S1 for detailed results).

Sample FY-80418, a volcanic breccia. Twenty-nine zircon grains, ranging from 100 to 200 μm in length with length-to-width ratios of 1:1 to 2:1, were analyzed. These grains displayed euhedral to subhedral shapes and distinct oscillatory zoning in the cathodoluminescence (CL) images (Figure 5a). The Th and U concentrations varied from 10 to 914 ppm and 3 to 987 ppm, respectively, with Th/U ratios between 0.38 and 2.97 (excluding one outlier at 0.38), suggesting a magmatic origin [58]. The $^{206}Pb/^{238}U$ ages of 18 concordant spots ranged from 368 to 394 Ma, producing a weighted mean age of 382.9 ± 3.4 Ma (MSWD = 3.0), corresponding to the late Middle Devonian (Figure 6a).

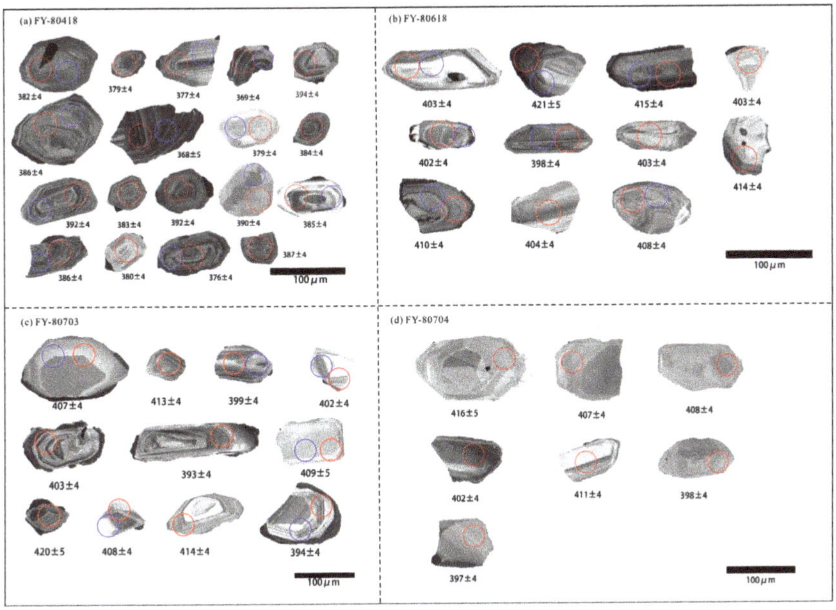

Figure 5. CL images of the zircons from the volcanic and volcaniclastic rocks in the Fuyun–Qinghe area (red circles represent U–Pb dating points while blue circles show Lu–Hf isotope points). (**a**) Sample FY-80418, (**b**) sample FY-80618, (**c**) sample FY-80703, (**d**) sample FY-80704.

For sample FY-80618, a basalt, 29 zircon grains were analyzed. These grains measured 50 to 150 μm in length, with length-to-width ratios of 1.5:1 to 2:1, and exhibited oscillatory zoning in the CL images (Figure 5b). The Th and U concentrations ranged from 51 to 1418 ppm and 83 to 1174 ppm, respectively, with Th/U ratios of 0.18 to 2.1 (mostly above 0.4). Eleven concordant spots yielded $^{206}Pb/^{238}U$ ages between 398 and 421 Ma, with a weighted mean age of 401.7 ± 4.7 Ma (MSWD = 2.6) (Figure 6b).

Sample FY-80703, an andesite, was analyzed using 15 zircon grains. The grains ranged from 50 to 200 μm in length, with length-to-width ratios of 1:1 to 3:1, and displayed oscillatory zoning in the CL images (Figure 5c). The Th and U concentrations varied from 57 to 277 ppm and 69 to 2236 ppm, respectively, with Th/U ratios of 0.09 to 1.26 (mostly exceeding 0.4, except for two points). Eleven concordant spots produced $^{206}Pb/^{238}U$ ages ranging from 393 to 420 Ma, yielding a weighted mean age of 405.1 ± 5.6 Ma (MSWD = 3.7), corresponding to the late Early Devonian (Figure 6c).

Lastly, sample FY-80704, also an andesite, was analyzed using 15 zircon grains. These grains measured 50 to 200 μm in length, with length-to-width ratios of 1:1 to 3:1, and exhibited oscillatory zoning in the CL images (Figure 5d). The Th and U concentrations ranged from 33 to 813 ppm and 56 to 758 ppm, respectively, with Th/U ratios of 0.44 to

1.26. Seven concordant spots yielded $^{206}Pb/^{238}U$ ages between 395 and 416 Ma, producing a weighted mean age of 404.8 ± 6.7 Ma (MSWD = 2.9) (Figure 6d).

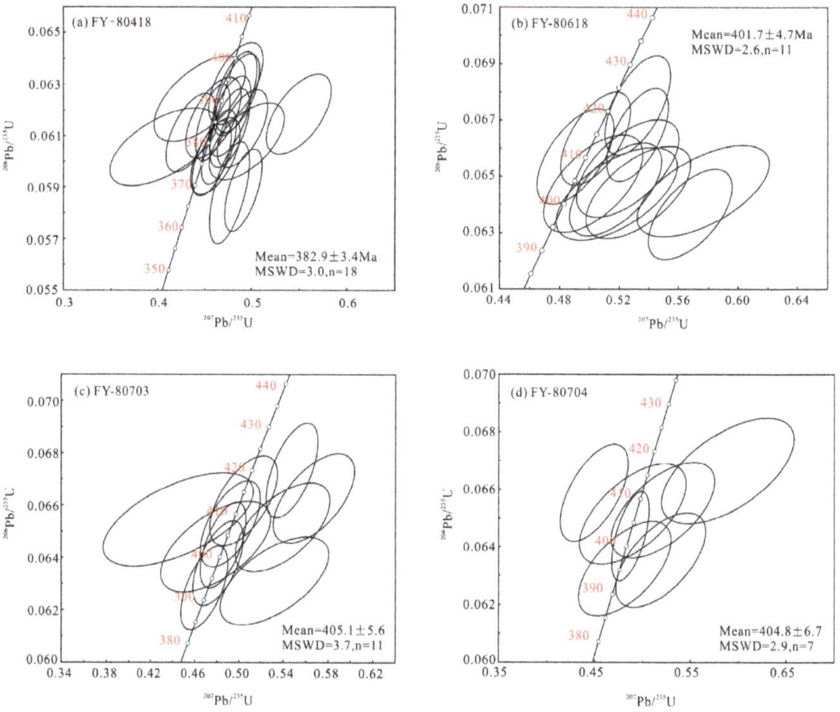

Figure 6. Zircon U–Pb concordia diagrams for the volcanic and volcaniclastic rocks examined in the Fuyun–Qinghe area. (**a**) Sample FY-80418, (**b**) sample FY-80618, (**c**) sample FY-80703, (**d**) sample FY-80704.

5.2. Major and Trace Elements

The major element data of the volcanic rocks are shown in Table S2. Among them, nine samples were collected from the Qiaoxiahala mining area and Kurt town on the north bank of the Ulungu River, and 18 samples were compiled from the literature [45,49]. All samples were Early–Middle Devonian intermediate to basic volcanic rocks. The samples were corrected for loss on ignition and volatile content, and the geochemical diagrams were constructed after normalizing the data to 100%.

The volcanic rocks showed SiO_2 levels in the range of 42.79~57.22% (n = 27), with $Na_2O + K_2O$ values of 0.52~7.74%; most samples had low K_2O and high Na_2O contents. Samples had CaO values of 3.59~11.5%, MgO values of 2.83~7.08% and Mg# values of 35.60~56.92. In a plot of $Na_2O + K_2O$ versus SiO_2 (Figure 7a), the samples were both subalkaline and alkaline series. Samples of the subalkaline series are mainly basaltic andesite and andesite, with little basalt. Samples of the alkaline series are mainly trachybasalt, basaltic trachyandesite, and trachyandesite. According to the recommendations of the IUGS, samples falling into trachybasalt, basaltic trachyandesite, and trachyandesite fields can be further divided based on whether their $Na_2O - 2 \geq K_2O$. Therefore, samples of trachybasalt, basaltic trachyandesite, and trachyandesite can be further divided into hawaiite, potassic trachybasalt, mugearite, shoshonite, and benmoreite. From the above, in combination with the characteristics of field occurrence, hand specimens, and petrography, the specific rock names are shown in Table S2. In the AFM diagram, the subalkaline samples are mostly calc-alkaline series (Figure 7b). Alkaline series sample points are present as the basalt and MA series in the SiO_2-MgO diagram, while most subalkaline samples are MA rocks, with a few HMA rocks (Figure 7c). In the A/CNK-A/NK diagram, all samples pertain to the metaluminous series rocks (Figure 7d). In the plot of K_2O versus SiO_2, the subalkaline samples belong to the low-K and middle-K CA series (Figure 7e). The alkaline samples (in the TAS diagram) in the SiO_2-(Na_2O + K_2O − CaO) diagram show characteristics of A and AC series, while the subalkaline samples (in the TAS diagram) show AC, CA, and C series features (Figure 7f).

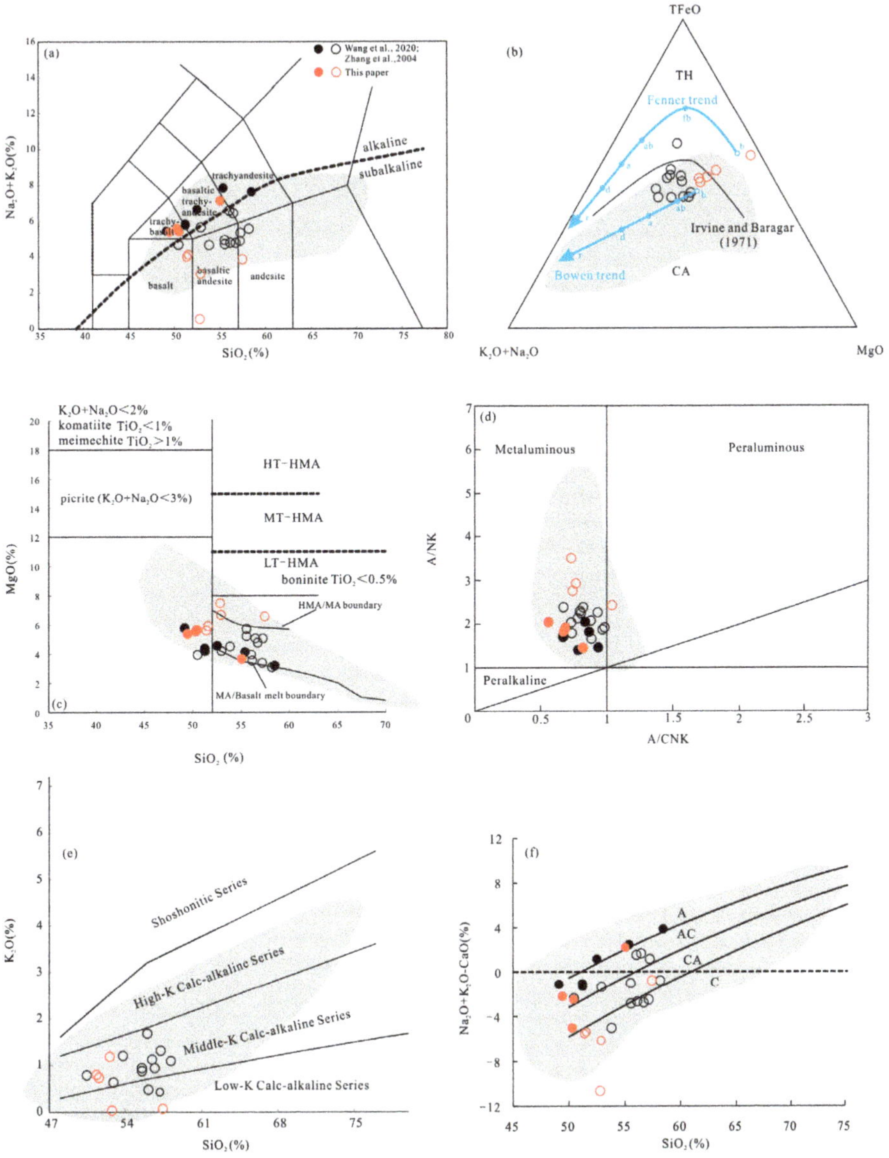

Figure 7. (**a**) Total alkali versus silica (TAS) diagram ([59]; the boundary line between the alkaline and subalkaline series is from [60]). (**b**) AFM diagram (the dashed separation line is from [60]; TH = tholeitic; CA = calc-alkaline. b = basalt; fb = ferro-basalt; ab = andesite basalt; a = andesite;

d = dacite; r = rhyolite.). (**c**) SiO_2 versus MgO diagram [61]. (**d**) A/CNK versus A/NK diagram [62]. (**e**) SiO_2 versus K_2O diagram [63]. (**f**) SiO_2 versus ($Na_2O + K_2O - CaO$) diagram ([64]; A = alkaline series; AC = alkali–calcic series; CA = calc-alkaline series; C = calcic series). (Data of shaded fields are from [26,65,66], while data of black circles are from [45,49], solid circles represent alkaline series, and hollow circles indicate subalkaline series).

In addition, there were special labeled rock types such as sub-boninite (FY-80704, SiO_2 = 54.36%, >52%, MgO = 6.24%, <8%, but higher than the boundary between HMA/MA, with 0.5% < TiO_2 = 0.57% < 1%, so we define it as sub-boninite), adakite (A01-2, etc., SiO_2 > 56%, $Al_2O_3 \geq 15\%$, MgO $\leq 6\%$, Sr ≥ 400 ppm, Y < 18 ppm), NEAB (FY-80618, etc., $Na_2O > K_2O$, TiO_2 > 1%, Nb/La > 0.5, Nb/U > 10), and HMA and MA (see above).

We have also compiled the geochemical data of Neogene–Quaternary samples from the Taitao Island and southern Patagonian plateaus at the Chile Triple Junction. These samples were believed to be members of the igneous rocks of the magmatic arc formed by the Chile Ridge. The shaded fields in Figure 7 represent the ophiolitic and volcanic suites generated by Neogene to Quaternary ridge subduction and near-trench magmatism at the Chile Triple Junction, as well as the basaltic rocks of the Southern Patagonian plateau, which resulted from arc and back-arc magmatism linked to the Chile Ridge subduction. Upon comparison, it is evident that all samples presented in this research point within the compositional ranges of rocks formed by ridge subduction at the Chile Triple Junction (Figure 7). This indicates that the aforementioned rocks and their characteristics align perfectly with those of the rocks formed by the Chile mid-ocean ridge subduction.

The trace element compositions of the volcanic rocks are presented in Table S3. In a primitive mantle-normalized spider diagram (Figure 8a), both alkaline and subalkaline series exhibit enrichment in large-ion lithophile elements (LILEs, e.g., Rb, Ba, U, and Pb) and depletion in high-field-strength elements (HFSEs, e.g., Nb, Ta, and Ti). These patterns likely reflect the influence of subducting slab fluids and/or melt influx, as well as the metasomatism of the mantle source [23]. The depletion of Nb, Ta, and Ti suggests the presence of residual minerals such as hornblende and/or Fe–Ti oxides (e.g., rutile and ilmenite). The trends of most samples closely resemble those of the island arc basalts (IABs), characterized by enrichment in light rare earth elements (LREEs) and depletion in heavy rare earth elements (HREEs). The ΣLREE values range from 41.54 to 297.46×10^{-6}, while ΣHREE values range from 8.93 to 42.09×10^{-6}, with $(La/Yb)_N$ ratios between 1.79 and 8.49. The δEu values range from 0.85 to 1.10. When plotted on chondrite-normalized rare earth element patterns (Figure 8b), the samples show slight positive or negative Eu anomalies, with a positive Sr anomaly (Figure 8a) indicating minimal plagioclase fractionation during crystallization.

5.3. Zircon Hf Isotopes

In situ Hf isotope analysis was conducted on selected zircon U–Pb dating spots from samples FY-80418, FY-80618, FY-80703, and FY-80704 (Table S4). The primary zircons from these samples exhibited initial $^{176}Hf/^{177}Hf$ ratios ranging from 0.282671 to 0.283067. The corresponding εHf(t) values varied between +4.79 and +19.63, while the two-stage model ages (TDM_2) spanned from 191 to 1060 Ma.

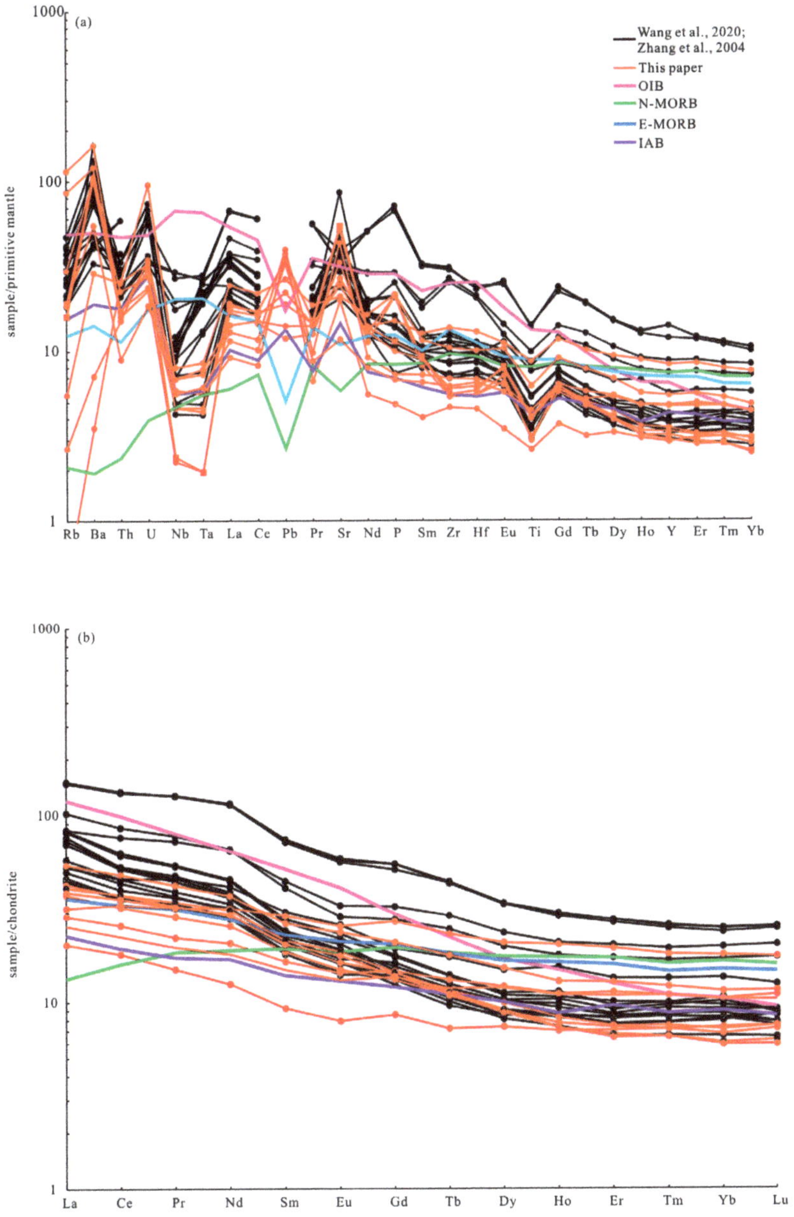

Figure 8. Primitive mantle-normalized trace element spider diagram (**a**) and chondrite-normalized REE pattern diagram (**b**) for the volcanic rocks in the Fuyun–Qinghe area. The chondrite and primitive mantle values are from [67] and [68], respectively. (Data of average ocean island basalt (OIB), normal mid-ocean ridge basalt (N-MORB), and enriched mid-ocean ridge basalt (E-MORB) are from [69], and the data of the average island arc basalt (IAB) are from [70]. Data of black lines are from [45,49]).

6. Discussion

6.1. Crustal Contamination and Fractional Crystallization

The majority of the Early–Middle Devonian samples displayed elevated Nb/Ta and Nb/Th ratios, surpassing those typically associated with crustal-derived magmas [71,72]. These samples also exhibited low Rb/Sr ratios (0.0003–0.1064) and high Ti/Y ratios (223–431), further supporting the absence of significant crustal contamination [73]. Additionally, substantial crustal contamination would typically result in negative anomalies for Sr, P, and U in primitive mantle-normalized trace element diagrams [68]; however, the Early–Middle Devonian volcanic rocks in this study showed positive anomalies for these elements (Figure 8a). Collectively, these findings indicate that the Early–Middle Devonian volcanic rocks in the Fuyun–Qinghe area did not undergo extensive crustal contamination.

Regarding fractional crystallization, there was an inapparent correlation between MgO and TiO_2 for both alkaline and subalkaline series rocks in the variation diagrams for major oxides and Cr, Ni, and MgO contents (Figure 9), and the correlations between MgO, TFe_2O_3, CaO, and CaO/Al_2O_3 of subalkaline samples (mostly CA rocks) were also not obvious. However, the alkaline series had obvious positive correlations. The aforementioned characteristics suggested that there had been no fractional crystallization in the CA series rocks; they might have undergone magmatic processes primarily characterized by varying degrees of partial melting or involved multiple magma sources. The relatively obvious correlations among some elements of the alkaline series rocks suggested a relatively consistent depth of magma sources, or they were subjected to fluid processes. Furthermore, a strong positive correlation was observed between Cr, Ni, and MgO, indicating the fractionation of the Cr-spinel (Figure 9).

6.2. Magma Sources and Petrogenesis

As mentioned earlier, the Early–Middle Devonian volcanic rock assemblages in the study area are hawaiite, mugearite, potassic trachybasalt, basaltic andesite, andesite, benmoreite, etc., which contain special labeled rocks such as adakite, sub-boninite, NEAB, HMA, and MA. Boninite forms through the high-degree partial melting of mantle peridotite under low-P and high-T conditions, facilitated by fluids enriched in incompatible elements derived from subducting oceanic crust [74]. In contrast, NEAB originates from the partial melting of a mantle wedge metasomatized by adakitic melts [75,76]. When subducting slab melts interact with mantle peridotite, they increase the Nb abundance and result in higher La/Yb ratios [2]. HMA, on the other hand, is generated by the partial melting of the mantle wedge above a subduction zone, driven by the release of H_2O from the subducted oceanic crust [61]. In the P_2O_5 versus TiO_2 and Nb/Yb versus Nb diagrams (Figure 10), samples of NEAB are present and fall within the slab melt metasomatism field. Similarly, in the primitive mantle-normalized Hf/Sm versus Ta/La diagram (Figure 11), some sample points align with fluid-related metasomatism or lie near the boundary between fluid- and melt-related metasomatism, indicating that the mantle peridotite was influenced by subducting fluids or melts. Furthermore, the subalkaline rocks are predominantly low-K to medium-K calc-alkaline series, characterized by a relatively high MgO content. Their enrichment in large-ion lithophile elements (LILEs) and depletion in high-field-strength elements (HFSEs) are consistent with the features of island arc magmatic rocks. In the primitive mantle-normalized trace element spider diagram (Figure 8a), the patterns of these samples closely resemble those of the island arc basalts (IABs). Consequently, it is evident that the Early–Middle Devonian volcanic rocks were formed in an arc environment associated with ocean subduction.

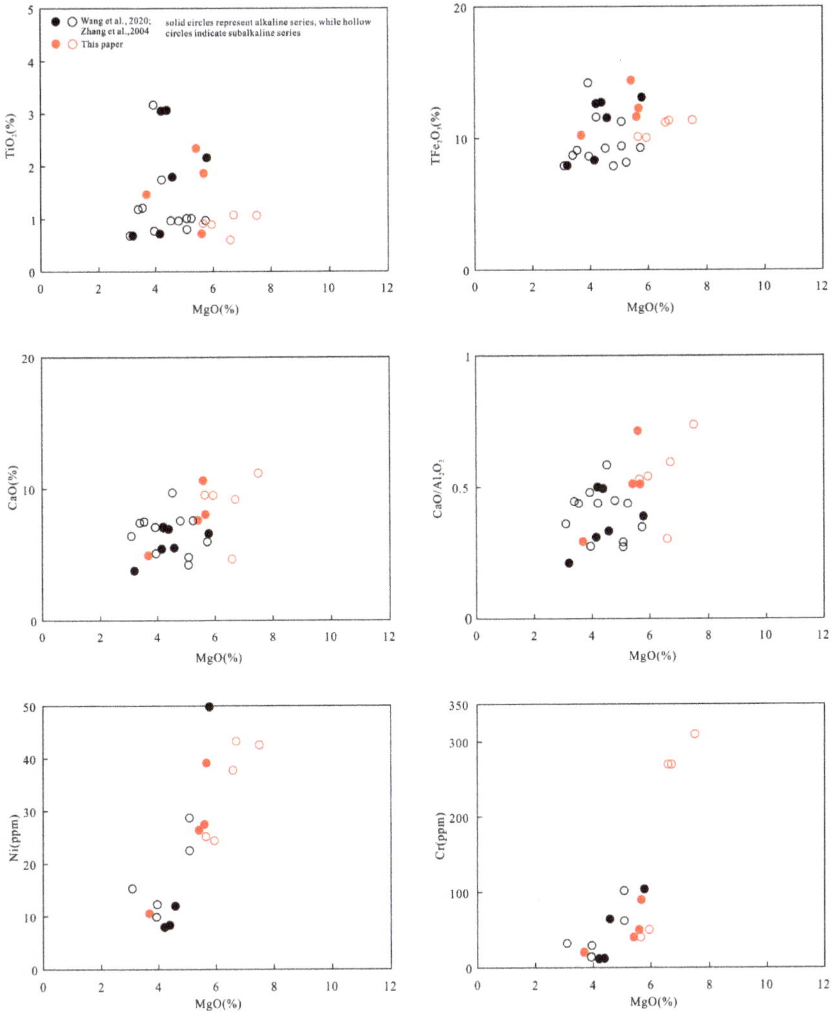

Figure 9. Variation diagrams for some major oxides and Cr and Ni versus MgO contents for the volcanic rocks in the Fuyun–Qinghe area. (Data of black circles are from [45,49]).

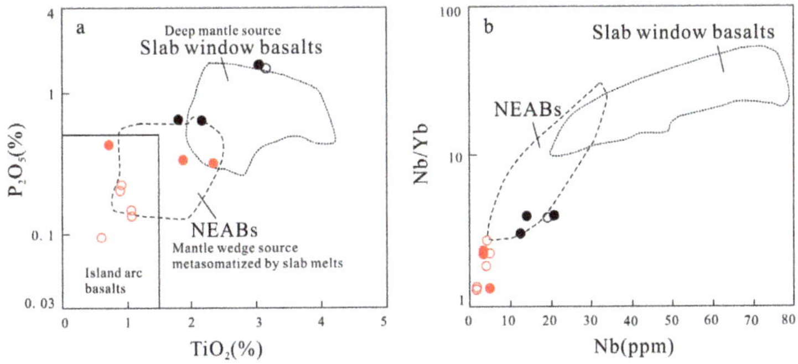

Figure 10. P_2O_5 versus TiO_2 (**a**) and Nb/Yb versus Nb (**b**) diagrams for the volcanic rocks in the Fuyun–Qinghe area [77] (the legends are the same as in Figure 7).

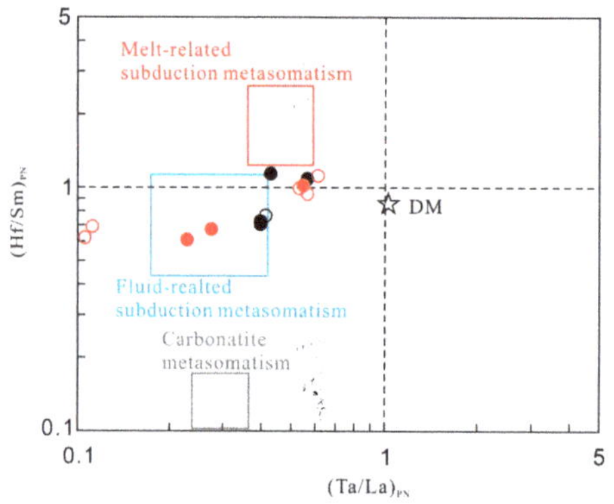

Figure 11. $(Hf/Sm)_{PN}$ versus $(Ta/La)_{PN}$ diagram for the volcanic rocks in the Fuyun–Qinghe area [78]. (PN, primitive mantle-normalized; DM, depleted mantle; the legends are the same as in Figure 7).

Experimental studies indicate that garnet preferentially retains heavy rare earth elements (HREEs) over middle rare earth elements during melting. As a result, melts derived from garnet–facies peridotite exhibit higher middle rare earth element/HREE ratios compared to those from spinel–facies peridotite [79]. The Sm/Yb ratios of basaltic rocks (including NEABs), HMAs, and sub-boninite from the Early–Middle Devonian volcanic rocks align with the melting curves of garnet–spinel mantle peridotite (Figure 12), suggesting that their precursor magma originated from 5–20% partial melting of enriched garnet–spinel lherzolite mantle. This is consistent with the relatively flat chondrite-normalized REE patterns, which are influenced by the presence of both spinel and garnet. Additionally, in a chondrite-normalized La/Sm versus Tb/Yb plot (Figure 13), basaltic samples and HMAs fall between the 5 and 10% partial melting curves of the primitive mantle and enriched mantle-20 fields [78], further supporting the idea that the precursor magma was generated by the partial melting of the enriched mantle. While the oceanic lithosphere typically exhibits depletion characteristics, the enriched mantle in this context likely reflects asthenospheric materials. This enrichment may have been triggered by the upwelling of asthenospheric materials through a slab window formed by the subduction of a spreading ridge.

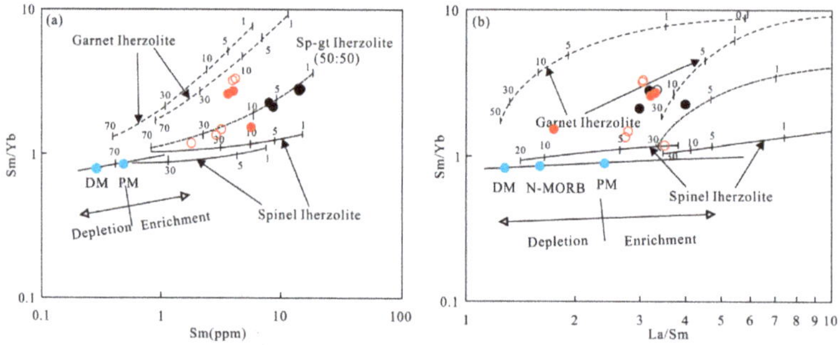

Figure 12. Sm/Yb versus Sm (**a**) and Sm/Yb versus La/Sm (**b**) diagrams for the volcanic rocks in the Fuyun–Qinghe area. The mantle array is defined by a depleted MORB mantle (DMM [80]) and a primitive mantle (PM [68]). The melting curves for spinel lherzolite ($Ol_{53} + Opx_{27} + Cpx_{17} + Sp_{11}$) and garnet peridotite ($Ol_{60} + Opx_{20} + Cpx_{10} + Gt_{10}$) with both DMM and PM compositions are from [81]. (PM, primitive mantle; N-MORB, normal mid-ocean ridge basalt; DM, depleted mantle; Numbers along lines represent the degree of the partial melting; the legends are the same as in Figure 7).

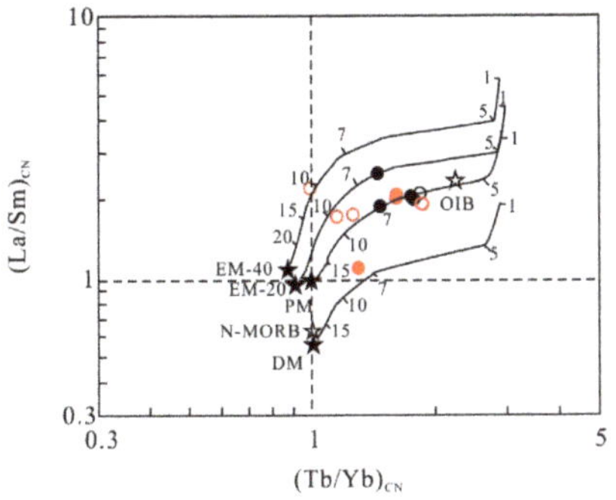

Figure 13. $(La/Sm)_{CN}$ versus $(Tb/Yb)_{CN}$ diagram showing the calculated compositions of basaltic magmas produced by variable degrees of decompression fractional melting of DM, PM, EM-20, and EM-40 mantle sources for the volcanic rocks in the southern margin of the Chinese Altay [78]. (CN, chondrite-normalized; EM-40, EM-20 are enriched mantle-40 and enriched mantle-20, the compositions of them are detailed in [78]; PM, primitive mantle; N-MORB, normal mid-ocean ridge basalt; DM, depleted mantle; OIB, ocean island basalt; the legends are the same as in Figure 7).

It is noteworthy that some of these volcanic rocks displayed geochemical characteristics similar to those of MORB, such as low K and Ti, and high Al concentrations and high Zr/Hf ratios with low La/Yb ratios. In the plot of P_2O_5 versus TiO_2, several sample points entered the deep mantle source region. In addition, our samples did not contain picrite, but picrite was also reported in the lower part of the Beitashan Formation in the southwest of Qinghe County by Zhang et al. [82]. Therefore, the Early–Middle Devonian rock assemblages in the Fuyun–Qinghe area also contained picrite. Picrite is the product of high-level melting of garnet peridotite metasomatized by fluid under high-temperature conditions [82]. Therefore, all the above rocks such as picrite, basalt, HMA, and their characteristics suggested that the asthenosphere mantle was another magma source.

Adakite and MA were also identified in the volcanic samples. Adakite primarily forms through two mechanisms: partial melting of young oceanic crust (less than 25 million years old) [83], due to its high heat flow units (HFUs, 2.8–8), or the partial melting of the thickened lower crust [84,85]. As previously noted, the study area was characterized by marine facies during the Early–Middle Devonian, eliminating the possibility of a thickened lower crust. Thus, the adakites in this region were likely generated by the subduction of young oceanic crust, which must have been younger than 25 Ma. Meanwhile, MA is thought to result from the interaction between magmas produced by the dehydration and melting of subducted oceanic crust and the overlying mantle wedge [86].

The discussions on fractional crystallization proved that the CA series rocks had different magma sources, and the εHf(t) and TDM_2 of zircon varied widely, ranging from +4.79 to +18.26 and 191 to 1060 Ma, respectively, which also suggests the existence of different magma sources.

In summary, we believe that there were at least three magma sources for the volcanic rocks in the Fuyun–Qinghe area: young and hot subducted oceanic crust, overlying mantle wedge, and asthenosphere mantle.

6.3. Tectonic Setting

The tectonic setting of the Chinese Altay during the Devonian period remains a subject of debate. Wang et al. [87] linked the abrupt rise in the heat flux and geothermal gradient during the Early Devonian to the activation of thermal subduction mechanisms. Meng et al. [88] proposed that the Devonian volcanic rocks in the Ashele mining district were generated by mantle plume upwelling and slab break-off. Other researchers have argued that the Chinese Altay was situated in an arc environment, though the specific type of magmatic arc is still contested. Proposed models include the island arc [20,21], continental marginal arc [19,89,90], back-arc basin [22,91], forearc extension [92], and ridge subduction [2,3,23,24,93–95].

As mentioned earlier, the rock assemblages and geochemical characteristics of the volcanic rocks from the Early–Middle Devonian pointed to the features of arc rocks, indicating that there should be a tectonic setting and crust–mantle structure of the subduction zone. Furthermore, four of the collected samples exhibited $(La/Yb)_{CN}$ values of 1.9, 2.64, 2.94, and 2.97 (ranging from 1.3 to 3.4, indicating enrichment in LREEs compared to N-MORB), and $(La/Nb)_{CN}$ values of 2.02, 2.04, 1.85, and 1.83 (ranging from 1.2 to 2.35, indicating a depletion in HFSEs relative to LILEs and LREEs). These findings indicated that they belonged to E-MORB with a subduction-related imprint [26]. The remaining samples displayed a significant enrichment in LREEs and LILEs, along with a depletion in HFSEs, characteristic of arc magmatic rocks ($(La/Yb)_{CN}$ ranging from 4.20 to 9.04; $(La/Nb)_{CN}$ ranging from 2.09 to 7.66). In the Ti/100-Zr-Y*3 and Hf/3-Th-Ta triangle diagrams (Figure 14), the sample points predominantly fell within the CAB and MORB, as well as the CAB and E-MORB domains, further underscoring the dual characteristics of the Early–Middle Devonian volcanic rocks.

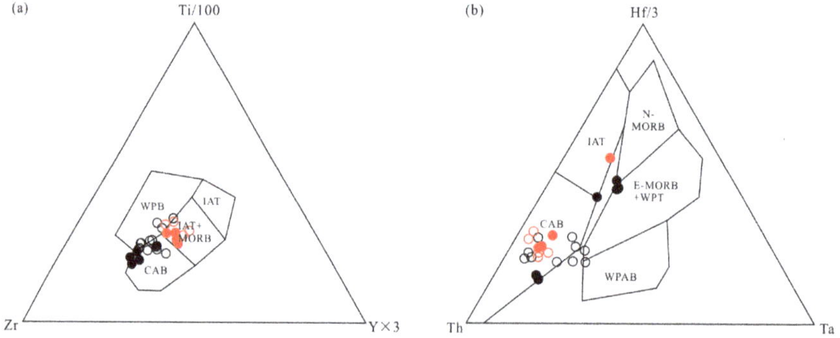

Figure 14. (**a**) Ti/100-Zr-Y*3 [96] and (**b**) Hf/3-Th-Ta diagrams for the volcanic rocks in the Fuyun–Qinghe area [97]. (CAB, calc-alkali basalt; IAT, island arc tholeiite; MORB, mid-ocean ridge basalt; WPB, within-plate basalt; N-MORB, normal mid-ocean ridge basalt; E-MORB, enriched mid-ocean ridge basalt; WPT, within-plate tholeiite; WPAB, within-plate alkali basalt; legends are the same as in Figure 7).

Numerous models have been proposed to explain the tectonic setting of igneous rocks exhibiting dual geochemical characteristics of both arc-like and MORB-like signatures. These include fore-arc basins, back-arc basins, decompression melting, asthenospheric upwelling due to slab break-off, and ridge subduction [3,17,22,23,26,95]. Fore-arc basalts are typically characterized by low LREE contents [74], with REE patterns resembling those of MORB. However, the samples in this study display LREE enrichment. Additionally, fore-arc basalts are generally associated with the initial stages of subduction, whereas subduction-related arc magmatism had already emerged during the Early Paleozoic in the Chinese Altay [46,98]. Furthermore, the presence of adakites in the Fuyun–Qinghe area during the same period suggests the partial melting of the young and hot oceanic crust [83], likely occurring on both sides of a slab window. The Kurt ophiolite on the southern margin of the Chinese Altay has been interpreted as a product of a back-arc basin [22]. However, distinctive rock types such as adakite, boninite, HMA, NEAB, picrite, and A-type granite also formed in this region during the Late Paleozoic [17,25,45]. Such diverse rock assemblages cannot be solely attributed to a cold and old back-arc basin. In the context of slab break-off, basaltic rocks typically exhibit affinities to both arc basalts and intraplate basalts, accompanied by rock assemblages such as intraplate basalts, A_2-type siliceous rocks, and contemporaneous bimodal volcanic rocks [99]. Consequently, the geochemical features and rock assemblages of the Early–Middle Devonian volcanic rocks in the Fuyun–Qinghe area do not align with a slab break-off tectonic setting.

Ridge subduction can induce ridge–trench interaction and trigger asthenosphere upwelling through slab windows, resulting in particular rock assemblages and geochemical characteristics, such as adakites, NEABs, HMAs, boninites, A-type granites, bimodal volcanic rocks, picrites, and so on [25,30]. The Taitao Peninsula and Taitao Ridge from the Chile Triple Junction produced the most complete magma products related to ridge subduction from the Neogene to the Quaternary, which provided a typical example and abundant references for studying oceanic ridge subduction in the Chinese Altay. Guivel et al. [26] summarized six major ridge and near-trench magmatic activities in the Taitao Peninsula and Taitao Ridge. The four segments of the active oceanic ridge in southern Chile (Seg.1 $\rightarrow$ Seg.4) were mainly of the MORB type, while the Taitao Ridge was mainly of the MORB type with subduction imprints. In other words, calc-alkaline igneous rocks originated from the asthenosphere below the slab window and the subduction slab at the edge of the slab window. The igneous rocks from the Taitao Peninsula possessed both MORB and MORB with subduction imprint characteristics. The coexistence of the MORB type with subduction imprints and arc CA series rocks in the petrological–geochemical types is uncommon but arises from the ridge–trench interaction during active ridge subduction, plate divergence, and convergence. Similarly, Le Moigne et al. [65] investigated two Pliocene to Pleistocene volcanic–sedimentary units on the Taitao Peninsula, categorizing them into those with MORB affinities, CA (calc-alkaline) affinities, and intermediate characteristics. They proposed that the varying metamorphic overprints resulted from the collision of two distinct ridge segments with the South American margin near the Taitao Peninsula. Additionally, the Bahia Barrientos Ophiolite (BBO), associated with the subduction of the Chile Ridge, is also present in the Taitao Peninsula. The tectonic emplacement of the ophiolites occurred around the same time as the subduction of the Chile Ridge at approximately 6 Ma, with the Taitao Ridge being interpreted as a nascent ophiolite [26]. Our samples included adakite, NEAB, HMA, MA, and sub-boninite. Some of these rocks exhibited dual characteristics resembling both MORB-like and arc-like features, with the majority being calc-alkaline series rocks, which aligned with the subduction-imprinted MORB and typical calc-alkaline igneous suite found in the Taitao Ridge [26]. Our samples also contained alkaline basalts related to a slab window, potentially indicating the presence of alkaline plateau basalts associated with a deep slab window generated after the cessation stage of arc volcanism (or closer to back-arc position) [66]. In the Mayinebo Ophiolitic Mélange belt in the southern part of the study area, previous research had determined the zircon U-Pb ages of gabbros and a pillow basalt within the belt. The gabbros were formed at 404.6 ± 2.9 Ma and 403 ± 5 Ma, while the pillow basalt was formed at 397 ± 6 Ma [100,101], whose dating results closely align with the ages presented in this study. They might represent near-trench ophiolites analogous to BBO. Thus, the volcanic rock assemblages and characteristics in the Fuyun–Qinghe area during the Early–Middle Devonian were quite similar to those of the Taitao Ridge and Taitao Peninsula. Combined with the arc rock assemblages and characteristics and the three magma sources mentioned above, it is speculated that the Fuyun–Qinghe area, the southern margin of the Chinese Altay, could be in a tectonic setting of PAO ridge subduction, while predecessors called the ocean the Junggar Ocean.

The Chinese Altay also featured the presence of Alaska-type mafic–ultramafic complexes and dikes from the Devonian period (Early and Late Devonian) [2,23], and the

Middle Devonian Beitashan Formation picrite was found in Qinghe County [82,102]. Furthermore, the Fuyun Supute area developed Middle Devonian Altay Formation bimodal volcanic rocks [103]. The Ashele Formation bimodal volcanic rocks, as well as the Early Devonian bimodal intrusive rocks in the Habahe area, are also significant [88]. The widespread occurrence of ~400 Ma granites with positive εHf(t) values, coupled with high-temperature, low-pressure metamorphism around 390 Ma [104], further supports the likelihood that the Chinese Altay was influenced by the PAO ridge subduction during the Devonian (see Figure 15).

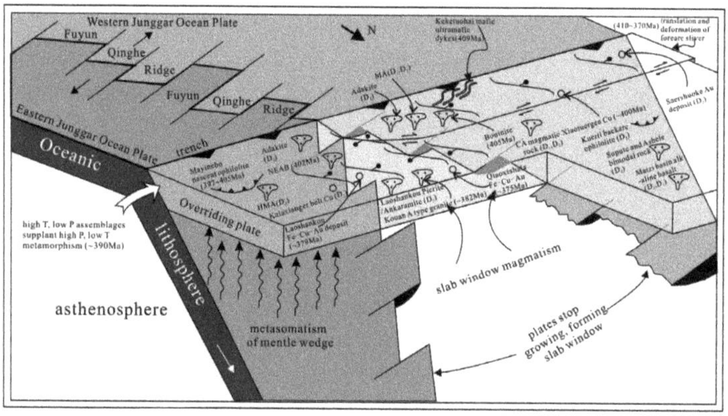

Figure 15. The spreading ridge subduction model and its magmatism in the Fuyun–Qinghe area (modified after [31]).

Therefore, this paper proposes the following model, based on the current distribution of rock types and potential tectonic positions, and the research by Sisson et al. [31]. The nature of the PAO slab is illustrated in Figure 15 The rock types and potential tectonic positions encompassed adakite, which was formed on both sides of the slab window, mafic–ultramafic complexes and dikes resulting from near-trench magmatism, as well as nascent serpentinite, HMA, boninite, picrite, MA, adakite, NEAB, and A-type granite, all derived from complex magmatism during normal arc subduction. Additionally, there were bimodal magmatic rocks formed by back-arc magmatism and alkaline basalt, among others. These represent the magmatic activities occurring in various positions within the subduction zones during this period, including young oceanic crust subduction, mid-ocean ridge subduction, near-trench magmatism, and back-arc magmatism. However, more in-depth and detailed studies are needed on the scope of the ridge and the matching of its components, such as the spatial distribution of ophiolite, bimodal volcanic rocks, and even alkaline series and calc-alkaline series rocks. This study combines the collected samples with previous geological data to preliminarily map the spatial distribution of the Devonian igneous rocks related to ridge subduction in the Fuyun–Qinghe area (Figure 15). In the northwest, the region features Supute bimodal volcanic rocks, Kuerti back-arc basin ophiolites, and Maizi alkaline basalts generated by back-arc magmatism [22,103,105]. Moving southward, the region exhibits near-trench magmatic activities and slab window processes, manifested by picrites, boninites, HMA, nascent ophiolites, and A-type granites in areas such as Qiaoxiahala–Laoshankou–Buergen [25,82,100,101,106–108]. The spatial distribution pattern of the aforementioned igneous rocks suggests that the northwest portion of the study area experienced back-arc magmatism during this geological interval, whereas the southeastern region was situated in proximity to the trench and concurrent development of a slab window. The distributional characteristics of adakite and NEAB may indicate the edges of the developing slab window. Notably, the inferred locations of the mid-ocean ridge magmatic activity, which are primarily based on the current spatial distribution of volcanic rocks, may not fully represent the configuration of magmatic processes during this period. This discrepancy arises from potential post-magmatic tectonic reworking and/or the incomplete preservation of volcanic records in the study area.

Based on our research, we posit that during the Early–Middle Devonian, the Junggar Ocean, a branch of the PAO in the Fuyun–Qinghe area, experienced ridge subduction (we called it the Fuyun–Qinghe Ridge). We inferred that the Fuyun–Qinghe Ridge separated the Junggar Ocean into three plates from the current evidence, namely the central Fuyun–Qinghe Ridge, the western Junggar Ocean Plate, and the eastern Junggar Ocean Plate (Figure 15). The oceanic plates mentioned above were juvenile plates that had recently been pulled out from the Fuyun–Qinghe Ridge, and the juvenile oceanic plate also subducted and penetrated beneath the Altay orogenic belt along with the Fuyun–Qinghe Ridge. Due to the asymmetric expansion of the oceanic ridge, the subduction of the expanding ridge is an inevitable outcome of the closure of an ocean [109]. The Devonian ridge subduction in the Chinese Altay was similar to the ridge subduction on the east coast of the Pacific Ocean today. The scale of the Pacific Ocean is gradually decreasing nowadays, and the Junggar Ocean also gradually disappeared during the Devonian. Although the Junggar Ocean had a certain scale, it had entered the climax period of transition from ocean to continent and had moved towards a stage of accelerated extinction.

Due to the development of transform faults along mid-ocean ridges, the migration of magmatic arcs formed by ridge subduction along transform faults occurs in the present-day Chile Triple Junction (Taitao Ridge) subduction zone, with an estimated migration rate of approximately 100 km/Ma. In the Ashele mining district of the northwestern Chinese Altay, high-Mg dacites and basalts dated at ~403 Ma were developed, which formed in a slab window setting associated with ridge subduction [24]. However, during the late Early Devonian to Middle Devonian, slab window-related magmatic activities were predominantly distributed in the Fuyun-Qinghe area. We preliminarily propose that during the transition between the Early and Middle Devonian, the ridge and slab window migrated from the Ashele area in the northwest (approximately 400 km from the Fuyun-Qinghe area) to the Fuyun-Qinghe region. However, the specific details of this process require further in-depth research.

7. Conclusions

1. LA-ICP-MS zircon U-Pb dating of the volcanic rocks in the Fuyun–Qinghe area revealed that the basalt from the Qiaoxiahala mining area formed at 401.7 ± 4.7 Ma, while the andesites from Kurt town yielded ages of 405.1 ± 5.6 Ma and 404.8 ± 6.7 Ma. The volcanic breccia from Kurt town was dated to 382.9 ± 3.4 Ma. These results indicate that the rocks were formed during the Early–Middle Devonian.

2. The Early–Middle Devonian volcanic rocks in the Fuyun–Qinghe area have both alkaline and subalkaline series rocks in the TAS diagram, with rock assemblages including hawaiite, mugearite, potassic trachybasalt, basaltic andesite, andesite, benmoreite, etc., which contained special magmatic arc rock assemblages such as adakite, sub-boninite, NEAB, picrite, HMA, and MA. They were enriched in LILEs and depleted in HFSEs, exhibiting similar geochemical characteristics to the magmatic arc rocks formed by the Taitao Ridge subduction in Chile.

3. The basaltic rocks (including NEAB), sub-boninite, and HMA originated from the partial melting of garnet–spinel mantle peridotite metasomatized by subducted fluids (melts), with some rocks likely incorporating asthenospheric materials. In contrast, the MA and adakite were likely produced by the partial melting of the young and hot subducting oceanic crust (slab) and subsequent interaction with the mantle wedge peridotite.

4. The rock assemblages and petrogeochemical characteristics of Fuyun-Qinghe samples closely resemble those of the Taitao Ridge and Taitao Peninsula magmatic arc by the Taitao Ridge subduction during the Neogene–Quaternary period at the Chile Triple Junction. Based on these similarities, we propose that the Fuyun–Qinghe area was likely situated in a magmatic arc setting formed by the PAO ridge subduction during the Early–Middle Devonian.

Supplementary Materials: The following supporting information can be downloaded at https://www.mdpi.com/article/10.3390/app15073736/s1. Table S1: LA–ICP–MS zircon U–Pb dating data for the volcanic rocks in the Fuyun–Qinghe area; Table S2: Major (%) element data for the volcanic rocks in the Fuyun–Qinghe area; Table S3: Trace (ppm) element data for the volcanic rocks in the Fuyun–Qinghe area; Table S4: Lu-Hf isotopic data for the volcanic rocks in the Fuyun–Qinghe area.

Author Contributions: Conceptualization, J.L. and C.L.; writing, J.L.; review and editing, C.L. and Q.L.; supervision, Z.L. and Y.L.; investigation, J.L., C.Z., X.G., X.Y. and M.W. All authors have read and agreed to the published version of the manuscript.

Funding: This study was supported by the China Geological Survey Project (DD20221645, DD20190370, 1212011121075) and the National Natural Science Foundation of China (Grants No. 92162213 and 40802020).

Institutional Review Board Statement: Not applicable.

Informed Consent Statement: Not applicable.

Data Availability Statement: Data are contained within the article and Supplementary Materials.

Acknowledgments: We extend our gratitude to Meng Dai from AGRS and Tongyang Zhao from the Geological Survey Institute of Xinjiang Uygur Autonomous Region, China, for their support in field work and sample collection, and gratefully acknowledge the Inner Mongolia Geological Survey Institute for complimentary providing zircon geochronological analyses and testing services. Special appreciation is extended to Liquan Xu and Jianwei Xiao from the institute for their professional support throughout the analytical work, including sample testing procedures, data interpretation, and technical consultation. We also appreciate the insightful comments provided by the reviewers.

Conflicts of Interest: Author Chenghao Zhou was employed by the company China National Nuclear Corporation. The remaining authors declare that the research was conducted in the absence of any commercial or financial relationships that could be construed as a potential conflict of interest.

References

1. Sengör, A.; Natal'in, B.; Burtman, V. Evolution of the Altaid tectonic collage and Paleozoic crustal growth in Asia. *Nature* **1993**, *364*, 299–307. [CrossRef]
2. Cai, K.; Sun, M.; Yuan, C.; Zhao, G.; Xiao, W.; Long, X. Keketuohai mafic–ultramafic complex in the Chinese Altai, NW China: Petrogenesis and geodynamic significance. *Chem. Geol.* **2012**, *294–295*, 26–41. [CrossRef]
3. Wang, Z.; Li, Y.; Yang, G.; Zhang, Y.; Li, H.; Zhao, Y.; Lindagato, P. Geochronology, geochemistry, and petrogenesis of the Kezijiaer gabbros, southern Chinese Altai: Evidence for ridge subduction. *Geol. J.* **2020**, *55*, 2254–2268. [CrossRef]
4. Yang, G.; Li, Y.; Tong, L.; Wang, Z.; Si, G.; Lindagato, P.; Zeng, R. Natural observations of subduction initiation: Implications for the geodynamic evolution of the Paleo-Asian Ocean. *Geosystems Geoenvironment* **2022**, *1*, 100009. [CrossRef]
5. Pan, G.; Lu, S.; Xiao, Q.; Zhang, K.; Yin, F.; Hao, G.; Luo, M.; Ren, F.; Yuan, S. Division of tectonic stages and tectonic evolution in China. *Earth Sci. Front.* **2016**, *23*, 10. [CrossRef]
6. Xiao, W.J. Altaids, continental growth and metallogeny. *Natl. Sci. Rev.* **2023**, *10*, nwad004. [CrossRef]
7. Gan, B.; Li, Z.; Song, Z.; Li, J. Middle Cambrian granites in the Dunhuang Block (NW China) mark the early subduction of the southernmost Paleo-Asian Ocean. *Lithos* **2020**, *372–373*, 105654. [CrossRef]
8. Wang, E.; Zhai, X.; Chen, W.; Wu, L.; Song, G.; Wang, Y.; Guo, Z.; Zhao, J.; Wang, J. Late Devonian A-Type Granites from the Beishan, Southern Central Asia Orogenic Belt: Implications for Closure of the Paleo-Asia Ocean. *Minerals* **2023**, *13*, 565. [CrossRef]

9. Niu, Y.; Shi, G.R.; Wang, J.; Liu, C.; Zhou, J.; Lu, J.; Song, B.; Xu, W. The closing of the southern branch of the Paleo-Asian Ocean: Constraints from sedimentary records in the southern Beishan Region of the Central Asian Orogenic Belt, NW China. *Mar. Pet. Geol.* **2021**, *124*, 104791. [CrossRef]

0. Liu, Q.; Zhao, G.; Han, Y.; Eizenhöfer, P.R.; Zhu, Y.; Hou, W.; Zhang, X. Timing of the final closure of the Paleo-Asian Ocean in the Alxa Terrane: Constraints from geochronology and geochemistry of Late Carboniferous to Permian gabbros and diorites. *Lithos* **2017**, *274–275*, 19–30. [CrossRef]

1. Weng, K.; Cao, J.; Karibovich, D.; Movlanov, J.; Chen, B.; Ma, Z. Zircon U-Pb ages in the Nuratau ophiolitic mélange in the southern Tianshan, Uzbekistan: Implication for the closure of Paleo-Asian Ocean. *China Geol.* **2024**, *7*, 365–368. [CrossRef]

2. Duan, F.; Zhi, Q.; Xiao, W.; Li, Y. Late Palaeozoic final subduction records of the southwestern Paleo-Asian Ocean revealed by ca. 286–283 Ma basaltic rocks in the West Junggar, NW China. *Int. Geol. Rev.* **2023**, *65*, 3128–3145. [CrossRef]

3. Liu, B.; Han, B.; Chen, J.; Ren, R.; Zheng, B.; Wang, Z.; Feng, L. Closure Time of the Junggar-Balkhash Ocean: Constraints From Late Paleozoic Volcano-Sedimentary Sequences in the Barleik Mountains, West Junggar, NW China. *Tectonics* **2017**, *36*, 2823–2845. [CrossRef]

4. Xiao, W.; Han, C.; Yuan, C.; Sun, M.; Lin, S.; Chen, H.; Li, Z.; Li, J.; Sun, S. Middle Cambrian to Permian subduction-related accretionary orogenesis of Northern Xinjiang, NW China: Implications for the tectonic evolution of central Asia. *J. Asian Earth Sci.* **2008**, *32*, 102–117. [CrossRef]

5. Song, D.; Xiao, W.; Ao, S.; Mao, Q.; Wan, B.; Zeng, H. Contemporaneous closure of the Paleo-Asian Ocean in the Middle-Late Triassic: A synthesis of new evidence and tectonic implications for the final assembly of Pangea. *Earth-Sci. Rev.* **2024**, *253*, 104771. [CrossRef]

6. Wang, H.; Xiao, W.; Li, R.; Chen, H.; Tan, Z.; Mao, Q.; Shi, M. High-grade complexes record the Late Permian-Middle Triassic arc metamorphism in the southernmost Altaids: Implications for the final closure of the Paleo-Asian Ocean. *Lithos* **2023**, *442–443*, 107054. [CrossRef]

7. Niu, H.; Sato, H.; Zhang, H.; Ito, J.I.; Yu, X.; Nagao, T.; Terada, K.; Zhang, Q. Juxtaposition of adakite, boninite, high-TiO_2 and low-TiO_2 basalts in the Devonian southern Altay, Xinjiang, NW China. *J. Asian Earth Sci.* **2006**, *28*, 439–456. [CrossRef]

8. Xu, Q.; Li, Y.; Yang, G.; Ning, W.; Tong, L.; Duan, F.; Wu, L.; Ren, P.; Xiao, W. Petrogenesis and tectonic setting of the Middle Devonian Beitashan Formation volcanic rocks in the northern East Junggar, NW China: Insights from geochemistry, zircon U–Pb dating, and Hf isotopes. *Geol. J.* **2019**, *55*, 1964–1983. [CrossRef]

9. Zheng, J.; Chai, F.; Yang, F. The 401–409 Ma Xiaodonggou granitic intrusion: Implications for understanding the Devonian Tectonics of the Northwest China Altai orogen. *Int. Geol. Rev.* **2015**, *58*, 540–555. [CrossRef]

0. Zhang, Z.; Zhou, G.; Kusky, T.M.; Yan, S.; Chen, B.; Zhao, L. Late Paleozoic volcanic record of the Eastern Junggar terrane, Xinjiang, Northwestern China: Major and trace element characteristics, Sr–Nd isotopic systematics and implications for tectonic evolution. *Gondwana Res.* **2009**, *16*, 201–215. [CrossRef]

1. Liang, P.; Chen, H.; Hollings, P.; Xiao, B.; Wu, C.; Bao, Z.; Cai, K. The Paleozoic tectonic evolution and metallogenesis of the northern margin of East Junggar, Central Asia Orogenic Belt: Geochronological and geochemical constraints from igneous rocks of the Qiaoxiahala Fe-Cu deposit. *J. Asian Earth Sci.* **2016**, *130*, 23–45. [CrossRef]

2. Xu, J.; Chen, F.; Yu, X.; Niu, H.; Zheng, Z.P. Kuerti ophiolite in Altay area of north Xinjiang: Magmatism of an ancient back-arc basin. *Acta Petrol. Mineral.* **2001**, *20*, 344–352.

3. Cai, K.; Sun, M.; Yuan, C.; Zhao, G.; Xiao, W.; Long, X.; Wu, F. Geochronological and geochemical study of mafic dykes from the northwest Chinese Altai: Implications for petrogenesis and tectonic evolution. *Gondwana Res.* **2010**, *18*, 638–652. [CrossRef]

4. Wu, Y.; Yang, F.; Liu, F.; Geng, X.; Li, Q.; Zheng, J. Petrogenesis and tectonic settings of volcanic rocks of the Ashele Cu–Zn deposit in southern Altay, Xinjiang, Northwest China: Insights from zircon U–Pb geochronology, geochemistry and Sr–Nd isotopes. *J. Asian Earth Sci.* **2015**, *112*, 60–73. [CrossRef]

5. Shen, X.; Zhang, H.; Wang, Q.; Wyman, D.A.; Yang, Y. Late Devonian–Early Permian A-type granites in the southern Altay Range, Northwest China: Petrogenesis and implications for tectonic setting of "A2-type" granites. *J. Asian Earth Sci.* **2011**, *42*, 986–1007. [CrossRef]

6. Guivel, C.; Lagabrielle, Y.; Bourgois, J.; Maury, R.C.; Fourcade, S.; Martin, H.; Arnaud, N. New geochemical constraints for the origin of ridge-subduction-related plutonic and volcanic suites from the Chile Triple Junction (Taitao Peninsula and Site 862, LEG ODP141 on the Taitao Ridge). *Tectonophysics* **1999**, *311*, 83–111. [CrossRef]

7. Lagabrielle, Y.; Guivel, C.; Maury, R.C.; Bourgors, J.; Fourcade, S.; Martin, H. Magmatic tectonic effects of high thermal regime at the site of active ridge subduction: The Chile Triple Junction model. *Tectonophysics* **1996**, *326*, 255–268. [CrossRef]

8. Cole, R.B.; Stewart, B.W. Continental margin volcanism at sites of spreading ridge subduction: Examples from southern Alaska and western California. *Tectonophysics* **2009**, *464*, 118–136. [CrossRef]

9. Liu, X.; Xiao, W.; Xiao, Y.; Liu, P.; Song, Y.; Huang, W.; Zhang, Z.; Liu, X. The major characteristics of ridge subduction in the Central Asian Orogenic Belt and their research advances. *Geochimica* **2025**, *54*, 3–23.

30. Shen, X.; Zhang, H.; Ma, L. Ridge Subduction and the Possible Evidences in Chinese Altay, Xinjiang. *Geotecton. Metallog.* **2010**, *34*, 181–195.
31. Sisson, V.B.; Pavlis, T.L.; Roeske, S.M.; Thorkelson, D.J. Introduction: An overview of ridge-trench interactions in modern and ancient settings. In *Geological Society of America Special Paper*; Geological Society of America: Boulder, CO, USA, 2003; Volume 371, pp. 1–18. [CrossRef]
32. Windley, B.F.; Kröner, A.; Guo, J.; Qu, G.; Li, Y.; Zhang, C. Neoproterozoic to Palaeozoic geology of the Altai orogen, NW China: New zircon age data and tectonic evolution. *J. Geol.* **2002**, *110*, 719–737.
33. Wang, T.; Jahn, B.; Kovach, V.P.; Tong, Y.; Wilde, S.A.; Hong, D.; Li, S.; Salnikova, E.B. Mesozoic intraplate granitic magmatism in the Altai accretionary orogen, NW China: Implications for the orogenic architecture and crustal growth. *Am. J. Sci.* **2014**, *314*, 1–42. [CrossRef]
34. XUARGSI (Xinjiang Uygur Autonomous Region Geological Survey Institute). *Regional Geological Annals of Xinjiang Uygur Autonomous Region*; Xinjiang Uygur Autonomous Region Geological Survey Institute: Urumqi, China, 2021.
35. Wang, T.; Tong, Y.; Jahn, B.; Zou, T.; Wang, Y.; Hong, D.; Han, B. SHRIMP U–Pb Zircon geochronology of the Altai No. 3 Pegmatite, NW China, and its implications for the origin and tectonic setting of the pegmatite. *Ore Geol. Rev.* **2007**, *32*, 325–336. [CrossRef]
36. Zhang, C.; Liu, L.; Santosh, M.; Luo, Q.; Zhang, X. Sediment recycling and crustal growth in the Central Asian Orogenic Belt: Evidence from Sr–Nd–Hf isotopes and trace elements in granitoids of the Chinese Altay. *Gondwana Res.* **2017**, *47*, 142–160. [CrossRef]
37. Liang, P.; Chen, H.; Wu, C.; Liu, Z. Geochemistry, geochronology and oxygen fugacity of volcanic and intrusive rocks from the Laoshankou Fe–Cu–Au deposit in the northern margin of East Junggar, NW China. *Earth Sci. Front.* **2018**, *25*, 96–118. [CrossRef]
38. Shen, X.; Zhang, H.; Wang, Q.; Ma, L.; Yang, Y. Early Silurian (~440 Ma) adakitic, andesitic and Nb-enriched basaltic lavas in the southern Altay Range, Northern Xinjiang (western China): Slab melting and implications for crustal growth in the Central Asian Orogenic Belt. *Lithos* **2014**, *206–207*, 234–251. [CrossRef]
39. Zhang, H.; Shen, X.; Ma, L.; Niu, H.; Yu, X. Geochronologyy of the Fuyun adakite, north Xinjiang and its constraint to the initiation of the Paleo-Asian Ocean subduction. *Acta Petrol. Sin.* **2008**, *24*, 1054–1058.
40. Dong, Z. Paleozoic Geological Evolution of the Chinese Altai Orogen in the Fuyun-Qinghe Region: Constraints on the Accretionary Orogenic Processes of the Paleo-Asian Ocean. Doctoral Dissertation, Northwest University, Xi'an, China, 2020.
41. Geng, X.; Yang, F.; Qin, J.; Wen, C. Devonian porphyry Cu deposits in NW China: The Xiaotuergen example. *Int. Geol. Rev.* **2016**, *59*, 1219–1233. [CrossRef]
42. Li, Q.; Lv, S.; Yang, F.; Geng, X.; Chai, F. Geological characteristics and genesis of the Laoshankou Fe–Cu–Au deposit in Junggar, Xinjiang, Central Asian Orogenic Belt. *Ore Geol. Rev.* **2015**, *68*, 59–78. [CrossRef]
43. Yang, T.N.; Li, J.Y.; Zhang, J.; Hou, K.J. The Altai-Mongolia terrane in the Central Asian Orogenic Belt (CAOB): A peri-Gondwana one? Evidence from zircon U–Pb, Hf isotopes and REE abundance. *Precambrian Res.* **2011**, *187*, 79–98. [CrossRef]
44. Chai, F.; Yang, F.; Liu, F.; Geng, X.; Lv, S.; Jiang, L.; Zang, M.; Chen, B. Geochronology and genesis of volcanic rocks in Beitashan Formation at the northern margin of the Junggar, Xinjiang. *Acta Petrol. Sin.* **2012**, *28*, 2183–2198.
45. Zhang, H.; Niu, H.; Sato, H.; Shan, Q.; Yu, X.; Ito, J.; Zhang, Q. Late Paleozoic adakite and Nb-enriched basalt from northern Xinjiang: Evidence for the southward subduction of the Paleo-Asian Ocean. *Geol. J. China Univ.* **2004**, *10*, 106–113.
46. Zhang, C.; Liu, D.; Zeng, J.; Jiang, S.; Luo, Q.; Kong, X.; Yang, W.; Liu, L. Nd-O-Hf isotopic decoupling in S-type granites: Implications for ridge subduction. *Lithos* **2019**, *332–333*, 261–273. [CrossRef]
47. Lv, S.; Yang, F.; Chai, F.; Zhang, X.; Jiang, L.; Liu, F.; Zhang, Z.; Geng, X.; Ouyang, L. Zircon U-Pb dating for intrusions in Laoshankou ore district in Northern margin of East Junggar and their significances. *Geol. Rev.* **2012**, *58*, 149–164.
48. Chu, F.; Zhao, T.; Zhang, J.; Zhu, Z.; Tang, Z.; Yan, X. Geochemical Characteristics and Geochronology of Sherbulake Ophiolite in Xinjiang. *Xinjiang Geol.* **2013**, *31*, 167–172.
49. Wang, Y.; Zhu, Z.; Li, X.; Li, P.; Zheng, F.; Gong, X. The SHRIMP Zircon U-Pb dating of the volcanic rocks of the Tuoranggekuduke Formation in the Early Devonian, Eastern Junggar and geological implications. *Xinjiang Geol.* **2020**, *38*, 161–167.
50. Wiedenbeck, M.; Allé, P.; Corfu, F.; Griffin, W.; Meier, M.; Oberli, F.; Vonquat, A.; Roddick, J.; Spiegel, W. Three natural zircon standards for U-Th-Pb, Lu-Hf, tracelement and ree analyse. *Geostand. Newsl.* **1995**, *19*, 1–23. [CrossRef]
51. Wiedenbeck, M.; Hanchar, J.; Peck, W.; Sylvester, P.; Valley, J.; Whitehouse, M.; Kronz, A.; Morishita, Y.; Nasdala, L.; Fiebig, J.; et al. Further characterisation of the 91500 zircon crystal. *Geostand. Geoanalytical Res.* **2004**, *28*, 9–39. [CrossRef]
52. Jackson, S.; Pearson, N.; Griffin, W.; Belousova, E. The application of laser ablation-inductively coupled plasma-mass spectrometry to in situ U–Pb zircon geochronology. *Chem. Geol.* **2004**, *211*, 47–69. [CrossRef]
53. Zhang, L.; Zhu, D.; Wang, Q.; Zhao, Z.; Liu, D.; Xie, J. Late Cretaceous Volcanic Rocks in the Sangri Area, Southern Lhasa Terrane, Tibet: Evidence for Oceanic Ridge Subduction. *Lithos* **2019**, *326–327*, 144–157. [CrossRef]
54. Anderson, T. Correction of common Lead in U–Pb analyses that do not report ^{204}Pb. *Chem. Geol.* **2002**, *192*, 59–79. [CrossRef]
55. Liu, Y.S.; Hu, Z.C.; Gao, S.; Günther, D.; Xu, J.; Gao, C.G.; Chen, H.H. In Situ Analysis of Major and Trace Elements of Anhydrous Minerals by LA–ICP–MS without Applying an Internal Standard. *Chem. Geol.* **2008**, *257*, 34–43. [CrossRef]

56. Ludwig, K.R. ISOPLOT 3: A geochronological toolkit for microsoft excel. *Berkeley Geochronol. Cent. Spec. Publ.* **2003**, *4*, 74.
57. Wu, F.; Yang, Y.; Xie, L.; Yang, J.; Xu, P. Hf isotopic compositions of the standard zircons and baddeleyites used in U–Pb geochronology. *Chem. Geol.* **2006**, *234*, 105–126. [CrossRef]
58. Wu, Y.; Zheng, Y. Zircon Genetic Mineralogy and Its Constraints on U–Pb Age Interpretation. *Chin. Sci. Bull.* **2004**, *49*, 1589–1604. [CrossRef]
59. Le Bas, M.; Le Maitre, R.; Streckeisen, A.; Zanettin, B. A chemical classification of volcanic rocks based on the total alkali–silica diagram. *J. Petrol.* **1986**, *27*, 745–750. [CrossRef]
60. Irvine, T.N.; Baragar, W.R.A. A Guide to the Chemical Classification of the Common Volcanic Rocks. *Can. J. Earth Sci.* **1971**, *8*, 523–548. [CrossRef]
61. Deng, J.; Flower, M.F.J.; Liu, C.; Mo, X.; Su, S.; Wu, Z. Nomeuclature, diagnosis and origin of high-magnesian andesites (HMA) and magnesian andesites (MA): A review from petrographic and experimental data. *Geochim. Cosmochim. Acta* **2009**, *73*, A279.
62. Maniar, P.D.; Piccoli, P.M. Tectonic Discrimination of Granitoids. *Geol. Soc. Am. Bull.* **1989**, *101*, 635–643. [CrossRef]
63. Peccerillo, R.; Taylor, S.R. Geochemistry of Eocene Calc-Alkaline Volcanic Rocks from the Kastamonu Area, Northern Turkey. *Contrib. Mineral. Petrol.* **1976**, *58*, 63–81. [CrossRef]
64. Frost, B.R.; Barnes, C.G.; Collins, W.J.; Arculus, R.J.; Ellis, D.J.; Frost, C.D. A Geochemical Classification for Granitoids Rocks. *J. Petrol.* **2001**, *42*, 2033–2048. [CrossRef]
65. Le Moigne, J.; Lagabrielle, Y.; Whitechurch, H.; Girardeau, J.; Bourois, J.; Maury, R. Petrology and geochemistry of the Ophiolitic and Volcanic Suites of the Taitao Peninsula–Chile Triple Junction Area. *J. S. Am. Earth Sci.* **1996**, *9*, 43–58.
66. Ramos, V.A.; Kay, S.M. Southern Patagonian plateau basalts and deformation: Backarc testimony of ridge collisions. *Tectonophysics* **1992**, *205*, 261–282.
67. Boynton, W.V. Geochemistry of the rare earth elements: Meteorite studies. In *Rare Earth Element Geochemistry*; Henderson, P., Ed.; Elsevier: Amsterdam, The Netherlands, 1984; pp. 63–114.
68. Sun, S.; McDonough, W.F. Chemical and isotopic systematics of ocneanic basalts: Implications for mantle composition and processes. In *Magmatism in Ocean Basins*; Saunders, A.D., Norry, M.J., Eds.; Geological Society of Special Publication: London, UK, 1989; pp. 313–345.
69. Niu, Y.; O'Hara, M.J. Origin of ocean island basalts: A new perspective from petrology, geochemistry, and mineral physics considerations. *J. Geophys. Res. Solid Earth* **2003**, *108*, 2209. [CrossRef]
70. Kelemen, P.B.; Hanghøj, K.; Greene, A.R. One View of the Geochemistry of Subduction-Related Magmatic Arcs, with an Emphasis on Primitive Andesite and Lower Crust. *Treatise Geochem. Second Ed.* **2014**, *4*, 749–806. [CrossRef]
71. Green, T.H. Significance of Nb/Ta as an indicator of geochemical processes in the crust–mantle system. *Chem. Geol.* **1995**, *120*, 347–359. [CrossRef]
72. Rudnick, R.L.; Gao, S. Composition of the continental crust. *Treatise Geochem.* **2003**, *3*, 1–64.
73. Han, B.; Wang, S.; Kagami, H. Trace element and Nd-Sr isotope constraints on origin of the Chifeng flood basalts, North China. *Chem. Geol.* **1999**, *155*, 187–199. [CrossRef]
74. Reagan, M.K.; Ishizuka, O.; Stern, R.J.; Kelley, K.A.; Ohara, Y.; Blichert-Toft, J.; Bloomer, S.H.; Cash, J.; Fryer, P.; Hanan, B.B.; et al. Fore-arc basalts and subduction initiation in the Izu-Bonin-Mariana system. *Geochem. Geophys. Geosystems* **2010**, *11*, Q03X12. [CrossRef]
75. Defant, M.J.; Kepezhinskas, P. Evidence suggests slab melting in arc magmas. *Eos Trans. Am. Geophys. Union* **2001**, *82*, 65–69. [CrossRef]
76. Zhang, T.; Deng, J.; Wang, M.; Li, C.; Zhang, L.; Sun, W. Geochemistry and genesis of the Nadun Nb-enriched arc basalt in the Duolong mineral district, western Tibet: Indication of ridge subduction. *Geosci. Front.* **2022**, *13*, 101283. [CrossRef]
77. Li, S.; Zhu, D.; Wang, Q.; Zhao, Z.; Zhang, L.; Liu, S.; Chang, Q.; Lu, Y.; Dai, J.; Zheng, Y. Slab-derived adakites and subslab asthenosphere-derived OIB-type rocks at 156 ± 2 Ma from the north of Gerze, central Tibet: Records of the Bangong–Nujiang oceanic ridge subduction during the Late Jurassic. *Lithos* **2016**, *262*, 456–469.
78. La Flèche, M.R.; Camiré, G.; Jenner, G.A. Geochemistry of post-Acadian, carboniferous continental intraplate basalts from the Maritimes Basin, Magdalen islands, Québec, Canada. *Chem. Geol.* **1998**, *148*, 115–136.
79. Shaw, J.E. Petrogenesis of the Largest Intraplate Volcanic Field on the Arabian Plate (Jordan): A Mixed Lithosphere-Asthenosphere Source Activated by Lithospheric Extension. *J. Petrol.* **2003**, *44*, 1657–1679.
80. McKenzie, D.; O'Nions, R.K. Partial melt distribution from inversion of rare earth element concentrations. *J. Petrol.* **1991**, *32*, 1021–1091.
81. Aldanmaz, E.; Pearce, J.A.; Thirlwall, M.F.; Mitchell, J.G. Petrogenetic evolution of late Cenozoic, post-collision volcanism in western Anatolia, Turkey. *J. Volcan. Geotherm. Res.* **2000**, *102*, 67–95.
82. Zhang, Z.; Yan, S.; Chen, B.; Zhou, G.; He, Y.; Chai, F.; He, L. Middle Devonian picrites of south margin of Altay Orogenic Belt and implications for tectonic setting and petrogenesis. *Earth Sci.* **2005**, *30*, 289–296.

83. Defant, M.J.; Drummond, M.S. Derivation of some modern arc magmas by melting of young subducted lithosphere. *Nature* **1990**, *347*, 662–665.
84. Zhang, Q.; Wang, Y.; Liu, H.; Wang, Y.; Li, Z. On the space-time distribution and geodynamic environments of adakites in China annex: Controversies over differing opinions for adakites in China. *Earth Sci. Front.* **2003**, *10*, 385–400.
85. Zhang, Q.; Xu, J.; Wang, Y.; Xiao, L.; Liu, H.; Wang, Y.L. Diversity of adakite. *Geol. Bull. China* **2004**, *23*, 959–965.
86. Deng, J.; Liu, C.; Feng, Y.; Xiao, Q.; Su, S.; Zhao, G.; Kong, W.; Cao, W. High magnesian andesitic/dioritic rocks (HMA) and magnesian andesitic/dioritic rocks (MA): Two igneous rock types related to oceanic subduction. *Geol. China* **2010**, *37*, 1112–1118.
87. Wang, Y.; Yuan, C.; Long, X.; Sun, M.; Xiao, W.; Zhao, G.; Cai, K.; Jiang, Y. Geochemistry, zircon U–Pb ages and Hf isotopes of the Paleozoic volcanic rocks in the northwestern Chinese Altai: Petrogenesis and tectonic implications. *J. Asian Earth Sci.* **2011**, *42*, 969–985. [CrossRef]
88. Meng, G.; Tang, H.; Liu, H.; Qin, J.; Wu, X.; Li, C. Devonian igneous rocks and tectono-magmatic evolution in the Ashele basin, Xinjiang. *Acta Geol. Sin.* **2022**, *96*, 445–464.
89. Wang, T.; Hong, D.; Jahn, B.M.; Tong, Y.; Wang, Y.; Han, B.; Wang, X. Timing, Petrogenesis, and Setting of Paleozoic Synorogenic Intrusions from the Altai Mountains, Northwest China: Implications for the Tectonic Evolution of an Accretionary Orogen. *J. Geol.* **2006**, *114*, 735–751. [CrossRef]
90. Chai, F.; Mao, J.; Dong, L.; Yang, F.; Liu, F.; Geng, X.; Zhang, Z. Geochronology of metarhyolites from the Kangbutiebao Formation in the Kelang basin, Altay Mountains, Xinjiang: Implications for the tectonic evolution and metallogeny. *Gondwana Res.* **2009**, *16*, 189–200. [CrossRef]
91. Cui, X.; Sun, M.; Zhao, G.; Yao, J.; Zhang, Y.; Han, Y.; Dai, L. A Devonian arc–back-arc basin system in the southern Chinese Altai: Constraints from geochemical and Sr-Nd-Pb isotopic data for meta-basaltic rocks. *Lithos* **2020**, *366–367*, 105540. [CrossRef]
92. Lv, Z.; Zhang, H.; Tang, Y.; Liu, Y.; Zhang, X. Petrogenesis of syn-orogenic rare metal pegmatites in the Chinese Altai: Evidences from geology, mineralogy, zircon U-Pb age and Hf isotope. *Ore Geol. Rev.* **2018**, *95*, 161–181. [CrossRef]
93. Shen, X.; Zhang, H.; Wang, Q.; Saha, A.; Ma, L.; Santosh, M. Zircon U–Pb geochronology and geochemistry of Devonian plagiogranites in the Kuerti area of southern Chinese Altay, northwest China: Petrogenesis and tectonic evolution of late Paleozoic ophiolites. *Geol. J.* **2017**, *53*, 1886–1905. [CrossRef]
94. Tang, G.; Wyman, D.A.; Wang, Q.; Li, J.; Li, Z.; Zhao, Z.; Sun, W. Asthenosphere–lithosphere interaction triggered by a slab window during ridge subduction: Trace element and Sr–Nd–Hf–Os isotopic evidence from Late Carboniferous tholeiites in the western Junggar area (NW China). *Earth Planet. Sci. Lett.* **2012**, *329–330*, 84–96. [CrossRef]
95. He, Y.; Sun, M.; Cai, K.; Xiao, W.; Zhao, G.; Long, X.; Li, P. Petrogenesis of the Devonian high-Mg rock association and its tectonic implication for the Chinese Altai orogenic belt, NW China. *J. Asian Earth Sci.* **2015**, *113*, 61–74. [CrossRef]
96. Pearce, J.A.; Cann, J.R. Tectonic setting of basic volcanic rocks determined using trace element analyses. *Earth Planet. Sci. Lett.* **1973**, *19*, 290–300. [CrossRef]
97. Wood, D.A.; Joron, J.L.; Treuil, M. A re-appraisal of the use of trace elements to classify and discriminate between magma series erupted in different tectonic settings. *Earth Planet. Sci. Lett.* **1979**, *45*, 326–336. [CrossRef]
98. Sun, M.; Yuan, C.; Xiao, W.; Long, X.; Xia, X.; Zhao, G.; Lin, S.; Wu, F.; Kröner, A. Zircon U–Pb and Hf isotopic study of gneissic rocks from the Chinese Altai: Progressive accretionary history in the early to middle Palaeozoic. *Chem. Geol.* **2008**, *247*, 352–383. [CrossRef]
99. Chen, Y.; Zhu, D.; Zhao, Z.; Meng, F.; Wang, Q.; Santosh, M.; Wang, L.; Dong, G.; Mo, X. Slab breakoff triggered ca. 113 Ma magmatism around Xainza area of the Lhasa Terrane, Tibet. *Gondwana Res.* **2014**, *26*, 449–463. [CrossRef]
100. Sun, X. The Material Composition, Structural Deformation and Geological Significance of Kobita Bea J Bai Bass Tao Paleozoic Accretionary Wedge, Qinghe County, Xinjiang, China. Master's Thesis, Chang'an University, Xi'an, China, 2013.
101. Liu, W. Zircons shrimp U-Pb dating of Mayinerbuo ophilit and its geological implications. *Xinjiang Geol.* **2011**, *29*, 385–388.
102. Zhang, Z.; Zhou, G.; Yan, S.; Chen, B.; He, Y.; Chai, F.; He, L. Geology and Geochemistry of the Late Paleozoic Volcanic Rocks of the South Margin of the Altai Mountains and Implications for Tectonic Evolution. *Acta Geol. Sin.* **2007**, *81*, 344–358.
103. Zhou, G.; Qin, J.; Zhang, Z.; Zhang, L.; Ying, L.; Ahemet, J.; Mao, W.; He, B.; Deng, J.; Cai, Y. The Discovery of Bimodal Volcanics in Supute Area of Fuyun County, Xinjiang, and Its Geological Significance. *Geol. Rev.* **2007**, *53*, 337–348.
104. Sun, M.; Long, X.; Cai, K.; Jiang, Y.; Wang, B.; Yuan, C.; Zhao, G.; Xiao, W.; Wu, F. Early Paleozoic ridge subduction in the Chinese Altai: Insight from the abrupt change in zircon Hf isotopic compositions. *Sci. China Ser. D Earth Sci.* **2009**, *52*, 1345–1348. [CrossRef]
105. Zhang, J.; Wang, J.; Ding, R. Petrology of marine volcanics and its relationship with ore-forming progess in the Maizi area, Altay. *Geol. Rev.* **1999**, *45*, 1116–1125.
106. Chen, Y.; Liu, D.; Wang, D.; Tang, Y.; Zhou, R.; Chen, Z. Discover and geological significance of picrite rocks in north Junggar, Xinjiang. *Geol. Bull. China* **2004**, *23*, 1059–1065.
107. Zhang, H.; Niu, H.; Yu, X.; Sato, H.; Ito, J.; Shan, Q. Geochemical characteristic of the shaerbulake boninites and their tectonic significance, Fuyun County, northern Xinjiang, China. *Geochimica* **2003**, *32*, 155–160.

08. Shi, Y.; Wang, Y.; Wang, J.; Wang, L.; Ding, R. Petrogenesis of ankaramite from Middle Devonian Beitashan Formation, northern margin of Junggar and its implication on Late Paleozoic ridge subduction. *Acta Petrol. Sin.* **2017**, *33*, 565–578.
09. Meneghini, F.; Kisters, A.; Buick, I.; Fagereng, Å. Fingerprints of late Neoproterozoic ridge subduction in the Pan-African Damara belt, Namibia. *Geology* **2014**, *42*, 903–906. [CrossRef]

applied sciences

MDPI

Article

A Thermodynamic Model for the Solubility of SO_2 in Multi-Ion Electrolyte Solutions and Its Applications

Baoyi Feng, Zequn Zhang, Mei Xu and Shide Mao *

School of Earth Sciences and Resources, China University of Geoscience, Beijing 100083, China; 2001230075@email.cugb.edu.cn (B.F.); 15822242726@163.com (Z.Z.); 3001230046@email.cugb.edu.cn (M.X.)
* Correspondence: maoshide@163.com

Abstract: A solubility model of SO_2 in multi-ion electrolyte solutions has been developed by the activity-fugacity relation at vapor-liquid equilibria. The fugacity coefficient of SO_2 in the vapor phase is calculated by the equation of state (EOS) of pure SO_2, and the activity coefficient of SO_2 in the liquid phase is calculated by the Pitzer activity coefficient theory. The model can reproduce the reliable solubility data of SO_2 in pure water and multi-ion electrolyte solutions (Na^+, K^+, Cl^-, SO_4^{2-}) within or close to experimental uncertainties. Although the second-order and third-order interaction parameters between SO_2 and Mg^{2+} and Ca^{2+} have been adopted by an approximation, the solubility model can also be extended to predict the SO_2 solubility in seawater. In addition, combining with the EOS of a CO_2-SO_2 fluid mixture, the model can be used to predict the solubility of a CO_2-SO_2 mixture in aqueous electrolyte solutions. The calculated results are consistent with experimental data, which indicates that the solubility model has certain predictive ability.

Keywords: SO_2; solubility model; electrolyte solutions; equation of state

1. Introduction

SO_2 in the atmosphere mainly comes from volcanic eruptions [1], burning fossil fuels in power plants [2], metal processing and smelting [3]. In the process of CO_2 capture and sequestration (CCS), SO_2 is also an impurity, and to sequester it with CO_2 in the deep saline aquifers can reduce the cost. Deep saline aquifers contain a variety of electrolyte ions (e.g., Na^+, K^+, Ca^+, Mg^{2+}, Cl^-, SO_4^{2-}). Therefore, it is essential to study the solubility of single gases and mixed gases in multi-ion electrolyte solutions. The solubility model of CO_2 in multi-ion electrolyte solutions has been reported in our previous studies [4,5], so this work focuses on the solubility model of SO_2 in multi-ion electrolyte solutions under different temperature-pressure (*T*-*P*) conditions. This research is of great practical significance for the co-storage of CO_2-SO_2 gas mixturlles in deep saline aquifers.

Due to the strong corrosive potential of SO_2, experimental solubility data for SO_2 in multi-ion electrolyte solutions are difficult to collect, especially at high temperatures and pressures. Therefore, predictive SO_2 solubility models are desired. In recent decades, many scholars have developed thermodynamic models to predict SO_2 solubility in multi-ion electrolyte solutions (Table 1).

Table 1. Thermodynamic models for calculating SO_2 solubilities in pure water, aqueous salt solutions or seawater.

References	Salt	T (K)	P (bar)	m_{salt} (mol/kg)
Hunger et al. [6]	H_2SO_4, Na_2SO_4	298–323	10^{-4}–0.1	0-saturation
Xia et al. [7]	NaCl, NH_4Cl	313–393	0–37	0–6
Rodríguez-Sevilla et al. [8]	seawater	278.15–318.15	0.976	unstated
Mondal [9]	water	293–333	0–0.01 *	0
Zimmermann et al. [10]	HCl, NaCl, HBr, NaBr	298	1.015	0–4.613
Shaw et al. [11]	water	298–313	0.2–3.6	0
Tan et al. [12]	water, NaCl	313–393	0–40	0–6
Cox et al. [13]	seawater, brine	298.15–323.15	unstated	0–1
Zhou et al. [14]	water	263.15–393.15	10–300	0
Saadallah et al. [15]	water	293–393	0–25	0

Notes: T, P and m_{salt} denote the temperature, pressure and molality of salt, respectively; * is the partial pressure of SO_2.

Hunger et al. [6] applied the methods from Meissner [16,17] and Bromley [18] to estimate activity coefficients, which can predict the solubility of SO_2 in Na_2SO_4 or H_2SO_4 solutions from 10^{-4} to 0.1 bar at 298 K and 323 K. Xia et al. [7] developed a phase equilibrium model to calculate SO_2 solubility in NaCl and NH_4Cl aqueous solutions, from 313 to 393 K at total pressures up to 37 bar, which is capable of representing most of the experimental data within experimental uncertainties. Rodríguez-Sevilla et al. [8] adopted the extended Debye-Hückel theory [19] and the Pitzer [20] ion-interaction model to evaluate the activity coefficients of ionic species, which can be used to predict the absorption equilibria of dilute SO_2 in seawater in the temperature range from 278.15 to 318.15 K with a pressure of up to 0.976 bar. Mondal [9] proposed a model for the solubility of SO_2 using Henry's Law in pure water with temperatures ranging between 293 and 333 K, with SO_2 partially pressurized up to 0.01 bar. Zimmermann et al. [10] developed a thermodynamic model with the Pitzer activity coefficient theory and the Sechenov's approach from Weisenberger and Schumpe [21], which can predict the SO_2 solubility in HCl, NaCl, HBr and NaBr electrolyte solutions. The applicable P-T-m range is 298 K, 1.015 bar, 0–4.61 kg/mol. The developed thermodynamic model by Zimmermann et al. [10] is in good agreement with experimental data. Shaw et al. [11] developed a solubility model of SO_2 in pure water with a valid T-P range of 298–313 K and 0.2–3.6 bar. The model is in good agreement with experimental data at lower pressures, but improvements are needed at higher pressures ($p > 1.5$ bar). Tan et al. [12] adopted the PC-SAFT/PMSA and CPA EOS to evaluate SO_2 solubility in pure water and NaCl electrolyte solutions up to 393 K and 40 bar and 6 mol/kg. Miri et al. [22] applied the SAFT1 model to estimate CO_2-SO_2 solubility in pure water, covering a valid T-P range of 303.15–373.15 K and 60–300 bar. Cox et al. [13] developed an equilibrium model by the Pitzer and Sechenov approach, which provides a reliable prediction of SO_2 solubility in seawater and brine at 298 K and 323 K with molality of electrolytes up to 1 mol/kg. Zhou et al. [14] proposed a PC-SAFT model to predict the SO_2 solubility in water, covering a valid T-P range of 263.15–393.15 K and 10–300 bar, which can accurately predict the SO_2 solubility in water above 10 bar and 273.15 K, but at lower temperatures and pressures, it cannot accurately calculate the SO_2 solubility. Saadallah et al. [15] used the Monte Carlo (MC) molecular simulation technique to predict the vapor-liquid and liquid-liquid equilibria including the solubility data of SO_2-H_2O mixture in the temperature range of 293–393 K and pressures up to 25 bar, and the results from the MC simulation are in agreement with experimental data at liquid-vapor and liquid-liquid equilibria.

As mentioned above, the models of SO_2 solubility in multi-ion electrolyte solutions are still lacking, and most of them either have lower accuracy or have a limited temperature-pressure-salinity range. Therefore, it is necessary to develop an accurate model for the solubility of SO_2 in multi-ion electrolyte solutions. In this work, we first review experimental data of SO_2 solubility in pure water and multi-ion electrolyte solutions. Then an accurate thermodynamic model is developed to predict the SO_2 solubility in water and multi-ion electrolyte solutions by the activity-fugacity relationship at vapor-liquid equilibria. Finally, the applications of the model are briefly discussed.

2. Review of Experimental Data for SO_2 Solubility in Pure Water and Aqueous Electrolyte Solutions

In this article, the solubility data of four different systems (SO_2-H_2O, SO_2-NaCl-H_2O, SO_2-KCl-H_2O and SO_2-Na_2SO_4-H_2O) have been found and compiled as shown in Table 2, where the information of each data set is provided, including the temperature, pressure, salinity and total number of data points. SO_2 solubility in water represents the total amount of SO_2 species dissolved in water, including SO_2(aq), HSO_3^- and SO_3^{2-}.

Table 2. SO_2 solubility measurements in pure water and aqueous single-salt solutions.

References	Solution	T (K)	P (bar)	Nd
Bunsen [23]	water	273.00–313.00	1^+	41
Sims [24]	water	281.20–323.20	1^+	22
Fox [25]	water	298.20–308.20	1^+	2
Simth and Parkhurst [26]	water	278.20–333.20	0.20–1.60	8
Hudson [27]	water	283.15–363.15	1^+	42
Sherwood [28]	water	273.15–323.15	0.00–1.00	109
Maass and Maass [29]	water	283.20–300.20	0.00–3.50	29
Campbell and Maass [30]	water	298.15–383.15	0.14–4.50	55
Morgan and Maass [31]	water	273.00–298.00	0.00–1.20	38
Conrad and Beuschlein [32]	water	298.20	0.40–1.00	6
Johnstone and Leppla [33]	water	298.20–323.20	0.00–0.20	16
Young [34]	water	373.00–423.00	1.70–7.20	22
Beuschlein and Simenson [35]	water	296.40–386.15	0.12–2.11	53
Parkinson [36]	water	263.00–294.00	0.00–0.06	42
Rabe and Harris [37]	water	303.15–353.15	0.09–1.20	43
Vosolsobe et al. [38]	water	293.20–333.20	0.04–0.34	35
Tokunaga [39]	water	283.15–313.15	1^+	4
Lavrova and Tudorovskaya [40]	water	299.15–363.15	1^+	5
Douabul and Riley [41]	water	278.97–303.25	1^+	6
Byerley et al. [42]	water	298.15–323.15	1^+	2
Rumpf and Maurer [43]	water	293.14–393.33	0.35–25.09	66
Siddiqi et al. [44]	water	291.05–294.45	0.02–0.03	50
Mondal [9]	water	293.00–333.00	0.02–0.21	20
Shaw et al. [11]	water	297.05–312.25	0.25–1.96	3
Cox et al. [13]	water	298.15–323.15	0.00–0.12	19
Zhou et al. [14]	water	273.15–373.15	10.00–300.00	45
Saadallah et al. [15]	water	293.15–353.15	0.00–12.00	45
	m_{NaCl} (mol/kg)			
Fox [25]	0.50–3.22	298.00–308.00	1^+	6
Millero et al. [45]	0.01–6.00	278.15–298.15	1^+	17
Xia et al. [7]	2.90–6.00	313.15–393.15	0.30–37.00	90
Zimmermann et al. [10]	0.49–4.28	298.15	0.00–0.30	37
Cox et al. [13]	1.00	298.15–323.15	0.03–0.12	19
	m_{KCl} (mol/kg)			
Fox [25]	0.00–0.05	298.00–308.00	1^+	12
Hudson [27]	0.50–4.40	298.00–308.00	1^+	43
Cox et al. [13]	0.05–1.00	298.15–323.15	1.00	19
	$m_{Na_2SO_4}$ (mol/kg)			

Table 2. *Cont.*

References	Solution	*T* (K)	*P* (bar)	Nd
Fox [25]	0.50–3.22	298.00–308.00	1	12
Hudson [27]	0.13–1.41	293.15–313.15	1^+	23
Rumpf and Maurer [46]	0.50–1.00	313.15–393.15	1.00–28.00	65

Notes: T, P and m_{salt} denote the temperature, pressure and molality of salt, respectively; Salt stands for NaCl, KCl and Na_2SO_4, respectively; Nd is the number of experimental data. 1^+ is the partial pressure of SO_2.

2.1. SO_2-H_2O Systems

Experimental data of SO_2 solubility in pure water are relatively extensive (Table 2). However, some of the solubility data are unreliable. Sherwood [28] measured the SO_2 solubility in water under relatively low pressure, but the data are quite different from the reliable data of others under the same conditions. Experimental data of SO_2 solubility from Johnstone and Leppla [33], Parkinson [36] and Vosolsobe et al. [38] were evaluated in the solubility data series book by Young [34]. Because their pressure is too low to be judged reasonably, experimental data of solubility have a large deviation. Experimental SO_2 solubility data of Siddiqi et al. [44], Mondal [9], Shaw et al. [11] and Cox et al. [13] are measured for the SO_2-N_2-H_2O system under extremely low pressures. The data yield big deviations when they are used in parameterization. Most of the experimental SO_2 solubility data of Morgan and Maass [31], Beuschlein and Simenson [35], Rabe and Harris [37], Zhou et al. [14] and Saadallah et al. [15] are consistent with one another. Figure 1 shows that the experimental SO_2 solubility data of Bunsen [23], Sims [24], Fox [25], Tokunaga [39], Lavrova and Tudorovskaya [40] and Byerley et al. [42] are consistent with each other when the partial pressure of SO_2 is 1 atm, but the data of Douabul and Riley [41] at 280 K yield a slight deviation. Therefore, all the data except the unreliable data mentioned above are used in the parameterization of this model, covering a wide *T*-*P* range of 273.15–393.15 K and 0.00–300.00 bar.

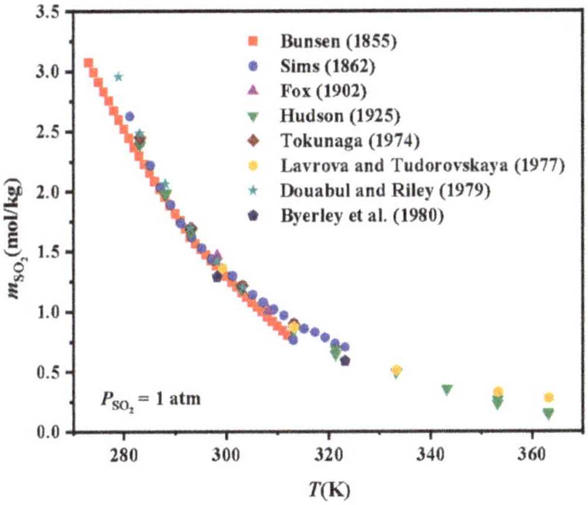

Figure 1. SO_2 solubility in pure water (experimental data) [23–25,27,39–42].

2.2. SO_2-NaCl-H_2O Systems

There are only five experimental data sets reported for SO_2-NaCl-H_2O systems (Table 2). The experimental SO_2 solubility data of Zimmermann et al. [10] and Cox et al. [13] are measured for SO_2-N_2-NaCl-H_2O systems under low pressures, and they are not used in the parameterization. The partially experimental SO_2 solubility data of Xia et al. [7] yield big deviations when they are used in parameterization. Therefore, the remaining reliable data are used to optimize the model parameters, covering a T-P-m_{NaCl} range of 278.15–393.15 K, 0.00–35.07 bar and 0–6 mol/kg.

2.3. SO_2-KCl-H_2O Systems

Experimental data of SO_2 solubility in aqueous KCl solutions are relatively scarce. Only 74 data points have been found for this type of system. Data from Cox et al. [13] for the SO_2-KCl-H_2O systems have the same problem as the SO_2-NaCl-H_2O systems. Due to the small amount of experimental data, the reliability of the experimental data cannot be determined. Therefore, the remaining two data sets are retained, covering a T-P-m_{KCl} range of 298.00–308.00 K, 1 bar and 0.00–4.40 mol/kg.

2.4. SO_2-Na_2SO_4-H_2O Systems

Only three data sets of SO_2 solubility have been reported for SO_2-Na_2SO_4-H_2O systems. When some SO_2 solubility data from Hudson [27] and Rumpf and Maurer [46] are added to optimize the model parameters, this yields big deviations. Therefore, the other reliable solubility data are used to optimize the model parameters, covering a T-P-$m_{Na_2SO_4}$ range of 298.00–393.15 K, 0.63–27.69 bar and 0.13–3.22 mol/kg.

3. The Thermodynamic Model for SO_2 Solubility in Multi-Ion Solutions and Parameterization

3.1. The Thermodynamic Model for SO_2 Solubility in Multi-Ion Solutions

SO_2 solubility in aqueous electrolyte solutions depends on the balance between the chemical potential in the liquid phase $\mu^l_{SO_2}$ and that in the vapor phase $\mu^v_{SO_2}$. The chemical potential can be written in terms of fugacity in the vapor phase and activity in the liquid phase:

$$\begin{aligned}\mu^l_{SO_2}(T,P,m_{salt}) &= \mu^{l(0)}_{SO_2}(T,P) + RT\ln\alpha_{SO_2}(T,P,m_{salt}) \\ &= \mu^{l(0)}_{SO_2}(T,P) + RT\ln m_{SO_2} + RT\ln\gamma_{SO_2}(T,P,m_{salt})\end{aligned} \tag{1}$$

$$\begin{aligned}\mu^v_{SO_2}(T,P,y) &= \mu^{v(0)}_{SO_2}(T) + RT\ln f_{SO_2}(T,P,y_{SO_2}) \\ &= \mu^{v(0)}_{SO_2}(T) + RT\ln y_{SO_2}P + RT\ln\varphi_{SO_2}(T,P,y_{SO_2})\end{aligned} \tag{2}$$

where $\mu^{l(0)}_{SO_2}$, the standard chemical potential of SO_2 in liquid, is defined as the chemical potential in a hypothetically ideal solution of unit molarity [47], and $\mu^{v(0)}_{SO_2}$, the standard chemical potential of SO_2 in vapor, is the hypothetical ideal gas chemical potential when the pressure is equal to 1 bar. T and P denote temperature and pressure, respectively. R is the molar gas constant. f_{SO_2} is the fugacity of SO_2 in the vapor phase, y_{SO_2} is the mole fraction of SO_2 in the vapor phase, φ_{SO_2} is the fugacity coefficient of SO_2. α_{SO_2} is the activity of SO_2, and m_{SO_2} and m_{salt} are the molality of SO_2 and salt, respectively. γ_{SO_2} is the activity coefficient of SO_2 in the liquid phase.

At phase equilibrium $\mu^l_{SO_2} = \mu^v_{SO_2}$, we obtain

$$\begin{aligned}\ln m_{SO_2} = {} & \ln y_{SO_2}P + \ln\varphi_{SO_2}(T,P,y_{SO_2}) \\ & -\frac{\mu^{l(0)}_{SO_2}(T,P) - \mu^{v(0)}_{SO_2}(T)}{RT} - \ln\gamma_{SO_2}(T,P,m_{salt})\end{aligned} \tag{3}$$

where $\mu^{v(0)}_{SO_2}$ can be assigned as zero for convenience in the parameterization because only the difference between these two chemical potentials is important. In this work, $\ln\varphi_{SO_2}$ can be calculated from EOS of pure SO_2 by Li [48] (see Appendix A), which will give rise to deviations, which can be canceled out in the later parameterization. y_{SO_2} is calculated by the following equation:

$$y_{SO_2} = \frac{P - P_{H_2O}^{salt}}{P} \tag{4}$$

where $P_{H_2O}^{salt}$ is the partial pressure of water in the vapor phase, which is approximated by Raoult's law, defined as

$$P_{H_2O}^{salt} = P_{H_2O}^{s} x_{H_2O} \tag{5}$$

where $P_{H_2O}^{s}$ is the saturated pressure of pure H_2O, which is calculated from the model of Wagner and Kretzschmar [49], and x_{H_2O} is the gas-free mole fraction of H_2O in the liquid phase.

$\ln \gamma_{SO_2}$ is expressed as a virial expansion of excess Gibbs energy [20].

$$\ln \gamma_{SO_2} = \sum_c 2\lambda_{SO_2-c} m_c + \sum_a 2\lambda_{SO_2-a} m_a + \sum_c \sum_a \xi_{SO_2\text{-a-c}} m_c m_a \tag{6}$$

where λ and ξ are second-order and third-order interaction parameters, respectively; c and a refer to cation and anion, respectively. γ_{SO_2} is not the real activity coefficient in aqueous salt solution. Substituting Equation (6) into Equation (3) yields

$$\begin{aligned} \ln m_{SO_2} = & \ln y_{SO_2} P + \ln \varphi_{SO_2} - \frac{\mu_{SO_2}^{l(0)}}{RT} \\ & - \sum_c 2\lambda_{SO_2-c} m_c - \sum_a 2\lambda_{SO_2-a} m_a - \sum_c \sum_a \xi_{SO_2\text{-a-c}} m_c m_a \end{aligned} \tag{7}$$

Following Pitzer et al. [50], we choose the following equation for the P-T dependence of λ's, ξ's and $\mu_{SO_2}^{l(0)}/RT$:

$$Par(T, P) = c_1 + c_2 T + \frac{c_3}{T} + c_4 PT \tag{8}$$

Equations (7) and (8) form the basis of the solubility model.

3.2. Parameterization

Since measurements can only be made in electronically neutral solutions, one of the parameters must be assigned arbitrarily [51–53]. $\lambda_{SO_2-Cl^-}$ is set to zero for convenience and remaining parameters are fit to the experimental solubility data. $\mu_{SO_2}^{l(0)}/RT$ is directly evaluated from the solubility data of SO_2 in pure water. $\lambda_{SO_2-Na^+}$ and $\xi_{SO_2-Na^+-Cl^-}$ are obtained by fitting the data to the reliable solubility data of SO_2 in aqueous NaCl solution. $\lambda_{SO_2-K^+}$ is determined by a linear regression to the reliable solubility data of SO_2 in aqueous KCl solutions reviewed above. $\lambda_{SO_2-SO_4^{2-}}$ is determined from the reliable solubility data of SO_2 in aqueous Na_2SO_4 solutions. Up to now, experimental data of SO_2 solubility in aqueous $MgCl_2$, $CaCl_2$ solutions have not been reported. Because SO_2 and CO_2 are the acid gases, $\lambda_{SO_2-Ca^{2+}}$, $\xi_{SO_2-Ca^{2+}-Cl^-}$, $\lambda_{SO_2-Mg^{2+}}$ and $\xi_{SO_2-Mg^{2+}-Cl^-}$ are approximated as $\lambda_{CO_2-Ca^{2+}}$, $\xi_{CO_2-Ca^{2+}-Cl^-}$, $\lambda_{CO_2-Mg^{2+}}$ and $\xi_{CO_2-Mg^{2+}-Cl^-}$ of Shi and Mao [5], respectively. Because ξ_{SO_2}-K^+-Cl^- and ξ_{SO_2}-Na^+-SO_4^{2-} have little effect on the SO_2 solubility, they are set as zero in this work. Table 3 lists the optimized parameters.

Table 3. Interaction parameters for SO_2 solubility.

T-P Coefficient	$\mu^{l(0)}_{SO_2}/RT$	$\lambda_{SO_2\text{-}Na^+}$	$\zeta_{SO_2\text{-}Na^+\text{-}Cl^-}$
c_1	0.13245076×10^{2}	$-0.11148943 \times 10^{1}$	0.2296582×10^{-4}
c_2	$-0.69839713 \times 10^{-4}$	$0.18138955 \times 10^{-4}$	
c_3	$-0.34267139 \times 10^{4}$	0.17449283×10^{3}	
c_4	$0.48022032 \times 10^{-7}$	$-0.69223494 \times 10^{-7}$	
	$\lambda_{SO_2\text{-}K^+}$	$\lambda_{SO_2\text{-}SO_4^{2-}}$	
c_1	0.14460352×10^{1}	0.46735219×10^{1}	
c_2	$-0.019673290 \times 10^{-5}$	$-0.079666020 \times 10^{-5}$	
c_3	$-0.27249399 \times 10^{3}$	$-0.68220367 \times 10^{3}$	
	* $\lambda_{SO_2\text{-}Mg^{2+}}$	* $\zeta_{SO_2\text{-}Mg^{2+}\text{-}Cl^-}$	
c_1	1.6528428×10^{-1}	-1.152724×10^{-2}	
c_2	1.4572839×10^{-4}		
	* $\lambda_{SO_2\text{-}Ca^{2+}}$	* $\zeta_{SO_2\text{-}Ca^{2+}\text{-}Cl^-}$	
c_1	-5.513825×10^{-1}	4.2222945×10^{-3}	
c_2	1.0502705×10^{-3}	$-4.44659191 \times 10^{-5}$	
c_3	1.3349619×10^{2}		
c_4	$-1.9361945 \times 10^{-4}$		
c_5	8.9529014×10^{-7}		
c_6	$-3.5780857 \times 10^{-10}$		

Note: * denotes that $\lambda_{SO_2-Ca^{2+}}$, $\zeta_{SO_2-Ca^{2+}-Cl^-}$, $\lambda_{SO_2-Mg^{2+}}$ and $\zeta_{SO_2-Mg^{2+}-Cl^-}$ are approximated as $\lambda_{CO_2-Ca^{2+}}$, $\zeta_{CO_2-Ca^{2+}-Cl^-}$, $\lambda_{CO_2-Mg^{2+}}$, and $\zeta_{CO_2-Mg^{2+}-Cl^-}$ [5], respectively.

4. Comparison with Experimental Data

Table 4 shows the average and maximal absolute deviations of the model from each data set for the SO_2 solubility in pure water and aqueous NaCl, KCl and Na_2SO_4 solutions. Figures 2–4 show the comparisons between the experimental results and model predictions.

Table 4. Calculated deviations of SO_2 solubilities in pure water and aqueous single-salt solutions.

References	*T* (K)	*P* (bar)	m_{salt} (mol/kg)	Nd	AAD (%)	MAD (%)
Bunsen [23]	273.00–313.00	1+	0	41	3.58	16.39
Sims [24]	281.20–323.20	1+	0	22	2.65	−6.77
Fox [25]	298.20–308.20	1+	0	2	3.62	−5.76
Simth and Parkhurst [26]	278.20–333.20	0.20–1.60	0	8	4.72	−7.39
Hudson [27]	283.15–363.15	1+	0	42	2.91	−5.43
Maass and Maass [29]	283.20–300.20	0.00–3.50	0	29	2.53	−5.67
Campbell and Maass [30]	298.15–383.15	0.14–4.50	0	55	6.95	−15.82
Morgan and Maass [31]	273.00–298.00	0.00–1.20	0	22	6.75	−16.90
Conrad and Beuschlein [32]	298.20	0.40–1.00	0	6	2.03	−3.03
Young [34]	373.00–423.00	1.70–7.20	0	9	6.30	13.72
Beuschlein and Simenson [35]	296.40–386.15	0.12–2.11	0	51	5.44	−17.55
Rabe and Harris [37]	303.15–353.15	0.09–1.20	0	39	4.79	−15.76
Tokunaga [39]	283.15–313.15	1+	0	4	3.78	−6.76
Lavrova and Tudorovskaya [40]	299.15–363.15	1+	0	5	2.04	3.48
Douabul and Riley [41]	278.97–303.25	1+	0	6	5.53	−10.35
Byerley, et al. [42]	298.15–323.15	1+	0	2	11.37	15.30
Rumpf and Maurer [43]	293.14–393.33	0.35–25.09	0	55	6.04	−17.19
Zhou et al. [14]	273.15–373.15	10.00–300.00	0	34	8.11	−21.46
Saadallah et al. [15]	293.15–353.15	0.00–12.00	0	36	6.98	−19.08
		m_{NaCl}(mol/kg)				
Fox [25]	298.00–308.00	1+	0.50–3.22	6	1.46	3.64
Millero et al. [45]	278.15–298.15	1+	0.01–6.00	17	4.52	−13.96
Xia, et al. [7]	313.15–393.15	0.30–37.00	2.90–6.00	57	5.33	−14.11
		m_{KCl}(mol/kg)				
Fox [25]	298.00–308.00	1+	0.00–0.05	12	11.53	−21.56
Hudson [27]	298.00–308.00	1+	0.50–4.40	43	1.28	−4.6

Table 4. *Cont.*

References	T (K)	P (bar)	m_{salt} (mol/kg)	Nd	AAD (%)	MAD (%)
		$m_{Na_2SO_4}$(mol/kg)				
Fox [25]	298.00–308.00	1	0.50–3.22	12	1.46	3.64
Hudson [27]	293.15–313.15	1^+	0.13–1.41	18	3.45	−5.73
Rumpf and Maurer [46]	313.15–393.15	1.00–28.00	0.50–1.00	35	6.18	18.07

Notes: T, P and m_{salt} denote the temperature, pressure and molality of salt, respectively; Nd is the number of experimental data; AAD is the average absolute deviation calculated from this model; MAD is the maximal absolute deviation calculated from this model. 1^+ is the partial pressure of SO_2.

4.1. SO_2-H_2O Systems

It can be seen in Table 4 and Figure 2 that the SO_2 solubility in pure water calculated from this model is in good agreement with most of the experimental data. Although there are large deviations for individual experimental data, such as Bunsen [23], Campbell and Maass [30], Morgan and Maass [31], Beuschlein and Simenson [35], Rabe and Harris [37], Byerley et al. [42], Rumpf and Maurer [43], Zhou et al. [14] and Saadallah et al. [15], the average relative deviation is 5.24%, which is within the experimental uncertainties of solubility (≈7%) [43].

Figure 2b–e shows the comparison between experimental data at 298.15–343.15 K and the calculated solubility curves from this model, Shaw et al. [11] and Saadallah et al. [15]. Figure 2b,c indicates that this solubility model is better than the model of Shaw et al. [11] at high pressures. This model can reproduce most of the experimental data at low pressures, but the model of Saadallah et al. [15] yields somewhat big deviations (Figure 2c–e). In all, the model can reproduce the solubility of SO_2 in water up to 423 K and 300 bar.

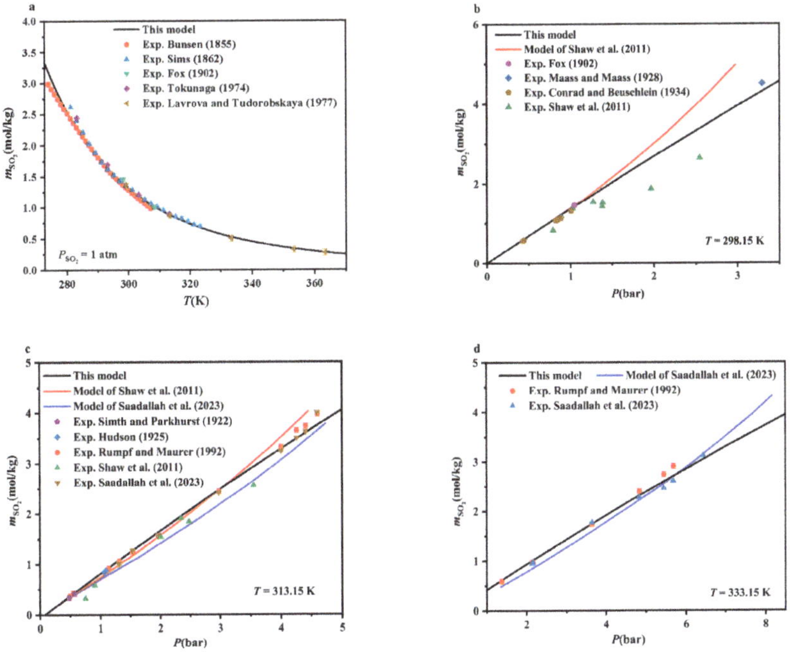

Figure 2. *Cont.*

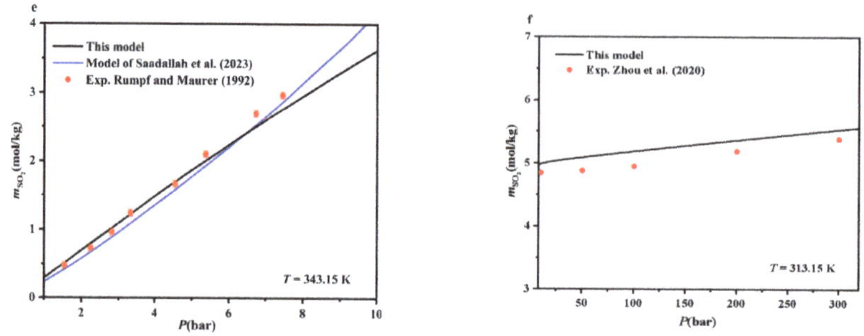

Figure 2. SO_2 solubilities in pure water (models vs. experimental data) [11,14,15,23–27,29,32,39,40,43]. (**a**) partial pressure of SO_2 = 1 atm; (**b**) T = 298.15 K; (**c**) T = 313.15 K at low pressures; (**d**) T = 333.15 K; (**e**) T = 343.15 K; (**f**) T = 313.15 K at high pressures. m_{SO_2}, T and P denote the solubility of SO_2, temperature and pressure, respectively.

4.2. SO_2-NaCl-H_2O Systems

Figure 3a–c shows that the SO_2 solubilities in aqueous NaCl solutions calculated from this model are in good agreement with the experimental data. The average absolute deviation of SO_2 solubility in an aqueous NaCl solution is 4.84%, within the experimental uncertainties of solubility. Figure 3 shows that the SO_2 solubility curve calculated from this model is consistent with the model of Tan et al. [12] at 393.15 K, but this model is more accurate than the model of Tan et al. [12] at 353.215 K and 373.15 K. The model can be applied in the T-P-m_{NaCl} range of 273–400 K, 0–50 bar and 0–6 mol/kg.

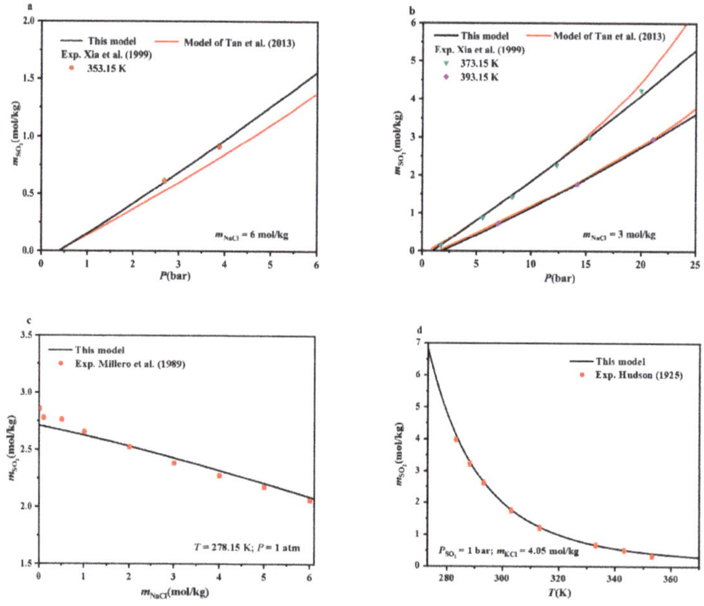

Figure 3. SO_2 solubilities in aqueous NaCl or KCl solutions (models vs. experimental data) [7,12,27,45]. (**a**) molality of NaCl = 6 mol/kg; (**b**) molality of NaCl = 3 mol/kg; (**c**) T = 278.15 K and P = 1 atm; (**d**) partial pressure of SO_2 = 1 atm and molality of KCl = 4.05 mol/kg. m_{SO_2}, T and P denote the solubility of SO_2, temperature and pressure, respectively. m_{NaCl} and m_{KCl} are the molality of NaCl and KCl.

4.3. SO_2-KCl-H_2O Systems

Figure 3d shows that the solubility curve of SO_2 in aqueous KCl solutions calculated from this model is in good agreement with the experimental data. The average absolute deviation of SO_2 solubility in an aqueous KCl solution is 3.51%, within the experimental uncertainties of solubility. The model can reproduce experimental solubility data from 298 to 308 K with a salinity of 0–4.5 mol/kg at one atmospheric pressure.

4.4. SO_2-Na_2SO_4-H_2O Systems

Figure 4 shows that the SO_2 solubilities in aqueous Na_2SO_4 solutions calculated from this model are in agreement with the experimental data. The average absolute deviation of SO_2 solubility in an aqueous Na_2SO_4 solution is 4.55%, which is within experimental uncertainties. On the whole, the model can approximately predict the solubility of SO_2 in aqueous Na_2SO_4 solutions.

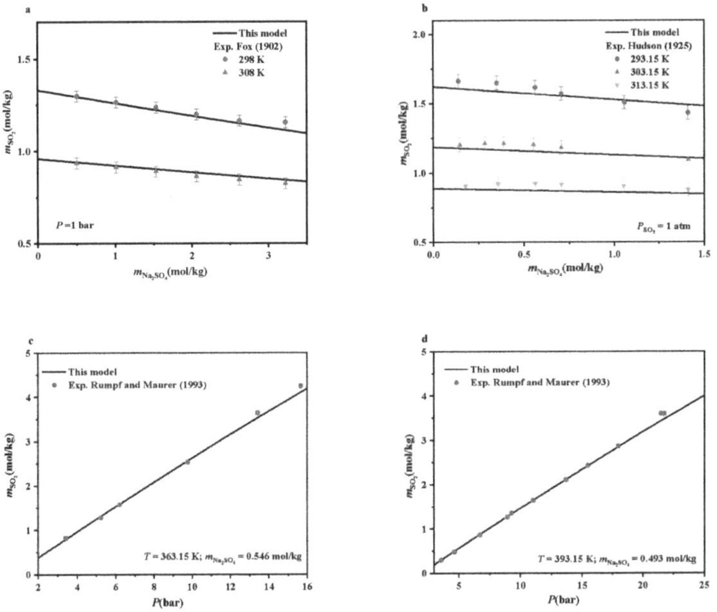

Figure 4. SO_2 solubilities in aqueous Na_2SO_4 solutions (models vs. experimental data), [25,27,43]. (**a**) $P = 1$ bar; (**b**) partial pressure of $SO_2 = 1$ atm; (**c**) $T = 363.15$ K and molality of $Na_2SO_4 = 0.546$ mol/kg; (**d**) $T = 393.15$ K and molality of $Na_2SO_4 = 0.493$ mol/kg. m_{SO_2}, T and P denote the solubility of SO_2, temperature and pressure, respectively. $m_{Na_2SO_4}$ is the molality of Na_2SO_4.

5. Applications of the Solubility Model

5.1. Prediction of SO_2 Solubility in Seawater

The solubility model of SO_2 in multi-ion electrolyte solutions can be used to predict the solubility of SO_2 in seawater. The average absolute deviation of our model from experimental SO_2 solubility in seawater is about 7.48% (Table 5). It can be seen in Figure 5 that the calculated results are a bit lower than the experimental data. It may be that the experimental solubility data of SO_2 in aqueous $MgCl_2$, $CaCl_2$ and K_2SO_4 solutions have not been found, so the second-order and third-order interaction parameters between SO_2 and Mg^{2+} and Ca^{2+} cannot be obtained, and they are only approximated as $\lambda_{CO_2-Ca^{2+}}$, $\xi_{CO_2-Ca^{2+}-Cl^-}$, $\lambda_{CO_2-Mg^{2+}}$ and $\xi_{CO_2-Mg^{2+}-Cl^-}$ from Shi and Mao [5].

Table 5. Calculated deviations of SO_2 solubilities in seawater.

System	Reference	*T* (K)	*P* (bar)	$m_{Seawater}$ (g/kg)	Nd	AAD (%)	MAD (%)
SO_2-Seawater	Douabul and Riley [41]	278.97–303.25	1^+	10.00–40.00	24	7.48	12.75

Notes: *T*, *P* and $m_{Seawater}$ denote the temperature, pressure and salinity of seawater, respectively; Nd is the number of experimental data; AAD is the average absolute deviation calculated from this model; MAD is the maximal absolute deviation calculated from this model. 1^+ is the partial pressure of SO_2.

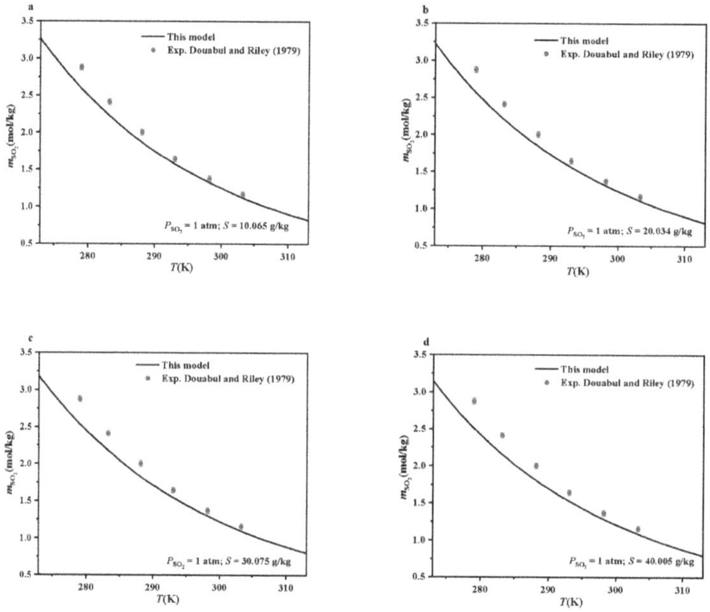

Figure 5. SO_2 solubilities in seawater (models vs. experimental data) [41]. (**a**) seawater salinity = 10.065 g/kg; (**b**) seawater salinity = 20.034 g/kg; (**c**) seawater salinity = 30.075 g/kg; (**d**) seawater salinity = 40.005 g/kg. m_{SO_2} is the solubility of SO_2, *T* is the temperature and P_{SO_2} is the partial pressure of SO_2.

5.2. Prediction of CO_2 Solubility of the CO_2-SO_2 Mixture in Pure Water

In this work, the fugacity-activity model can be used to predict the solubility of CO_2-SO_2 mixed gases in multi-ionic electrolyte solutions. If a fugacity coefficient of CO_2 or SO_2 is calculated from the EOS of CO_2-SO_2 of Li [48], combining the solubility model of CO_2 developed by Shi and Mao [5] and the solubility model of SO_2 here, we can calculate the solubility of CO_2 and SO_2 in multi-ion electrolyte solutions. However, only experimental CO_2-SO_2 solubility data in pure water have been reported [54]. Figure 6 shows the calculated solubilities of CO_2 of CO_2-SO_2 mixtures in pure water under different *T-P* conditions. It can be seen in Figure 6 that the CO_2 solubilities of CO_2-SO_2 mixtures in pure water increase with the increase of pressure at the same temperature and mole fraction of SO_2, but decrease with the increase of SO_2 at the same *T-P* conditions. The calculated solubilities of CO_2 with a 0.1 mole fraction of SO_2 are in agreement with the experimental data of Zhou [54] (Figure 6a,c,d).

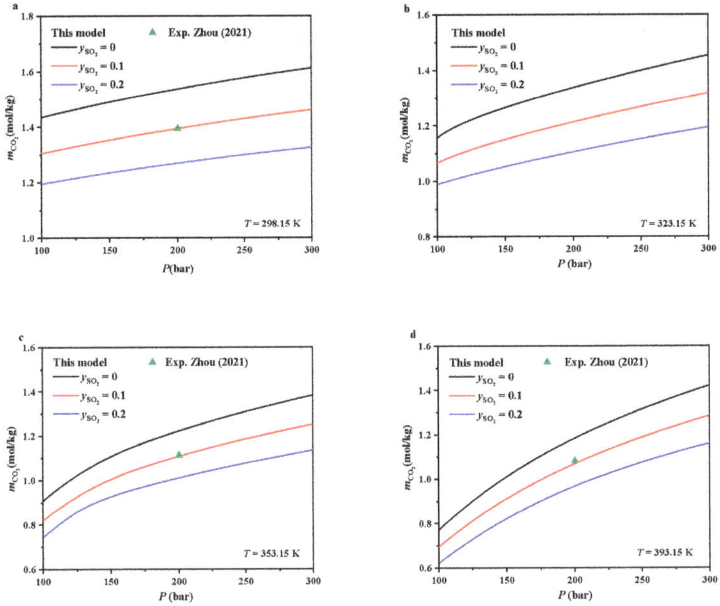

Figure 6. CO_2 Solubilities of CO_2-SO_2 mixtures in pure water (models vs. experimental data) [54]. (**a**) T = 298.15 K; (**b**) T = 323.15 K; (**c**) T = 353.15 K; (**d**) T = 393.15 K. m_{CO_2}, T and P denote the solubility of SO_2, temperature and pressure, respectively. y_{SO_2} is the mole fraction of SO_2 in vapor phase.

6. Conclusions

A SO_2 solubility model has been developed based on the EOS of pure SO_2 and the Pitzer activity coefficient theory. This model can accurately reproduce most of the experimental data with or close to experimental accuracy. Although the second-order and third-order interaction parameters between SO_2 and Mg^{2+} and Ca^{2+} have been adopted by an approximation, based on the electrolyte solution theory of Pitzer [20], this model can also be used to predict SO_2 solubility in seawater. Combined with the EOS of CO_2-SO_2 fluid mixture, the developed model can approximately predict the solubility of CO_2-SO_2 mixed gases in more complex brines.

It should be noted that the experimental data on the solubility of SO_2 and CO_2-SO_2 mixtures in aqueous $CaCl_2$, K_2SO_4 and $MgCl_2$ solutions are still lacking. Experimental studies can be focused on the solubility of these systems in the future.

Author Contributions: Methodology, writing—original draft preparation, validation and visualization: B.F.; methodology, software, formal analysis, investigation, data curation, writing—original draft preparation, validation and visualization: Z.Z.; methodology, writing—review and editing and visualization: M.X.; conceptualization, methodology, formal analysis, investigation, resources, writing—review and editing, supervision, project administration, funding acquisition and validation: S.M. All authors have read and agreed to the published version of the manuscript.

Funding: This research was funded by the National Natural Science Foundation of China grant (Numbers 42073056 and 42430813) and APC was funded by the National Natural Science Foundation of China.

Data Availability Statement: The original contributions presented in this study are included in the article. Further inquiries can be directed to the corresponding author.

Acknowledgments: We thank two anonymous reviewers for their detailed and helpful comments, which improved greatly the quality of the manuscript.

Conflicts of Interest: The authors declare no conflicts of interest.

Abbreviations

List of Symbols

m_i	molality (mol/kg) of component *I* in liquid phase
P	total pressure, that is $P_{SO_2} + P_{H_2O}$ in bar
Par	parameter
R	universal gas constant (83.14472 bar·cm^3·mol^{-1}·K^{-1})
T	absolute temperature in Kelvin
x_i	mole fraction of component *I* in liquid phase
y_i	mole fraction of component *I* in vapor phase
Greek letters	
α	activity
φ	fugacity coefficient
γ	activity coefficient
μ	chemical potential
ρ	density
$\lambda_{SO_2\text{-ion}}$	interaction parameter
$\xi_{SO_2\text{-cation-anion}}$	interaction parameter
Subscripts	
a	anion
c	cation
salt	salt solution
Superscripts	
v	vapor
l	liquid
(0)	standard state

Appendix A. The EOS of Pure SO_2

The dimensionless Helmholtz free energy α_{SO_2} of pure SO_2 is divided into two parts:

$$\alpha_{SO_2} = \alpha^0_{SO_2} + \alpha^r_{SO_2} \quad \text{(A1)}$$

where $\alpha^0_{SO_2}$ is the ideal-gas part of the Helmholtz free energy of SO_2, $\alpha^r_{SO_2}$ is the residual part of the Helmholtz free energy of SO_2.

$$\alpha^r_{SO_2} = \sum_{m=1}^{6} a_m \delta^{i_m} \tau^{j_m} + \sum_{m=7}^{14} a_m \delta^{i_m} \tau^{j_m} e^{-\delta^{k_m}} \quad \text{(A2)}$$

where $\delta = \rho/\rho_c$, $\tau = T_c/T$, ρ_c and T_c are the critical density and temperature of SO_2, ρ is the density, and T is the temperature. The exponents (i_m, j_m, k_m) come from Sun and Ely [55], and the values of parameter a_m are obtained by fitting it to thermodynamic data of SO_2 (Table A1).

The natural logarithm of the fugacity coefficient of pure SO_2 ($\ln \varphi$) is calculated from the following equation:

$$\ln \varphi = \frac{P}{\rho RT} - 1 - \ln\left(\frac{P}{\rho RT}\right) + \alpha^r_{SO_2} \quad \text{(A3)}$$

Table A1. Parameter values of EOS of pure SO_2.

m	a_m	i_m	j_m	k_m
1	$-1.6617272 \times 10^{-1}$	1	1.5	0
2	9.3106614×10^{-1}	1	0.25	0
3	-1.9592419×10^{0}	1	1.25	0
4	6.1581150×10^{-2}	3	0.25	0
5	1.7701346×10^{-4}	7	0.875	0
6	$-5.9686148 \times 10^{-4}$	2	1.375	0
7	1.7455469×10^{-3}	1	0	1
8	2.1284330×10^{-1}	1	2.375	1
9	5.0982105×10^{-1}	2	2	1
10	1.0299522×10^{-2}	5	2.125	1
11	$-2.5336956 \times 10^{-1}$	1	3.5	2
12	$-5.4155817 \times 10^{-2}$	1	6.5	2
13	$-5.9551762 \times 10^{-2}$	4	4.75	2
14	$-1.6326078 \times 10^{-2}$	2	12.5	3

References

1. Guo, J.; Li, F.; Zhang, Z.; Zhang, H.; Li, J.; Wu, K.; He, W. Retrieval of volcanic SO_2 emission rate. *J. Atmos. Environ. Opt.* **2024**, *19*, 98–110.
2. Wang, S.; Sun, G. Analysis of the relationship among thermal power Generation, coal-fired consumption and sulfur dioxide emission in China during 1991–2007. *Resour. Sci.* **2010**, *32*, 1230–1235.
3. Krotkov, N.A.; McClure, B.; Dickerson, R.R.; Carn, S.A.; Li, C.; Bhartia, P.K.; Yang, K.; Krueger, A.J.; Li, Z.; Levelt, P.F. Validation of SO_2 retrievals from the Ozone Monitoring Instrument over NE China. *J. Geophy. Res.-Atmos.* **2008**, *113*. [CrossRef]
4. Mao, S.; Zhang, D.; Li, Y.; Liu, N. An improved model for calculating CO_2 solubility in aqueous NaCl solutions and the application to CO_2-H_2O-NaCl fluid inclusions. *Chem. Geol.* **2013**, *347*, 43–58.
5. Shi, X.; Mao, S. An improved model for CO_2 solubility in aqueous electrolyte solution containing Na^+, K^+, Mg^{2+}, Ca^{2+}, Cl^- and $SO_4{}^{2-}$ under conditions of CO_2 capture and sequestration. *Chem. Geol.* **2017**, *463*, 12–28.
6. Hunger, T.; Lapicque, F.; Storck, A. Thermodynamic equilibrium of diluted SO_2 absorption into Na_2SO_4 or H_2SO_4 electrolyte solutions. *J. Chem. Eng. Data* **1990**, *35*, 453–463.
7. Xia, J.; Rumpf, B.; Maurer, G. The solubility of sulfur dioxide in aqueous solutions of sodium chloride and ammonium chloride in the temperature range from 313 K to 393 K at pressures up to 3.7 MPa: Experimental results and comparison with correlations. *Fluid Phase Equilib.* **1999**, *165*, 99–119.
8. Rodríguez-Sevilla, J.; Álvarez, M.; Díaz, M.C.; Marrero, M.C. Absorption equilibria of dilute SO_2 in seawater. *J. Chem. Eng. Data* **2004**, *49*, 1710–1716.
9. Mondal, M.K. Experimental determination of dissociation constant, Henry's constant, heat of reactions, SO_2 absorbed and gas bubble–liquid interfacial area for dilute sulphur dioxide absorption into water. *Fluid Phase Equilib.* **2007**, *253*, 98–107. [CrossRef]
10. Zimmermann, K.; Pasel, C.; Luckas, M.; Herbell, J.-D. Solubility of sulphur dioxide in aqueous electrolyte solutions at higher ionic strengths-Chloride and bromide containing systems. *Fluid Phase Equilib.* **2009**, *279*, 105–114. [CrossRef]
11. Shaw, A.C.; Romero, M.A.; Elder, R.H.; Ewan, B.C.R.; Allen, R.W.K. Measurements of the solubility of sulphur dioxide in water for the sulphur family of thermochemical cycles. *Int. J. Hydrogen Energy* **2011**, *36*, 4749–4756.
12. Tan, S.P.; Yao, Y.; Piri, M. Modeling the solubility of SO_2 + CO_2 mixtures in brine at elevated pressures and temperatures. *Ind. Eng. Chem. Res.* **2013**, *52*, 10864–10872.
13. Cox, C.; Pasel, C.; Luckas, M.; Bathen, D. Absorption of SO_2 in different electrolyte solutions, seawater and brine. *Fluid Phase Equilib.* **2015**, *402*, 89–101.
14. Zhou, Q.; Guo, H.; Yang, P.; Wang, Z. Solubility of SO_2 in water from 263.15 to 393.15 K and from 10 to 300 bar: Quantitative Raman spectroscopic measurements and PC-SAFT prediction. *Ind. Eng. Chem. Res.* **2020**, *59*, 12855–12861.
15. Saadallah, K.; Lachet, V.; Creton, B.; Caumon, M.-C.; Randi, A.; Sterpenich, J. Solubility study of binary systems containing sulfur dioxide and water: A combination of Raman spectroscopy and Monte Carlo molecular simulation. *Fluid Phase Equilib.* **2023**, *574*, 113901.
16. Meissner, H.P.; Kusik, C.L. Activity coefficients of strong electrolytes in multicomponent aqueous solutions. *AIChE J.* **1972**, *18*, 294–298.
17. Meissner, H.P.; Tester, J.W. Activity coefficients of strong electrolytes in aqueous solutions. *Ind. Eng. Chem. Process Des. Dev.* **1972**, *11*, 128–133.
18. Bromley, L.A. Thermodynamic properties of strong electrolytes in aqueous solution. *AICHE J.* **1973**, *19*, 313–320.

19. Bromley, L.A. Approximate individual ion values of β (or B) in extended Debye-Hückel theory for uni-univalent aqueous solutions at 298.15 K. *J. Chem. Thermodyn.* **1972**, *4*, 669–673. [CrossRef]
20. Pitzer, K.S. Thermodynamics of electrolytes. I. Theoretical basis and general equations. *J. Phys. Chem.* **1973**, *77*, 268–277.
21. Weisenberger, S.; Schumpe, D.A. Estimation of gas solubilities in salt solutions at temperatures from 273 K to 363 K. *AIChE J.* **1996**, *42*, 298–300. [CrossRef]
22. Miri, R.; Aagaard, P.; Hellevang, H. Examination of CO_2–SO_2 solubility in water by SAFT1. Implications for CO_2 transport and storage. *J. Phys. Chem. B* **2014**, *118*, 10214–10223. [CrossRef] [PubMed]
23. Bunsen, R. Ueber das Gesetz der Gasabsorption. *Justus Liebigs Ann. Chem.* **1855**, *93*, 1–50. [CrossRef]
24. Sims, T.H. Contributions to the knowledge of the laws of gas-absorption. *Q. J. Chem. Soc. Lond.* **1862**, *14*, 1–22. [CrossRef]
25. Fox, C.J.J. Über die löslichkeit des schwefeldioxyds in wässerigen salzlösungen und seine wechselwirkung mit den salzen. *Z. Phys. Chem.* **1902**, *41*, 458–482. [CrossRef]
26. Simth, W.T.; Parkhurst, R.B. The solubility of sulfur dioxide in suspensions of calcium and magnesium hydroxides. *J. Am. Chem. Soc.* **1922**, *44*, 1918–1927. [CrossRef]
27. Hudson, J.C. The solubility of sulphur dioxide in water and in aqueous solutions of potassium chloride and sodium sulphate. *J. Chem. Soc.* **1925**, *127*, 1332–1347. [CrossRef]
28. Sherwood, T.K. Solubilities of sulfur dioxide and ammonia in water. *Ind. Eng. Chem. Res.* **1925**, *17*, 745–747. [CrossRef]
29. Maass, C.E.; Maass, O. Sulfur dioxide and its aqueous solutions. I. Analytical methods, vapor density and vapor pressure of sulfur dioxide. Vapor pressure and concentrations of the solutions. *J. Am. Chem. Soc.* **1928**, *50*, 1352–1368. [CrossRef]
30. Campbell, W.B.; Maass, O. Equilibria in sulphur dioxide solutions. *Can. J. Res.* **1930**, *2*, 42–64. [CrossRef]
31. Morgan, O.M.; Maass, O. An investigation of the equilibria existing in gas-water systems forming electrolytes. *Can. J. Res.* **1931**, *5*, 162–199.
32. Conrad, F.H.; Beuschlein, W.L. Some equilibrium relations in the system calcium oxide-sulfur dioxide-water (acid region) at pressures below atmospheric. *J. Am. Chem. Soc.* **1934**, *56*, 2554–2562. [CrossRef]
33. Johnstone, H.F.; Leppla, P.W. The solubility of sulfur dioxide at low partial pressures. The Ionization constant and heat of Ionization of sulfurous acid. *J. Am. Chem. Soc.* **1934**, *56*, 2233–2238. [CrossRef]
34. Young, C.L. *Sulfur Dioxide, Chlorine, Fluorine and Chlorine Oxides, Solubility Data Series*; Pergamon Press: Oxford, UK, 1983.
35. Beuschlein, W.L.; Simenson, L.O. Solubility of sulfur dioxide in water. *J. Am. Chem. Soc.* **1940**, *62*, 610–612. [CrossRef]
36. Parkinson, R.V. The solubility of sulfur dioxide in water and sulfuric acid. *Tappi J.* **1956**, *39*, 517–519.
37. Rabe, A.E.; Harris, J.F. Vapor liquid equilibrium data for the binary system sulfur dioxide and water. *J. Chem. Eng. Data* **1963**, *8*, 333–336.
38. Vosolsobe, J.; Simecek, A.; Michalek, J. Rozpusnost kyslíku siričiného ve vodě. *Chem. Prum.* **1965**, *15*, 401.
39. Tokunaga, J. Solubilities of sulfur dioxide in aqueous alcohol solutions. *J. Chem. Eng. Data* **1974**, *19*, 162–165. [CrossRef]
40. Lavrova, E.M.; Tudorovskaya, G.L. Solubility of sulfur dioxide in aqueous hydrochloric acid solutions. *J. Appl. Chem. USSR* **1977**, *50*, 1102–1106.
41. Douabul, A.; Riley, J. Solubility of sulfur dioxide in distilled water and decarbonated seawater. *J. Chem. Eng. Data* **1979**, *24*, 274–276. [CrossRef]
42. Byerley, J.J.; Rempel, G.L.; Thang, V. Solubility of sulfur dioxide in water-acetonitrile solutions. *J. Chem. Eng. Data* **1980**, *25*, 55–56. [CrossRef]
43. Rumpf, B.; Maurer, G. Solubilities of hydrogen cyanide and sulfur dioxide in water at temperatures from 293.15 to 413.15 K and pressure up to 2.5 MPa. *Fluid Phase Equilib.* **1992**, *81*, 241–260. [CrossRef]
44. Siddiqi, M.A.; Krissmann, J.; Petersgerth, P.; Luckas, M.; Luckas, K. Spectrophotometric measurement of the vapor-liquid equilibris of (sulphur dioxide + water). *J. Chem. Thermodyn.* **1996**, *28*, 685–700. [CrossRef]
45. Millero, F.J.; Hershey, J.P.; Johnson, G. The solubility of SO_2 of and the dissociation of H_2SO_3 in NaCl solutions. *J. Atmos. Chem.* **1989**, *8*, 377–389. [CrossRef]
46. Rumpf, B.; Maurer, G. Solubility of sulfur dioxide in aqueous solutions of sodium and ammonium sulfate at temperatures from 313.15 K to 393.15 K and pressures up to 3.5 MPa. *Fluid Phase Equilib.* **1993**, *91*, 113–131.
47. Denbigh, K. *The Principles of Chemical Equilibrium*; Cambridge University Press: Cambridge, UK, 1971.
48. Li, X. The Prediction of *PVTx* and Vapor-Liquid Equilibrium (VLE) Properties of C-H-O-N-S Fluid Mixtures. Master's Thesis, China University of Geosciences, Beijing, China, 2017.
49. Wagner, W.; Kretzschmar, H. IAPWS industrial formulation 1997 for the thermodynamic properties of water and steam. In *International Steam Tables: Properties of Water Steam Based on the Industrial Formulation IAPWS-IF97*; Springer Vieweg: Berlin/Heidelberg, Germany, 2008; pp. 7–150.
50. Pitzer, K.S.; Peiper, J.C.; Busey, R.H. Thermodynamic properties of aqueous sodium chloride solutions. *J. Phys. Chem. Ref. Data* **1984**, *13*, 1–102.

51. Duan, Z.; Sun, R. An improved model calculating CO_2 solubility in pure water and aqueous NaCl solutions from 273 to 533 K and from 0 to 2000 bar. *Chem. Geol.* **2003**, *193*, 257–271.
52. Duan, Z.; Sun, R.; Liu, R.; Zhu, C. Accurate thermodynamic model for the calculation of H_2S solubility in pure water and brines. *Energy Fuels* **2007**, *21*, 2056–2065.
53. Harvie, C.E.; Møller, N.; Weare, J.H. The prediction of mineral solubilities in natural waters: The Na-K-Mg-Ca-H-Cl-SO_4-OH-HCO_3-CO_3-CO_2-H_2O system to high ionic strengths at 25 °C. *Geochim. Cosmochim. Acta* **1984**, *48*, 723–751.
54. Zhou, Q. Quantitative Raman Spectroscopic Investigation on Non-Equilibrium Dissolution and Water-Rock Reaction of SO_2-CO_2 Co-Injection into Pore Space. Ph.D. Thesis, China University of Geosciences, Wuhan, China, 2021.
55. Sun, L.; Ely, J.F. Universal equation of state for engineering application: Algorithm and application to non-polar and polar fluids. *Fluid Phase Equilib.* **2004**, *222–223*, 107–118.

applied sciences

MDPI

Article

Impacts of Holocene Sea Level Rise and the Opening of the Qiongzhou Strait on the Provenance of Sediments in the Beibu Gulf, South China Sea

Zhenang Cui [1] and Yueming Hou [2,*]

[1] Sanya Geology Institute of South China Sea, Guangzhou Marine Geological Survey, China Geological Survey, Sanya 572025, China; cuizhenang@163.com
[2] Institute for Marine Petroleum Geology, Guangzhou Marine Geological Survey, China Geological Survey, Guangzhou 510075, China
* Correspondence: hymcza@163.com

Abstract: The opening of the Qiongzhou Strait during the Holocene was a significant geological event in the Beibu Gulf, profoundly influencing sediment provenance and ocean circulation systems. Due to the scarcity of geological records documenting this event, the understanding of regional Holocene sedimentary evolution has been constrained. To investigate the impact of this event on sediment provenance and ocean currents in the Beibu Gulf, geochemical analyses were conducted on sediment core SO-31 retrieved from the South China Sea. The sediments in core SO-31 were stratigraphically divided into three units based on vertical geochemical profiles, reflecting changes in sea level and shifts in sediment provenance within the study area. The Th/Cr vs. Th/Sc scatter plot for core SO-31 indicate that sedimentary materials primarily originated from the Red River during 11,400–7700 a BP, and a significant change in provenance occurred in the study region around 7700 a BP, characterized by increased contributions from the Qiongzhou Strait and decreased contributions from the Red River. This suggests that the opening of the Qiongzhou Strait significantly influenced the sediment supply to the central Beibu Gulf around 7700 a BP. These findings provide critical geochemical evidence for studying the Qiongzhou Strait opening event and enhance our understanding of Holocene sedimentary evolution and "source–sink" transitions in the Beibu Gulf.

Keywords: geochemistry; sediment provenance; Beibu Gulf; South China Sea; Holocene

1. Introduction

Marine sediments, serving as carriers of environmental evolution information, are extensively utilized in paleoclimate and paleoenvironmental studies [1–5]. In recent years, the Beibu Gulf, situated in the northwest of the South China Sea, has emerged as a focal area for marine geological research owing to its unique geographical position and topographical conditions [6–9]. Since the last deglaciation, the Beibu Gulf has undergone significant changes in sedimentary environments due to the continuous rise in global sea levels. Notably, the opening of the Qiongzhou Strait during the early Holocene significantly altered the sedimentary system within the Beibu Gulf [10,11]. To elucidate regional paleoenvironmental and paleoclimatic changes, numerous studies have been conducted on the marine records of the Beibu Gulf, establishing modern sediment source–sink processes and Quaternary paleoclimate evolution based on mineralogical and geochemical analyses of Beibu Gulf sediments [12–17]. However, compared to extensive research on modern sediment source–sink processes, studies on sediment provenance transitions since the Holocene remain relatively limited [18,19], particularly concerning significant geological events such as the opening of the Qiongzhou Strait, which have profoundly impacted the source–sink dynamics in this region.

The Qiongzhou Strait, as a critical channel for material exchange between the Beibu Gulf and the eastern sea area, directly influenced the sediment supply and circulation patterns within the Beibu Gulf [20,21]. Previous studies have documented the thick Holocene sediments deposited at the adjacent region of western mouth of the Qiongzhou Strait [22], with contributions from the Qiongzhou Strait also observed in the middle and southern area of the Beibu Gulf [18,19]. Despite its importance, the geological events associated with the opening of the Qiongzhou Strait have not received adequate attention. Consequently, understanding the provenance supply to the Beibu Gulf since the opening of the Qiongzhou Strait remains unclear, and there is a lack of geological evidence effectively reflecting the timing and evolution of this event, which limits our comprehension of Holocene sedimentary evolution and the "source–sink" process. This study provides geochemical characteristics of the sediments based on their major and trace element compositions, deciphering the sedimentary provenance and depositional environment of core SO-31 in the Beibu Gulf, elucidating changes in provenance contributions from the Qiongzhou Strait and other major sediment sources to the central region of the Beibu Gulf. The research findings will provide critical geochemical evidence for a better understanding of the formation and evolution of the Qiongzhou Strait, as well as valuable insights into the paleoenvironmental changes in the Beibu Gulf.

2. Materials and Methods

A gravity core, SO-31, was obtained in the central part of the Beibu Gulf, South China Sea, at coordinates 107°22′19″ E, 18°37′40″ N, by the Chinese–German joint cruise of R/V Sonne 219 in 2011. This sedimentary profile is 738 cm in length and was recovered at a water depth of 75 m. The core site is ~210 km from the Red River mouth in Vietnam and about 150 km from the Changhua River mouth on Hainan Island (Figure 1). The SO-31 core, situated in the central region of the Beibu Gulf, is susceptible to environmental events in its surroundings, which may influence its sedimentary environment and could be recorded in the core sediments. Moreover, no significant depositional discontinuities were observed within the vertical profile of SO-31, suggesting continuous recording of the depositional history since the Holocene epoch. Consequently, this sedimentary column serves as an excellent representative for studying regional sedimentary environmental evolution.

The core was cut into two halves, photographed, described, and then subsampled for grain-size and geochemistry analysis. The core was sampled at 3 cm intervals, and the outer rims of the samples were removed to avoid contamination, yielding a total of 246 samples. Additionally, 16 surface sediment samples were collected from the Qiongzhou east (8 samples) and Hainan west nearshore (8 samples) by box sampler using the China R/V "Fendou 5" in 2012 (Figure 1), and the upper 5 cm of each sample was used for analysis of grain size and elements.

All samples were stored in polyethylene bags and kept frozen at −20 °C until chemical analysis in the laboratory at Guangzhou Marine Geological Survey (GMGS, China). Grain size of the bulk samples was measured by laser particle size analyzer (Mastersizer 2000), after removing organic matter and biogenic carbonate from the samples with 10% H_2O_2 and 1 N HCl, respectively. Prior to the geochemical analysis, the bulk samples were desalted with distilled water and dried at 60 °C in a clean oven, and then ground to 200 mesh (aperture 0.075 mm) or less in an agate mortar. The 10 major elements (Si, Al, Fe, Mg, Mn, Ca, Na, K, Ti, and P) and 12 trace elements (Sc, V, Cr, Co, Ni, Cu, Zn, Ga, Sr, Ba, Pb, and Zr) were measured by Inductively Coupled Plasma-Optical Emission Spectrometry (ICP-OES, 4300 DV).

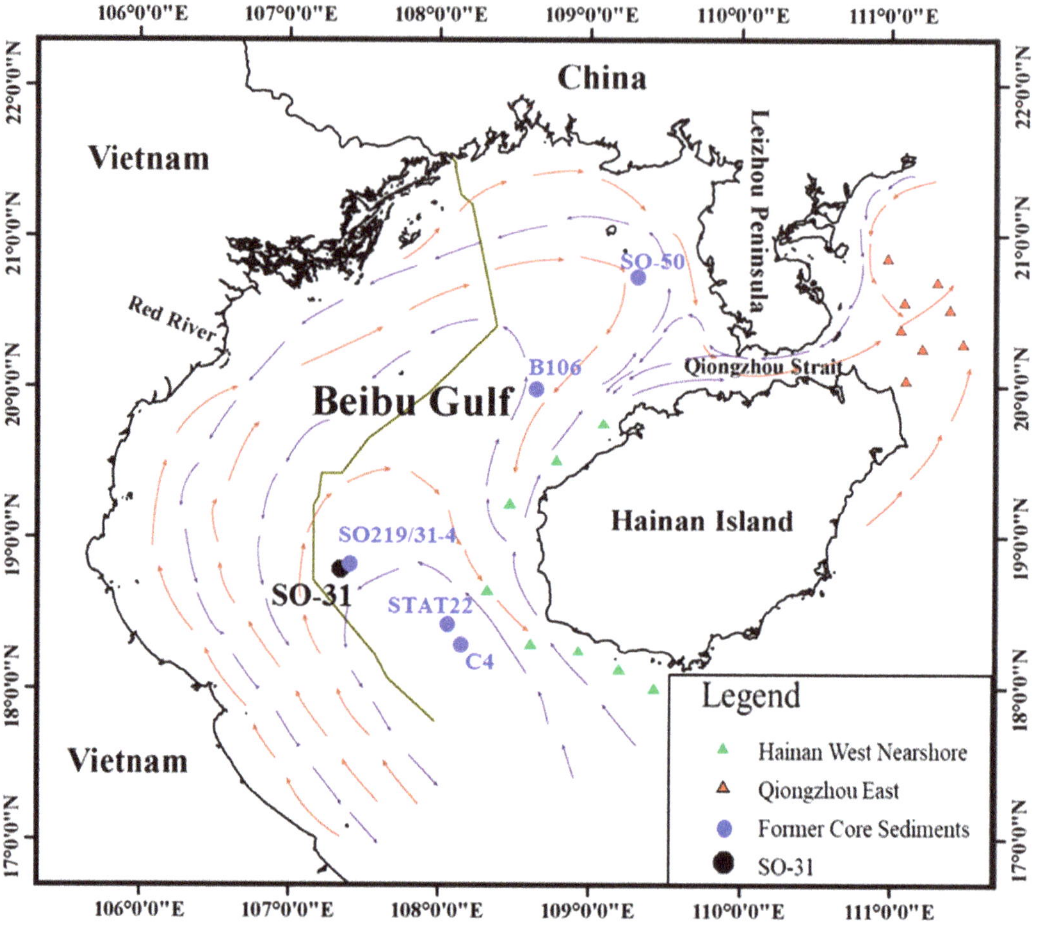

Figure 1. Geographical setting of the Beibu Gulf and location of the gravity core SO-31. The black triangles and black dots show the surficial sediments from Qiongzhou east and Hainan west nearshore, respectively. The yellow line indicates the boundary between Vietnam and The People's Republic of China. The arrows indicate the main current direction for the SW (red color)/NE (blue color) monsoon.

To analyze the contents of rare earth elements (REEs), about 100 mg of each sample was first digested by 1 mL 50% HNO_3 and 3 mL HF in a tightly closed Teflon vessel on a hotplate at about 160 °C for 48 h. After evaporation to dryness, 1 mL of 55% $HClO_4$ was added to the vessel, which was heated again to about 160 °C until the complete evaporation of the acid. After the sample cooled down to room temperature, 1.5 mL of 50% HNO_3 was added, and the sample was again heated at 160 °C for 12 h and then cooled to room temperature. Afterwards, the solution was moved to a measuring cylinder (including the solutions produced by rinsing the vessel with 50% HNO_3 to concentrate the remaining sample in the vessel), diluted with 10% HNO_3, and measured with an Inductively Coupled Plasma Mass Spectrometer (ICP-MS, X Series 2).

We conducted repeated analysis of several samples and standard sample (GSD-9) analysis to ensure testing accuracy and precision (relative error < 2%). The analysis of grain size and all the elements were done in the Laboratory of Guangzhou Marine Geology Survey (GMGS), Guangzhou, China. The trace elements, including REEs, from the sediments of the Qiongzhou Strait and the Hainan west nearshore were also measured at GMGS using same methods and same accuracy as stated for the SO-31 core sediments.

Nine samples composed of picked foraminifera shells, bivalve shells, and snail fossils were selected for Accelerator Mass Spectrometry (AMS) ^{14}C radiocarbon dating according to the species of shell organisms and depth in the core sediments. When selecting dating samples, bivalves or snail fossils that are easy to find were prioritized. If no bivalves and snail fossils were available at the relevant depth horizon, sediments were collected to select other dating samples like foraminifera shells. This work was completed by the State Key Laboratory of Nuclear Physics and Technology, Peking University, and calibrated with the software Cal. 8.2.0.

3. Results

3.1. Chronology Framework and Sedimentary Rates

The chronology for core SO-31 was based on the AMS ^{14}C dating data with a composite age–depth model (Table 1 and Figure 2). The indication of the core had a 2σ basal age of ca. 12,674–13,026 calibrated years before present (cal. yr. BP), which corresponded to a median sedimentation rate of 0.057 cm per year (cm/yr). According to the age–depth model illustrated in Figure 2, the sedimentation rate of this studied core ranged from 0.012 cm/yr to 0.140 cm/yr. Intervals of increased sedimentation accumulation rate (exceeding 0.057 cm/yr) were observed during two periods dated to cal. 12,828–10,667 cal. yr. BP and 5137 cal. yr. BP to the present, whereas the lowest-rate period (below 0.02 cm/yr) occurred ca. 10,667–5137 cal. yr. BP (Figure 2). The sediment dating data of SO-31 are largely consistent with the findings reported by Zhang et al. for SO219/31-4 [13], likely due to the proximity of the two sites (less than 500 m apart). Minor discrepancies of datasets between the SO-31 and SO219/31-4 can be attributed to variations in the specific locations and depths of the sampled sediments.

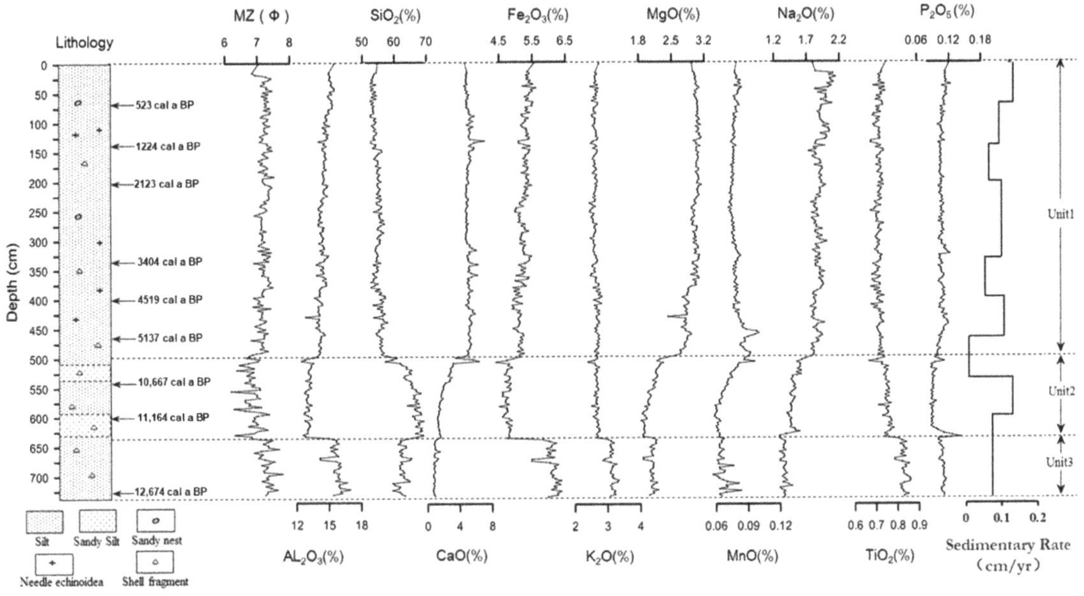

Figure 2. Geological column and down-core variations in mean grain size (MZ) and major elements of the core SO-31 sediments show the change in the sedimentary environment. Dashed horizontal lines indicate the boundaries between depositional units (Unit 1, Unit 2, and Unit 3). The nine ages marked with arrows in geologic column were obtained by the AMS ^{14}C dating.

Table 1. AMS ^{14}C dates from the SO-31 sediment core.

No.	Depth (cm)	Source	Conventional ^{14}C Age, Year	Cal. Year BP (2σ)
1	73	Bivalve shell	940 ± 30	382–655
2	142	Spiral Shell	1680 ± 30	1060–1366
3	205	Bivalve shell	2460 ± 30	1939–2307
4	337	Foraminifera	3510 ± 30	3217–3580
5	403	Bivalve shell	4370 ± 30	4320–4736
6	469	Foraminifera	4860 ± 30	4912–5318
7	538	Bivalve shell	9710 ± 40	10,442–10,937
8	601	Bivalve shell	10,060 ± 50	10,908–11,389
9	733	Foraminifera	11,350 ± 50	12,674–13,026

3.2. Characteristics of Major and Trace Elements

The major and trace element compositions, together with the mean grain size (MZ), for the core sediments are given in Tables 2 and 3. The mean grain size ranges from 6.18Φ to 7.76Φ, with the average of 7.14Φ, showing that silt-sized or finer grains dominate the core sediments, and the study area had a relatively calm sedimentary environment. Among the major elements measured for the entire profile, SiO_2 is a dominant component, averaging 57.72% in the sediment, followed by Al_2O_3, averaging 14.43%. The two elements amount to an average of more than 70%, indicating that the sediment is dominated by silicate and aluminosilicate. The CaO content is also relatively high, with an average of 4.0%, which is slightly lower than the average of the coastal area of the Hainan Island [23,24], but far higher than the sediment content of the Red River [25–27], indicating that bicarbonate deposition is relatively active in this area. Content The content of Al_2O_3 was slightly higher than the average of shallow sea sediments [18], but lower than that of the Red River [25–27], which may be related to the dilution of biocalcinous detritus in the study area, which may be due to the quiet mode of sedimentation. The contents of Fe_2O_3, K_2O, MgO, Na_2O, TiO_2, P_2O_5, and MnO in the sediments of SO-31 are close to the average values of the sediments from the shallow sea of the South China Sea, and the contents of most trace elements such as Co, Cu, Ni, Zn, V, Ba, Sc, Ga, Pb, Cr, and La in the core sediments are slightly higher than the average of the shallow area of the South China Sea [18], which may be caused by the difference in the sedimentary environment and material sources between the Beibu Gulf and other areas in the South China Sea. In general, geochemical composition similar to that of the upper crust (UCC) [28] indicates that the surrounding continent is the main source of sediments in this area.

The SO-31 core sediments can be divided into 3 sedimentary units: Unit 1, Unit 2, and Unit 3, in ascending order, according to the vertical variation in grain size and geochemistry (Figures 2 and 3). In Unit 1 (0–501 cm, 0–7700 a BP), the content of SiO_2 in sediments showed a gradual increase with depth; conversely, the contents of Al_2O_3, Fe_2O_3, and Na_2O showed a trend of decreasing with depth. At 501 cm depth, an obvious change in the contents of most of major elements (such as SiO_2, Al_2O_3, Fe_2O_3, P_2O_5, MgO, CaO, and Na_2O) and trace elements (such as Zn, Cu, Gd, and Sr) can be observed simultaneously. In Unit 2 (501–640 cm, 7700–11,600 a BP), SiO_2 content is slightly higher than the average, and the contents of K_2O and TiO_2 are close to that of the average. In Unit 3 (640–738 cm, 11,600–12,800 a BP), compared with Unit 2, the contents of SiO2 in Unit 3 decreased rapidly, and conversely, the contents of Al_2O_3, Fe_2O_3, K_2O, TiO_2, Co, Cu, Ni, Zn, V, Sc, Ga, Pb, Cr, and Zr increased significantly. Similarly to the major and trace elements, the rare earth elements (REEs and LREEs/HREEs) and the grain size of the core sediments also show similar variation in the vertical direction. The abrupt changes at depths of 501 cm and 640 cm indicate that the sedimentary environment has undergone significant changes.

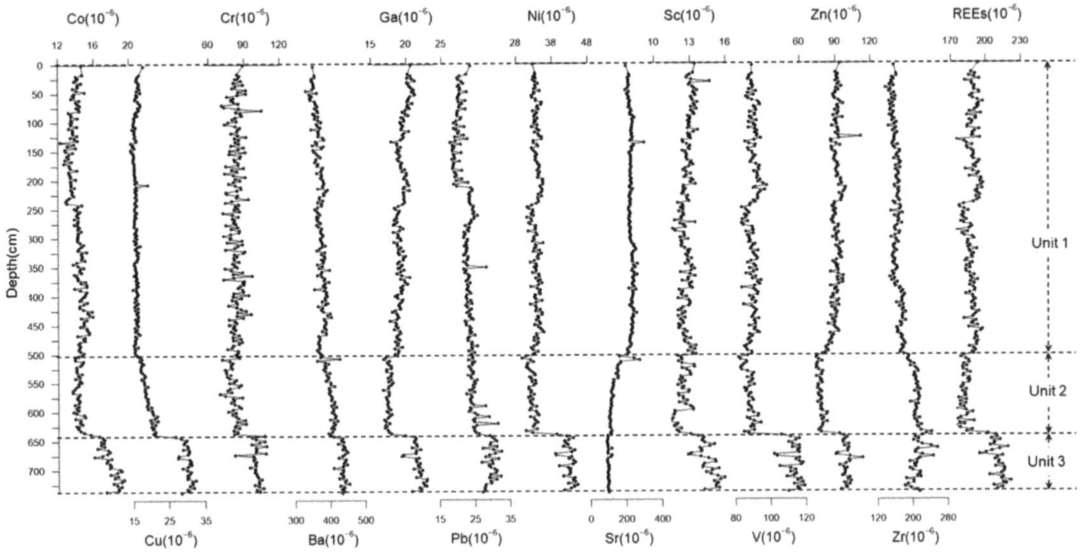

Figure 3. Down-core variations of trace elements and rare earth elements of the core SO-31 sediments show the change in the sedimentary environment. Dashed horizontal lines indicate the boundaries between depositional units (Unit 1, Unit 2, and Unit 3).

Table 2. Concentrations of major (%) and trace (10^{-6}) elements and mean grain size (Φ) in SO-31 sediments and the surficial sediments; n is the sample number.

		MZ	SiO_2	Al_2O_3	Fe_2O_3	CaO	MgO	K_2O	Na_2O	MnO	P_2O_5	TiO_2	Co	Ni	Cr	Sr	Zr	Sc	V	Ga	Ba	Rb	Nb	Th
All samples (n = 246)	Min	6.18	51.8	12.4	4.35	0.68	1.84	2.39	1.28	0.06	0.08	0.67	12.3	29.5	69.0	90.5	139	11.0	81.7	16.6	333	118	10.0	11.6
	Max	7.76	69.0	16.9	6.36	7.12	3.14	3.34	2.14	0.10	0.14	0.86	19.4	45.0	109	303	256	15.9	119	22.8	454	160	24.6	17.5
	Mean	7.14	57.7	14.4	5.26	4.00	2.68	2.71	1.74	0.08	0.10	0.73	14.7	34.6	85.1	193	178	12.9	92.8	19.2	385	136	20.2	15.3
Unit 1 (n = 171)	Min	6.66	51.8	12.4	4.35	3.20	2.19	2.39	1.55	0.07	0.09	0.67	12.3	29.5	70.9	163	139	11.6	81.7	16.7	333	126	17.2	13.6
	Max	7.51	61.7	15.6	5.61	7.12	3.14	2.84	2.14	0.10	0.12	0.74	15.9	35.8	105	303	192	14.7	98.1	21.3	427	147	23.3	17.5
	Mean	7.18	54.8	14.5	5.23	5.13	2.95	2.64	1.89	0.08	0.11	0.71	14.2	33.6	83.3	229	165	12.8	90.0	19.3	372	136	19.8	15.9
Unit 2 (n = 43)	Min	6.18	62.2	12.6	4.50	1.12	1.84	2.52	1.28	0.06	0.08	0.71	13.6	30.5	69.0	95.4	194	11.0	82.8	16.6	381	118	10.0	11.6
	Max	7.35	69.0	15.0	5.80	3.18	2.25	3.00	1.61	0.08	0.14	0.81	16.6	39.3	107	168	243	13.3	106	19.7	421	141	23.0	14.8
	Mean	6.83	66.0	13.1	4.76	1.86	2.02	2.68	1.47	0.07	0.09	0.75	14.3	32.5	81.3	124	210	12.3	88.6	17.4	402	126	19.7	13.3
Unit 3 (n = 32)	Min	6.88	59.3	14.1	5.46	0.68	1.88	2.86	1.28	0.06	0.10	0.78	16.0	38.8	81.9	90.5	182	12.8	102	19.2	410	137	18.0	13.4
	Max	7.76	65.1	16.9	6.36	1.51	2.21	3.34	1.47	0.08	0.11	0.86	19.4	45.0	109	119	256	15.9	119	22.8	454	160	24.6	16.6
	Mean	7.29	62.1	15.7	6.08	0.86	2.08	3.13	1.35	0.07	0.11	0.83	17.9	42.5	99.9	98.5	210	14.5	113	21.5	436	151	22.8	15.2
Qiongzhou East (n = 8)	Min	6.12	55.3	10.9	4.46	1.80	1.66	2.03	1.09	0.07	0.10	0.62	10.7	24.3	51.1	111	165	11.5	66	13.2	265	84	15.2	13.3
	Max	7.22	68.7	16.5	6.37	6.46	2.23	2.44	1.88	0.09	0.15	0.97	15.9	36.2	81.0	241	260	14.0	110	21.2	348	142	23.1	38.4
	Mean	6.84	60.6	13.7	5.27	3.39	1.98	2.28	1.58	0.08	0.12	0.79	13.2	29.9	71.7	152	205	12.6	85	17.3	327	118	20.2	25.5
Hainan West Nearshore (n = 8)	Min	4.50	54.3	7.95	2.73	3.02	1.47	1.97	0.95	0.06	0.08	0.50	6.88	14.9	36	174	73	9.0	40	8.91	284	104	12.4	11.1
	Max	7.64	69.3	17.0	6.11	10.6	2.50	2.71	2.01	0.08	0.13	0.88	15.4	35.3	88	469	340	13.8	101	21.5	370	165	21.6	23.1
	Mean	6.16	60.1	12.1	4.72	5.63	1.97	2.31	1.46	0.07	0.10	0.65	11.6	25.5	65	263	206	11.4	70	15.2	307	130	16.9	17.4
Red River (mean) [1]		5.52	44.2	11.6	11.2	9.89	11.4	1.68	3.03	0.18	0.81	2.26	5.05	-	21.1	23.5	16.8	15.1	128	-	564	147	25.3	18.0
UCC [2]		6.00	61.7	15.0	6.17	5.39	3.67	2.58	3.18	0.09	0.17	0.67	10.0	20.0	35.0	350	190	11.0	60.0	17.0	550	112	25	10.5

[1] Data from Tong et al., 2007 [25]; [2] Data from Taylor et al., 1985 [28].

Table 3. Concentrations of REEs (10^{-6}) and REE characteristic parameters in SO-31 sediments and the surficial sediments; n is the sample number.

		La	Ce	Pr	Nd	Sm	Eu	Gd	Tb	Dy	Ho	Er	Tm	Yb	Lu	REEs	LREE [3]/HREE [4]	δCe [5]	δEu [6]
All samples (n = 246)	Min	36.6	72.8	8.29	30.3	5.59	1.09	4.84	0.75	4.46	0.84	2.34	0.36	2.32	0.36	172.8	8.5	0.92	0.59
	Max	46.7	92.1	10.6	39.2	7.67	1.49	6.66	1.03	6.89	1.20	3.44	0.54	3.44	0.56	221.0	9.8	0.98	0.68
	Mean	40.2	81.2	9.12	33.5	6.29	1.22	5.49	0.84	5.08	0.95	2.72	0.42	2.66	0.42	190.1	9.3	0.95	0.64
Unit 1 (n = 171)	Min	36.6	74.6	8.29	30.3	5.59	1.09	4.96	0.75	4.46	0.84	2.34	0.36	2.36	0.36	173.7	9.1	0.93	0.59
	Max	41.9	85.4	9.46	34.8	6.83	1.27	5.77	0.90	5.42	1.01	2.88	0.45	2.82	0.44	197.6	9.8	0.98	0.67
	Mean	39.7	81.1	9.01	33.1	6.17	1.19	5.41	0.82	4.96	0.92	2.65	0.41	2.60	0.40	188.4	9.4	0.96	0.63
Unit 2 (n = 43)	Min	36.7	72.8	8.42	30.8	5.76	1.14	4.84	0.77	4.61	0.88	2.38	0.37	2.32	0.36	173.0	8.8	0.92	0.62
	Max	42.9	84.8	9.77	35.4	6.82	1.36	5.95	0.89	5.56	1.08	2.94	0.47	2.94	0.45	201.3	9.5	0.95	0.68
	Mean	38.8	76.9	8.79	32.5	6.11	1.20	5.30	0.81	4.88	0.93	2.58	0.40	2.54	0.39	182.1	9.2	0.93	0.65
Unit 3 (n = 32)	Min	40.8	81.0	9.40	35.0	6.78	1.27	5.70	0.90	5.56	1.04	2.97	0.46	2.90	0.45	194.4	8.5	0.92	0.62
	Max	46.7	92.1	10.6	39.2	7.67	1.49	6.66	1.03	6.89	1.20	3.44	0.54	3.44	0.56	221.0	8.9	0.93	0.67
	Mean	44.4	87.8	10.1	37.4	7.13	1.40	6.22	0.97	5.99	1.13	3.23	0.50	3.17	0.50	210.1	8.7	0.92	0.65
Qiongzhou East (n = 8)	Min	33.7	70.6	7.69	29.4	5.43	1.08	4.66	0.74	4.21	0.77	2.17	0.32	2.00	0.31	173.0	8.5	0.88	0.62
	Max	46.2	95.3	10.5	39.7	7.47	1.54	6.16	1.00	6.06	1.16	3.36	0.51	3.05	0.49	163.3	9.0	1.12	0.68
	Mean	39.5	82.5	9.11	33.8	6.48	1.28	5.49	0.85	5.07	0.94	2.74	0.41	2.57	0.40	222.6	9.7	1.05	0.66
Hainan West Nearshore (n = 8)	Min	25.2	58.2	5.83	21.6	3.81	0.73	3.37	0.48	2.79	0.50	1.60	0.25	1.46	0.27	191.2	9.4	0.96	0.60
	Max	47.2	97.9	10.9	39.7	7.44	1.41	6.21	0.97	5.85	1.08	3.23	0.49	3.18	0.53	126.1	8.7	1.06	0.65
	Mean	40.6	82.6	9.17	33.54	6.25	1.21	5.48	0.85	4.97	0.91	2.74	0.42	2.69	0.43	225.5	10.7	1.03	0.63
Red River (mean) [1]		52.03	112	11.5	43.3	7.73	1.67	7.99	1.03	5.67	1.09	3.15	0.43	2.85	0.40	250.5	10.1	1.07	0.65
UCC [2]		36.9	73.8	8.01	30.1	5.58	1.11	4.86	0.78	4.55	0.86	2.43	0.39	2.53	0.39	172.3	9.3	1.10	0.65

[1] Data from Tong et al., 2007 [25]; [2] Data from Taylor et al., 1985 [28]; [3] LREE = La + Ce + Pr + Nd + Sm + Eu; [4] HREE = Gd + Tb + Dy + Ho + Er + Tm + Yb + Lu; [5] $\delta Ce = 2Ce_N/(La_N + Pr_N)$; [6] $\delta Eu = 2Eu_N/(Sm_N + Gd_N)$; Ce_N, La_N, Pr_N, Eu_N, Sm_N, and Gd_N are the normalized values of chondrites measured for the corresponding elements.

4. Discussion

4.1. Potential Sources of the Sediments

The Beibu Gulf is a semi-closed gulf located northwest of the SCS and connected with the SCS through the southern end of the gulf and Qiongzhou Strait, which is located between Hainan Island and the Leizhou Peninsula (Figure 1). The Red River (Song Hong River) provides the major riverine discharge into the gulf, along with some smaller coastal rivers [29,30]. Discharge from the Pearl River, about 400 km to the northeast, may reach the gulf through Qiongzhou Strait [31]. Additionally, coastal erosion also is a significant source of sediments; amounts of suspended particulate matter (SPM) from the west area of Hainan Island can even be transported to the center of the Beibu Gulf [18]. Geologically, Cambrian to recent strata are outcropped in the land area around the Beibu Gulf. The granitic rocks, as the dominant rocks, occupy more than 60% of the area west of Hainan Island [32]; the basic igneous rocks of the Tertiary and Quaternary periods compose the basement of the Leizhou Peninsula; and Paleozoic sandstone and shale are widely exposed in the coastal areas of the Guangxi province, People's Republic of China [33]. In most of the Beibu Gulf surrounding land areas, the solid rocks are unevenly mantled by Quaternary alluvium, regolith, or deeply weathered rocks [33]. The Red River drainage system has considerable differences in its lithologies, including metamorphic rocks, limestone, granitoid rocks, felsic rocks, and ultramafic rocks [25]. The sediments derived from these different rocks are intensively mixed within the Red River estuary. The variation in rock types surrounding the Beibu Gulf inevitably resulted in differences in material composition and geochemical characteristics across these source areas, and this variability offers a valuable opportunity to trace and identify the different origins of sedimentary materials in the study area.

4.2. Identification of the Sediment Provenance

Previous studies have revealed that REEs, La, Sc, Ti, Co, Hf, Th, and Nb are especially useful for monitoring source-area composition [34–38]. In order to further study the sediment provenance, element ratios of Si/Al, Rb/K, Zr/Ti, Th/Sc, La/Co, δCe, and so on were selected for identification of sediment sources of SO-31 (Figure 4), given that the ratios of these elements can better offset the grain size and carbonate dilution effect. Additionally, these elements show distinct differences during magmatic differentiation and can be more effective in discriminating their provenance [39,40]. It is evident from Figure 4 that the vertical distribution of element ratios aligns well with the distribution of individual element contents (Figures 2 and 3). Notably, at a depth of 250 cm, significant variations are observed in indicators such as Ni/Co, Th/Sc, Th/Cr, and δCe. These changes further subdivide Unit 1 into two sub-units, namely, Unit 1A and Unit 1B, reflecting a substantial alteration in sediment sources within this region.

Th/Sc and Th/Cr are commonly utilized as indicators for differentiating sediment sources [41,42]. Tong (2007) demonstrated that Th/Cr and Th/Sc effectively distinguish sediments originating from the Pearl River, Red River, and Mekong River, which are the primary input rivers around the Beibu Gulf [25]. In the study area, Th/Sc and Th/Cr also exhibit marked differences among the three main source end-members: the Red River, the eastern part of the Qiongzhou Strait, and the western part of Hainan Island. Furthermore, the vertical variations in Th/Sc and Th/Cr correspond well with the unit division of SO-31, confirming their utility as effective indicators for analyzing sediment sources of SO-31. The scatter plots of ratios of Th/Sc and Th/Cr show that the three sedimentary units had different distribution characteristics. The scatter plots in Unit 1 (including Unit 1A and Unit 1B) had a similar distance with that of the Red River, the west nearshore of the Hainan Island, and the east side of the Qiongzhou Strait, indicating that the three possible source areas may have similar contributions to the sediments in the sea area. Compared with Unit

1, the scatter plots in Unit 2 moved gradually away from the east side of the Qiongzhou Strait and coastal sediments of the Hainan Island, but closer to the Red River, indicating the increasing contributions of the Red River runoff. Unit 3 is closer to the Red River than Unit 2 and Unit 1, which indicates that the Red River was the main source area for the core sediments in this unit.

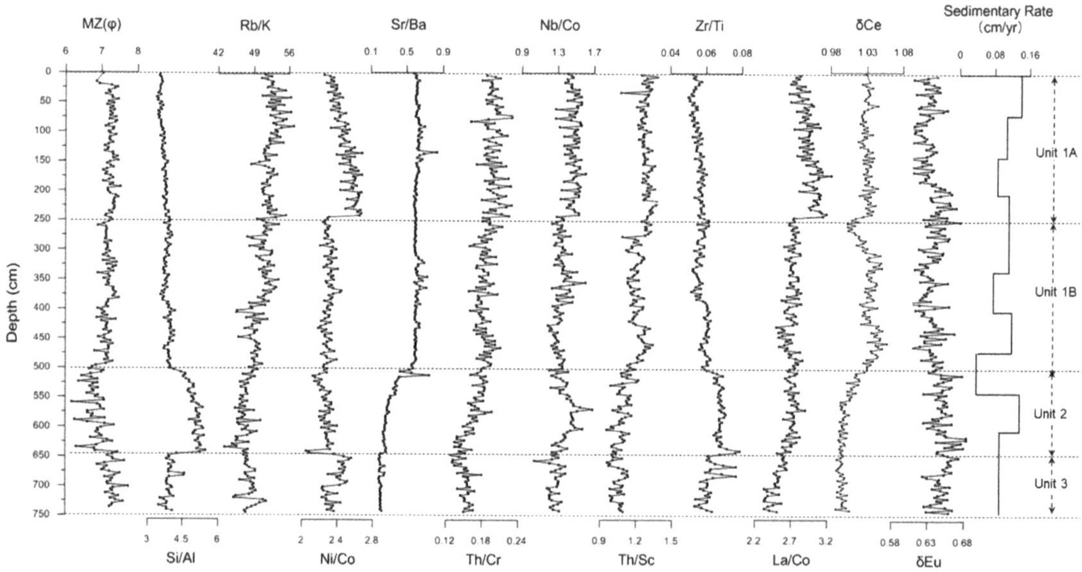

Figure 4. Down-core variations of selected indexes of the core SO-31 sediments show the changes in the source and the environment. Dashed horizontal lines indicate the boundaries between depositional units.

The provenance analysis results of the SO-31 sediments differ from the STAT22 sediments reported by Cui (2016) [18]. Geographically, the core STAT22 is situated on the eastern slope of the Beibu Gulf (Figure 1), and its sediments are predominantly influenced by sources from the western part of Hainan Island. Due to the "central depression" topography of the Beibu Gulf, sediments from the western part of the gulf, including those from the Red River, are less likely to be transported to the STAT22 location under the influence of the bottom current. In contrast, SO-31 is located in the central area of the Beibu Gulf, where materials from the surrounding area of the Beibu Gulf, including the Red River and the western part of the Beibu Gulf, can more easily reach this location. Therefore, the sediments at SO-31 provide a more comprehensive record of the provenance changes in the Gulf. In addition, it should also be noted that the scatter plot of Th/Sc vs. Th/Cr (Figure 5) shows that the sediments from the three units in SO-31 are not completely located among the Red River, the western nearshore of Hainan Island, and the eastern Qiongzhou Strait end-member, suggesting the presence of other significant sediment sources influencing this region, especially the western coastal area of the Beibu Gulf.

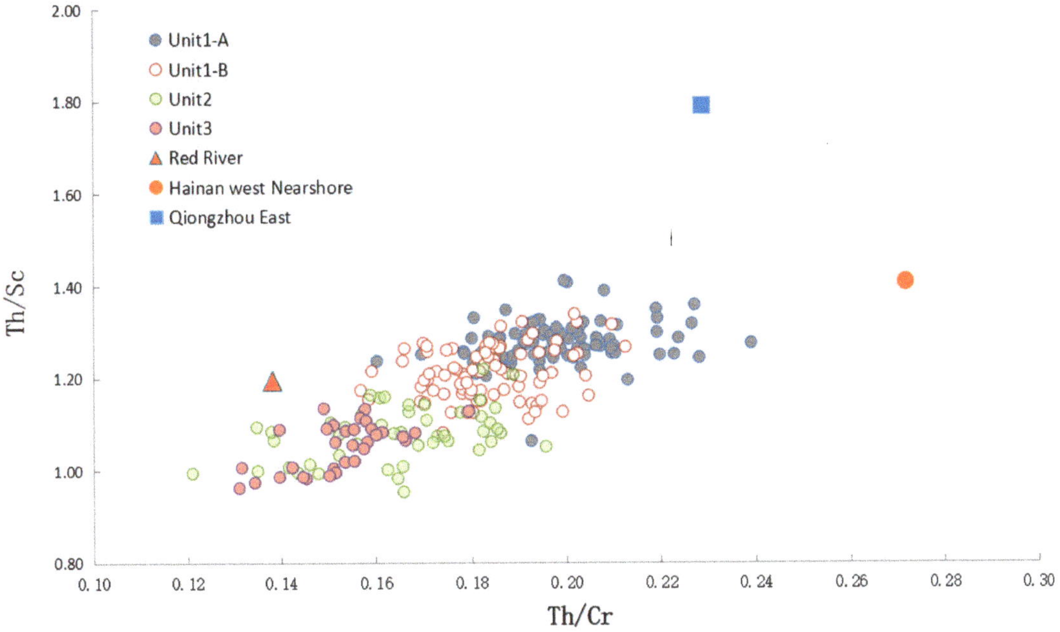

Figure 5. Discrimination plot of Th/Cr vs. Th/Sc for SO-31 core sediments.

4.3. Depositional Sedimentation History and Environmental Changes

During the Holocene, with the regional sea level rise associated with climate change, the Beibu Gulf experienced a series of drastic environmental changes. Especially, the Qiongzhou Strait opening event at the end of the Early Holocene [43,44] had a great impact on the hydrodynamic environment and material input of the Beibu Gulf [45,46].

At the beginning of the Holocene (12,800~11,600 a BP), the Beibu Gulf was in the period of relatively low sea level, and the regional sea level was ~60 m lower than in the present [47,48]. During a comparable period (12,900–11,700 a BP), Zhang et al. discovered intermittent occurrences of diatoms mainly composed of typical coastal species and nearshore planktonic species in the adjacent columnar sediment SO219/31-4, suggesting that it was in a fluvial or shallow-water sedimentary environment [13], this is consistent with the low Sr/Ba ratio (<0.25) found in the Unit 3 sediments in the present study. In the period, the Red River, being the largest river discharging into the Beibu Gulf, significantly influenced sedimentation in the region. It extensively incised the continental shelf, extending its channel southward even to the study area [49–52], potentially allowing direct accumulation of Red River sediments on the mid-outer shelf. Additionally, smaller rivers such as the Changhua, Ma, and Nanliu Rivers also contributed to regional sediment sources, though the Red River remained the predominant contributor. In contrast, the eastern side of the Qiongzhou Strait did not serve as a major sediment source during this period due to the strait's closure [43,44].

Between 11,600–7700 a BP, the location of SO-31 progressively moved farther from the Red River outlet and surrounding terrestrial areas as regional sea levels rose, which is supported by the reduced abundance of shell remains and CaO, alongside an increase in Sr/Ba ratios in Unit 2 sediments compared to those in Unit 3. This stage corresponds to stage 2 and stage 3 (11,700–9500 a BP and 9500–7200 a BP) as defined by Zhang et al. based on diatom assemblages. Unfortunately, this only illustrated the lowest sedimentation rate observed in the 9500–7200 stage and was attributed to the slowing of sea level rise, while the overall low sedimentation rate in the two stages failed to provide a detailed explanation [13]. Throughout this interval, the Red River continued to dominate sediment supply to the study area (Figure 5). Previous studies indicate that sea levels in the Beibu Gulf were rising rapidly in this period [53,54], and as sea levels increased, sediment deposition gradually shifted away from river mouths, including the Red River and other coastal rivers, leading to a decrease in sediment input and overall sedimentation rates in the region. Notably, at the end of Unit 2 (depth of 501 cm), there was a marked change in sedimentation rates and provenance, suggesting the potential west-to-east connection of the Qiongzhou Strait. Previous studies have shown that during the 11,600–7700 a BP period, the sea level in the Beibu Gulf was still rising rapidly. With the continuous rise in sea level, the sediment area gradually moved away from the estuary area of the Red River and other rivers entering the sea, and the provenance input of the Red River and other rivers entering the sea decreased, resulting in the overall decrease in the sedimentation rate in the area. It is worth noting that at the end of this unit (501 cm, 7700 a BP), there was a significant change in deposition rate and provenance in the area, which may indicate that the Qiongzhou Strait had been connected from east to west. With a significant influx of sediments from the eastern part of the Qiongzhou Strait into the Beibu Gulf, these materials were subsequently transported to the study area by ocean currents. Consequently, the contribution rate of the Red River to the provenance of the area was diminished, leading to a relative reduction in the Red River's influence on the regional provenance (Figure 5). Collectively, these observations suggest that the Qiongzhou Strait likely opened around 7700 a BP; this is consistent with Zhao et al. and Yao et al.'s assumptions [43,44].

From 7700–0 a BP, sediments originating from the eastern Qiongzhou Strait became one of the primary contributors to the core area sediments, as indicated by the proximity of Unit 1 sediments to Qiongzhou East sediments in Th/Sc vs. Th/Cr plots (Figure 5). Unit 1 sediments are positioned centrally among the three provenance end-members (the Red River, the western Hainan Island, and the eastern Qiongzhou Strait), indicating a mixed sedimentary product derived from all three sources (Figure 5); this aligns with previous interpretations by Dou (2012) and Cui (2015) [55,56]. Compared to Unit 1B, the scatter plot of Unit 1A (2800–0 a BP) in Th/Sc vs. Th/Cr shows closer affinity to the western Hainan Island and Qiongzhou Strait. This stage also corresponds to stages 4–8 (7200–0 a BP), as defined by Zhang et al. However, their focus was primarily on the increase in the sedimentation rate during Stage 6, which they attributed to sea-level drop and abnormal hydrodynamic conditions, while the reasons for the overall increase in the sedimentation rate across stages 4–8 remain unaddressed [13]. According to Shi (2007) [57], sea levels in the Beibu Gulf exhibited a downward trend during this period (2800–0 a BP). The continuous decline in sea level facilitated greater sediment transport to the study area, enhancing local sedimentation rates. Meanwhile, the Qiongzhou Strait channel underwent progressive incision and widening due to tidal currents, increasing sediment input from the east into the Beibu Gulf and, correspondingly, elevating sediment contribution rates in the region.

Several prevailing perspectives exist regarding the formation and evolution of the Qiongzhou Strait, particularly concerning its opening time during the Holocene. Zhao et al. conducted a comprehensive analysis of regional geological, geophysical, and paleontological data, concluding that the east–west waterway of the Qiongzhou Strait formed between 10.5 and 7.1 ka BP [43]. Yao et al., based on relative sea level change curves and sediment thickness data from the northwest South China Sea, determined that the east–west waterway of Qiongzhou Strait primarily developed between 11.0 and 8.5 ka BP, with full penetration occurring at approximately 8.5 ka BP [44]. Chen et al., using grain size analysis and AMS ^{14}C dating of core sediments from both sides of the strait, estimated the penetration time of the Strait to be around 8.0 ka BP [11]. In this study, the significant increase in provenance contribution from Qiongzhou Strait, observed at 7.7 ka BP in core SO-31, aligns well with the aforementioned conclusion despite minor temporal discrepancies. These results further corroborate that the Qiongzhou Strait opened in the early Holocene, approximately 8.0–7.7 ka BP. Additionally, Xu et al. also observed a significant increase in deposition rates at the C4 and B106 core sediment stations in the eastern Beibu Gulf around 7.0 ka BP [20], whereas no such notable increase was observed in the SO-31 sediments; this discrepancy suggests that the increased deposition rates in the two stations should not be attributed to the opening events of the Qiongzhou Strait.

5. Conclusions

(1) The changes in provenance and sedimentary environment within the Beibu Gulf of the South China Sea were primarily driven by regional sea level rise during the Holocene. Additionally, the opening of the Qiongzhou Strait in the early Holocene had a significant impact on the provenance supply to the Beibu Gulf.

(2) Sediments in core SO-31 were stratigraphically divided into three units based on vertical variations in the abundance of major and trace elements, reflecting fluctuations in sea level and shifts in sediment provenance within the study area.

(3) The Th/Cr vs. Th/Sc scatter plot for sediments from core SO-31 indicates that the contribution from the Qiongzhou Strait increased markedly around 7700 a BP, suggesting that the opening event of the Qiongzhou Strait occurred during this period and significantly influenced the source supply to the Beibu Gulf.

(4) The Th/Cr vs. Th/Sc scatter plot shows that sediments from core SO-31 do not fully fall between the three source end elements. This suggests potential biases in identifying major provenance areas in the region. Therefore, additional provenance areas should be considered, especially the western coast of the Beibu Gulf.

Author Contributions: Z.C.: conceptualization, methodology, validation, formal analysis, investigation, data curation, writing—original draft preparation. Y.H.: software, visualization, writing—review and editing. All authors have read and agreed to the published version of the manuscript.

Funding: This research was funded by both geological survey projects of the China Geological Survey (1212010914027, DD20221725, DD20230404, DD20240090) and BMBF in Germany (03F0607A).

Institutional Review Board Statement: Not applicable.

Informed Consent Statement: Not applicable.

Data Availability Statement: The raw data supporting the conclusions of this article will be made available by the authors on request.

Conflicts of Interest: The authors declare no conflicts of interest.

References

1. Sobrino, M.C.; Parada, C.A.; Cartelle, V.; Carreño, N.M.; Delgado, C.; Cazás, N.; Lázaro, I.; Gil, S.G. Sediment recycling during the Holocene marine transgression in Ría de Vigo (NW Iberia): Multiproxy evidence and environmental implications. *Quaternary Sci. Rev.* **2024**, *344*, 109006. [CrossRef]
2. Limoges, A.; Ribeiro, S.; Nieuwenhove, N.V.; Jackson, R.; Juggins, S.; Crosta, X.; Weckström, K. Marine diatoms record Late Holocene regime shifts in the Pikialasorsuaq ecosystem. *Global Change Biol.* **2023**, *23*, 6503–6516. [CrossRef] [PubMed]
3. Razjigaeva, N.G.; Ganzey, L.A.; Grebennikova, T.A.; Mokhova, L.M.; Degterev, A.V.; Ezhkin, A.K.; Rybin, A.V.; Arslanov, K.A.; Maksimov, F.E.; Petrov, A. The Records of Environmental Changes in Lacustrine-Swamp Sequences within the Mountain Area of Iturup Island since the Late Glacial Period. *Russ. J. Pac. Geol.* **2022**, *16*, 116–130. [CrossRef]
4. Peterson, D.C.; Phipps, B.J. Accommodation space controls on incised-valley sediment accumulation rates during the Holocene marine transgression (0–11ka) in Grays Harbor, a large meso-tidal estuary, Washington, USA. *Mar. Geol.* **2016**, *38*, 1–16. [CrossRef]

5. Thong, G.T.; Watitemsu, I.; Temsulemba, W. Geochemical evidences on the origin and paleo-oceanic depositional setting of the Naga Hills Ophiolite cherts, North East India. *Geol. J.* **2022**, *57*, 3114–3134. [CrossRef]
6. Andreas, G.; Jan, H. Relative sea-level changes induced by glacial isostatic adjustment and sediment loads in the Beibu Gulf, South China Sea. *Oceanologia* **2023**, *65*, 249–259.
7. Van, N.S.; Van, N.K.; Viet, B.D.; Duy, C.T. Biostratigraphy and palaeoenvironments of early Pleistocene deposits in the southern part of the Gulf of Tonkin, Vietnam. *Quatern Int.* **2021**, *604*, 1–15.
8. Liu, F.; Zhang, R.; Li, H.; Liu, H.; Yan, A.; Han, M.; Yu, K. Distribution and adsorption-desorption of organophosphate esters from land to sea in the sediments of the Beibu Gulf, South China Sea: Impact of seagoing river input. *Stoten* **2024**, *917*, 170359. [CrossRef]
9. Xu, G.; Liang, D.; Jia, L.; Xiao, Y.; Wen, L.; Sang, M.; Li, S.; Chen, B.; Fu, C.; Du, J.; et al. Early-Middle Holocene beach rocks at the western Qiongzhou Strait and their significance for paleo-sea level reconstruction. *Acta Geol. Sin.* **2023**, *97*, 2386–2398.
10. Xu, D.; Ge, Q.; Han, X.; Deng, T.; Xiao, T. Driving force of marine sedimentary source to sink; A case study from Beibu Gulf. *J. Mar. Sci.* **2022**, *40*, 17–32. (In Chinese)
11. Chen, L.; Zhang, Y.; Li, T.; Yang, W.; Chen, J. Sedimentary Environment and Its Evolution of Qiongzhou Strait and Nearby Seas since Last Ten Thousand Years. *J. Earth Sci. China* **2014**, *39*, 696–704.
12. Liang, D.Y.; Wu, S.Z.; Xu, G.Q.; Xia, C.J.; Gao, F.L.; Lin, Y.H.; Jia, L.Y. Paleoenvironmental changes in the coastal zone of the northwest South China Sea during the last 13 kyr. *Sci. Rep.* **2023**, *13*, 13540. [CrossRef]
13. Zhang, J.P.; Michal, T.; Andrzej, W.; Xia, Z.; Li, C. A fossil diatom-based reconstruction of sea-level changes for the Late Pleistocene and Holocene period in the NW South China Sea. *Oceanologia* **2023**, *65*, 211–229. [CrossRef]
14. Liang, D.; Xu, G.; Gao, F.; Wen, L.; Jia, L.; Liu, L.; Jiao, D.; Yang, C.; Wu, S. Holocene sediment source analysis and paleoclimatic significance of core KZK01 from the eastern part of the Beibu Gulf. *Front. Earth Sci.* **2023**, *11*, 1192206. [CrossRef]
15. Li, Y.; Fan, T.; Wang, A.; Zeng, J.; Lv, Y.; Zhang, M.; Liu, D. The Sediments in the Beibu Gulf Reveal Dramatic Paleoenvironmental Changes and Climate Events over the Past 20,000 Years. *J. Mar. Sci. Eng.* **2024**, *12*, 615. [CrossRef]
16. Guan, Y.; Chen, Y.; Sun, X.; Xu, L.; Xu, D.; Zhu, Z.; He, W. The Clay Mineralogy and Geochemistry of Sediments in the Beibu Gulf, South China Sea: A Record of the Holocene Sedimentary Environmental Change. *J. Mar. Sci. Eng.* **2023**, *11*, 1463. [CrossRef]
17. Li, M.; Ouyang, T.; Roberts, P.A.; Heslop, D.; Zhu, Z.; Zhao, X.; Qiu, Y. Influence of Sea Level Change and Centennial East Asian Monsoon Variations on Northern South China Sea Sediments Over the Past 36 kyr. *Geochem. Geophys. Geosyst.* **2018**, *19*, 1674–1689.
18. Cui, Z.; Schulz-Bull, E.; Hou, Y.; Xia, Z.; Waniek, J. Geochemical Characteristics and Provenance of Holocene Sediments (Core STAT22) in the Beibu Gulf, South China Sea. *J. Coastal Res.* **2016**, *32*, 1105–1115. [CrossRef]
19. Zhang, A.; Chen, M.; Gan, H.; Chen, Q.; Lan, B.; Fang, Q. Geochemical characteristics and sediment provenance of Core SO-50 sediments in the Beibu Gulf. *Hai Yang Xue Bao* **2018**, *10*, 107–117. (In Chinese)
20. Xu, D.; Li, J.; Chu, F.; Long, J.; Ye, L.; Han, X. The Response of Sedimentary Records in Eastern BeibuGulf to the Last Postglacial Transgression and Circulation. *Earth Sci. J. China Univ. Geosci.* **2013**, *38* (Suppl. S1), 70–82.
21. Chen, B.; Shi, M. Advances in Study of Beibu Gulf Circulation. *Guangxi Sci.* **2019**, *26*, 595–603. (In Chinese)
22. Ni, Y.; Endler, R.; Xia, Z.; Endler, M.; Harff, J.; Gan, H.; Schulz-Bull, D.; Waniek, J. The butterfly delta system of Qiongzhou Strait: Morphology, seismic stratigraphy and sedimentation. *Mar. Geol.* **2014**, *355*, 361–368. [CrossRef]
23. Xu, D.; Long, J.P.; Qian, J. The modern sedimentation rate and the distribution character of 7 cores in Hainan Island offshore. *J. Mar. Sci.* **2008**, *26*, 9–17. (In Chinese)
24. Cui, Z.; Lin, J.; Gan, H.; Liu, W.; Zhang, L. Geochemical characteristics of the surface sediments in the eastern Beibu Gulf, South China Sea. *Mar. Sci.* **2015**, *39*, 103–111. (In Chinese)
25. Tong, S. Element Geochemistry for Surface Sediments of Pearl, Red and Mekong Rivers. Master's Thesis, Tongji University, Shanghai, China, 2007. (In Chinese).
26. Tong, S.; Liu, Z.; PhonLe, K.; Huang, W. Chemical Weathering in the Red River Basin: Records of Major and Trace Elemental Geochemistry. *Bull. Mineral. Petrol. Geochem.* **2006**, *3*, 218–225. (In Chinese)
27. Taylor, S.; Mclennan, S. *The Continental Crust: Its Composition and Evolution: An Examination of the Geochemical Record Preserved in Sedimentary Rocks*; Blackwell Scientific Publications: Oxford, UK, 1985; p. 312.
28. Wetzela, A.; Unverricht, D. Sediment dynamics of estuarine Holocene incised-valley fill deposits recorded by Siphonichnus (ancient Red River, Gulf of Tonkin). *Palaeogeogr. Palaeocl.* **2020**, *560*, 110041. [CrossRef]
29. Hori, K.; Tanabe, S.; Saito, Y.; Haruyama, S.; Nguyen, V.; Kitamura, A. Delta initiation and Holocene sea-level change: Example from the Song Hong (Red River) delta, Vietnam. *Sediment Geol.* **2004**, *164*, 237–249. [CrossRef]
30. Wei, L.; Wang, C.; Cui, H.; Jia, J.; Zeng, L.; Su, M. Provenance of sediments in the northern shelf of the South China Sea: Insights from Pb isotopes in detrital K-feldspar. *Mar. Geol.* **2025**, *480*, 107448. [CrossRef]

31. Zhao, J.; Zhang, L.; Wang, X.; Ge, J.; Xu, M.; Chen, W.; Luo, M.; Liang, Q.; Yu, Q.; Luo, S.; et al. Characteristics of Paleontological Communities in Surface Sediments of the Southern South China Sea and Their Paleoclimatic and Paleoenvironmental Significance. *J. Earth Sci.* **2024**, *35*, 144–154. [CrossRef]
32. Xu, D. The Research of Sedimentary Geochemistry of the Hainan Island Nearshore Sea Area. Master's Thesis, The Second Institute of Oceanography, Hangzhou, China, 2007. (In Chinese).
33. Feng, W.K.; Xue, W.J.; Yang, D.Y. *The Geological Environment of Late Quaternary in the Northern South China Sea*; Guangdong Science and Technology Publishing House: Guangzhou, China, 1988; p. 261. (In Chinese)
34. Hu, M.; Zhu, Y.; Chen, W.; Yu, B.; Zhang, P.; Hu, C.; Jin, R. Investigating the Element Geochemical Behavior and Provenance of Surface Sediments in the Offshore Area of Sierra Leone, Africa: Insights from Major and Trace Elements. *Water* **2024**, *23*, 3540. [CrossRef]
35. Paola, G.; Bartolomeo, S.; Federica, B.; Chiara, P.; Beatrice, M.; Roberta, S.; Enzo, G.; Matteo, P.; Alessandro, L.; David, C. Sources and trends of trace elements and polycyclic aromatic hydrocarbons in a shallow lake in the Mediterranean area from sediment archives of the Anthropocene. *Environ. Sci. Pollut. Res.* **2022**, *29*, 85815–85828.
36. Ahmad, Z.; Singh, P. REE geochemistry of core sediments of Cauvery delta, India for provenance studies. *Acta Geochim.* **2020**, *39*, 512–526. [CrossRef]
37. Jung, H.; Lim, D.; Jeong, D.; Xu, Z.; Li, T. Discrimination of sediment provenance in the Yellow Sea: Secondary grain-size effect and REE proxy. *J. Asian Earth Sci.* **2016**, *123*, 78–84. [CrossRef]
38. Lim, D.J.; Hoi, S.; Choi, J.Y. REE partitioning in riverine sediments around the Yellow Sea and its importance in shelf sediment provenance. *Mar. Geol.* **2014**, *357*, 12–24. [CrossRef]
39. Brijraj, K.; AL-Mikhlafi, A.; Kaura, P. Geochemistry of Mansar Lake sediments, Jammu, India: Implication for source-area weathering, provenance, and tectonic setting. *J. Asian Earth Sci.* **2006**, *26*, 649–668.
40. Wronkiewicz, D.J.; Condie, K.C. Geochemistry of Archean shales from the Witwatersrand Supergroup, South Africa: Source-area weathering and provenance. *Geochim. Cosmochim. Acta* **1987**, *51*, 2401–2416. [CrossRef]
41. Guo, Y.; Li, C.; Wang, C.; Xu, J.; Jin, C.; Yang, S. Sediment Routing and Anthropogenic Impact in the Huanghe River Catchment, China: An Investigation Using Nd Isotopes of River Sediments. *Water Resour. Res.* **2021**, *57*, e2020WR028444. [CrossRef]
42. Li, Q.; Liu, S.; Han, B.; Wang, Y.; Dang, Q. Geochemical characteristics of the metapelites from the Xingxingxia group in the Eastern Segment of the Central Tianshan: Implications for the provenance and paleoweathering. *Sci. China Ser. D Earth Sci.* **2005**, *48*, 1637–1648. [CrossRef]
43. Zhao, H.; Wang, L.; Yuan, J. Origin and time of Qiongzhou Strait. *Mar. Geol. Quat. Geol.* **2007**, *27*, 33–40. (In Chinese)
44. Yao, Y.; Harff, J.; Meyer, M.; Zhan, W. Reconstruction of paleocoastlines for the northwestern South China Sea since the Last Glacial Maximum. *Sci. China Earth Sci.* **2009**, *52*, 1127–1136. [CrossRef]
45. Shi, M.; Chen, C.; Xu, Q.; Lin, H.; Liu, G.; Wnag, H.; Wang, F.; Yan, J. The Role of Qiongzhou Strait in the Seasonal Variation of the South China Sea Circulation. circulation. *J. Phys. Oceanogr.* **2002**, *32*, 103–121. (In Chinese) [CrossRef]
46. Tang, D.; Kawamura, H.; Lee, M.A.; Dien, T.V. Seasonal and spatial distribution of chlorophyll-α concentrations and water conditions in the Gulf of Tonkin, South China Sea. *Remote Sens. Environ.* **2003**, *85*, 475–483. [CrossRef]
47. Zhao, H.; Liu, Z.; Zhao, Y. Sea level and East Asian monsoon influenced chemical weathering records in the southern South China Sea over the past 21 ka. *Global Planet. Change* **2024**, *232*, 104324. [CrossRef]
48. Thanakorn, J.; Liu, Z. East Asian monsoon and sea-level controls on clay mineral variations in the southern South China Sea since the Last Glacial Maximum. *Quatern Int.* **2021**, *592*, 1–11.
49. Yang, H.; Qiao, L.; Liu, X. Shoreline evolution of the Red River Delta in the recent 30 years. *Mar. Geol. Quanternary Geol.* **2020**, *40*, 48–57. (In Chinese)
50. Yu, Y.; Mu, Y. The new institutional arrangements for fisheries management in Beibu Gulf. *Mar. Policy* **2006**, *30*, 249–260. [CrossRef]
51. Borges, J.; Huh, Y. Petrography and chemistry of the bed sediments of the Red River in China and Vietnam: Provenance and chemical weathering. *Sediment. Geol.* **2007**, *194*, 155–168. [CrossRef]
52. Zhou, G.; Cao, X.; Xia, J.; Wang, S.; Song, Z. A dramatic marine environment change in the Beibu Gulf of the South China Sea around 3.2 kyr BP. *Lithosphere* **2022**, *2022*(Special 9), 2632112.
53. Tanabe, S.; Hori, K.; Saito, Y.; Haruyama, S.; Vu, V.P.; Kitamura, A. Song Hong (Red River) delta evolution related to millennium-scale Holocene sea-level changes. *Quatern. Sci. Rev.* **2003**, *22*, 2345–2361. [CrossRef]
54. Hanebuth, T.J.J.; Voris, H.K.; Yokoyama, Y.; Saito, Y.; Okuno, J. Formation and fate of sedimentary depocentres on South- east Asia's Sunda Shelf over the past sea-level cycle and biogeographic implications. *Earth Sci. Rev.* **2011**, *104*, 92–110. [CrossRef]
55. Dou, Y.; Li, J.; Li, Y. Rare earth element compositions and provenance implication of surface sediments in the eastern Beibu Gulf. *Geochimica* **2012**, *41*, 147–157. (In Chinese)

56. Cui, Z.; Gan, H.; Liu, W.; Zhang, L. Geochemical characteristics of major elements and provenance implication of surface sediments in the eastern Beibu gulf. *Comput. Tech. Geophys. Geochem. Explor.* **2015**, *37*, 522–531. (In Chinese)
57. Shi, X.; Yu, K.; Chen, G. Progress in researches on sea level changes in South China Sea since Mid-Holocene. *Mar. Geol. Quat. Geol.* **2007**, *27*, 121–132.

applied sciences

MDPI

Article

Prediction of Selenium-Enriched Crop Zones in Xiaoyan Town Using Fuzzy Logic and Machine Learning Approaches

Jiacheng Li [1], Shuyun Xie [1,*], Wenbing Yang [2], Weihang Zhou [1], Emmanuel John M. Carranza [3], Weiji Wen [1] and Hongtao Shi [1]

1 State Key Laboratory of Geological Processes and Mineral Resources (GPMR), Faculty of Earth Sciences, China University of Geosciences, Wuhan 430074, China; lijiacheng_5315@163.com (J.L.); zwh15671032170@126.com (W.Z.); cugwenweiji@163.com (W.W.); 18779268469@163.com (H.S.)

2 Eighth Geological Brigade of Hubei, Xiangyang 441002, China; ywb-88@163.com

3 Department of Geology, University of the Free State, Bloemfontein 9301, South Africa; ejmcarranza@gmail.com

* Correspondence: tinaxie@cug.edu.cn

Abstract: Selenium-rich foods play a crucial role in human health and hold significant economic value for agricultural products. However, many regions in China are experiencing selenium deficiency, which has led to an increased demand for Se-rich agricultural products. This study focused on Nanzhang County, a key area within the "Organic Valley" of Hubei Province, China. We employed fuzzy weights-of-evidence, backpropagation neural network, and support vector regression models to predict optimal planting zones for Selenium-rich crops. A comparative analysis indicated that the backpropagation neural network model provided the highest prediction accuracy (R^2 = 0.77), identifying Selenium-rich crop zones covering 68.87% of the aera, where Selenium-rich crops made up 86.67% of all samples. Notably, the backpropagation neural network yielded excellent performance for rice and rapeseed, with R^2 values of 0.95 and 0.99, respectively. The findings also indicate that the Selenium content in crops is closely linked to Selenium levels in the soil and is significantly influenced by synergistic and antagonistic interactions with other elements. This study provides scientific support for the cultivation of selenium-rich crops. It plays a positive role in promoting the development of the local selenium-rich industry and the sustainable utilization of soil selenium resources.

Keywords: selenium-enriched crop zones; fuzzy weights-of-evidence; machine learning; geospatial prediction; cultivation recommendation

1. Introduction

Selenium (Se) is one of the essential trace elements for the human body [1,2]. It exerts biological functions such as antioxidation, anti-inflammation, anticancer activity, immune enhancement, and anti-aging, primarily through its role as a cofactor in selenoproteins, particularly glutathione peroxidase, which protects cells from oxidative damage by detoxifying lipid peroxides [3–5]. Prolonged Se deficiency in humans can lead to well-known endemic diseases, such as Keshan disease and Kashin-Beck disease, by impairing cardiovascular function and increasing the risk of myocardial infarction [6,7]. China is a country with soils severely deficient in Se, with an uneven distribution of Se resources, resulting in varying degrees of Se deficiency across most regions [8,9]. The primary source of Se for the human body is the consumption of Se-rich products [10–12]. Thus, Se-rich foods are not only essential for meeting daily Se intake but are also increasingly in demand due to heightened health awareness, leading to expanding market opportunities and rising economic value [9,13].

Located in Xiangyang City, Hubei Province, Nanzhang County is one of the three core towns of Xiangyang's "Organic Valley" [14]. Situated in the transitional zone between the Jing Mountains and the Jianghan Plain, Nanzhang County enjoys natural ecological resources nurtured by a northern subtropical monsoon climate [15], providing unique conditions for the development of organic Se-rich agricultural products. Delineating optimized planting zones for Se-rich crops in Nanzhang County and employing precision agricultural practices to improve the quality and yield of local Se-rich products will enhance their market competitiveness and promote the development of regional specialty agriculture.

Moreover, scientifically planning Se-rich crop planting zones can ensure the sustainable utilization of soil resources, fostering the long-term growth of the Se-rich product industry [16,17]. Concurrently, by identifying areas suitable for Se-low crop cultivation, measures such as soil improvement, crop variety selection, precision fertilization, and irrigation can be implemented to effectively develop Se-deficient soils, enhance Se content in crops, and significantly alleviate Se deficiency in certain regions, potentially driving the growth of the Se-rich product industry [18,19].

In previous studies, multivariate linear regression (MLR) models have been mainly used to predict the Se crop content [20,21]. Geochemical data reveal that the relationship between Se crop content and soil properties is characterized by nonlinearity and emergent behaviors, where complex interactions between various factors give rise to outcomes that cannot be predicted simply by analyzing the individual components [22], limiting the applicability of traditional linear regression methods in this field.

In recent years, the rapid development of fuzzy weights-of-evidence (FWofE) models and machine learning (ML) techniques has provided effective solutions to address these challenges. The FWofE model can quantitatively analyze the spatial relationships between the distribution of Se-rich crops and environmental influencing factors, enabling accurate predictions of Se-rich crop distribution. Comparative studies of ML approaches and multiple regression methods have demonstrated that ML techniques achieve higher accuracy and applicability in predicting Se crop content [21,23]. Unlike the MLR method, ML methods such as backpropagation neural network (BP-NN) and support vector regression (SVR) can capture nonlinear relationships, emergent behaviors, and high-dimensional interactions among variables [24,25]. In summary, compared to the MLR approach, FWofE, BP-NN, and SVR are particularly advantageous for modeling the intricate and nonlinear relationships between Se crop content and soil physicochemical properties, thereby supporting more precise predictions.

This study aimed to analyze the geochemical data of crops and their corresponding root soils in Xiaoyan Town, Nanzhang County (China), to identify key geochemical characteristics of root soils strongly correlated with Se crop content. The FWofE, BP-NN, and SVR methods were employed to predict the distribution of Se-rich crops and to delineate optimal planting zones within the study area. Additionally, SHAP (SHapley Additive exPlanations) and PDPs (Partial Development Plots) were applied to provide model interpretability, adding a novel perspective often overlooked in similar studies. Ultimately, Se-rich optimal planting zones for wheat, rice, rapeseed, and tea were delineated, providing planting recommendations and scientific support for the development of Se-rich agriculture in the region.

2. Study Area and Data

2.1. Study Area

The Nanzhang County is located within 111°26′–112°9′ E and 31°13′–32°1′ N, with a total area of 3859 km^2. The main exposed strata in the region include the Silurian,

Permian, Triassic, Cretaceous, Paleogene, and Quaternary, among which the Triassic is the most extensive and constitutes the primary strata in the study area. The region's soil types include paddy soil, purple soil, calcareous soil, and yellow-brown earth. The spatial distribution of calcareous soil largely coincides with that of the Triassic, making it the most widely distributed soil type. As a result, the study area experiences severe rocky desertification [26].

2.2. Sample Data

The collection and geochemical analysis of soil and crop samples were conducted by the Eighth Geological Brigade of the Hubei Geological Survey. Sampling and element analysis followed the standards outlined in the "Specifications for Soil Geochemical Survey (DZ/T 0145-2017)" and the "Specifications for Geochemical Land Quality (DZ/T 0295-2016)" [27,28]. The specific processes for soil and crop sample collection and analysis are detailed below:

2.2.1. Soil Samples

In Xiaoyan Town, Nanzhang County, surface soil samples were collected using a combination of grid and polygonal sampling based on a predetermined sampling map. A handheld GPS device was used to pinpoint sampling locations, with an average sampling density of 4.10 points/km^2. Surface soil samples and duplicate samples were collected from the 0 to 20 cm soil layer, removing debris such as weeds, roots, stones, bricks, and fertilizer clumps. To enhance representativeness, a primary sampling point was selected as the fixed location, and 3–5 subsamples were collected within a 50 m radius and combined into a single sample.

Soil sample processing included drying, sieving, packing, completing sample submission forms, assigning sample IDs, and submitting samples for analysis. Surface soil samples were air-dried, sieved through a 20-mesh sieve, and sent to the laboratory for analysis. In total, 29 elements were analyzed, including Se, Cd, Hg, As, Pb, Cu, Zn, Cr, Ni, Mo, Mn, B, N, Co, I, F, Sr, Ge, P, Corg, S, Cl, Fe, Na, K, Ca, Mg, Si, and Al. Besides, pH was measured as well.

2.2.2. Crop Samples

Crop samples were primarily collected during the harvest season. Grain samples were taken from multiple points within the sampling site and mixed in equal proportions to create a composite sample, and the dry weight of the final sample was maintained at no less than 200 g. For fresh samples with 80–90% water content, the fresh weight was estimated to be 5–10 times the dry weight. Field-collected fresh samples had a minimum weight of 2000 g. After threshing and homogenization, grain samples were divided into subsamples using the grid method and quartering method, resulting in a final sample weight of approximately 250 g for small grains and 500 g for larger grains.

In total, wheat, rice, rapeseed, tea leaf, tobacco leaf, potato, and corn samples were collected, along with corresponding root soil samples. The geochemical analysis for crop samples included 21 elements: K, Fe, S, Ca, P, Mg, Cu, Zn, Mn, F, Ni, Se, Mo, Co, Cr, As, Cd, Pb, Hg, Si, and Sr. The root soil samples were analyzed for 20 elements, including K, Fe, S, Ca, P, Mg, Cu, Zn, Mn, F, Ni, Se, Mo, Co, Cr, As, Cd, Pb, Hg, and Si; besides, pH was measured as well.

2.3. Se Soil Content

An iterative method was applied to remove outliers in the original Se soil data that exceeded $X + 3S$ or fell below $X - 3S$ (where X and S are the mean and standard deviation of the original Se soil data, respectively). This process ensured that all data values fell within the range of $X + 3S$ and $X - 3S$, forming a background dataset [29]. Geochemical parameters such as mean, median, standard deviation, and coefficient of variation (CV) were calculated for Se soil content before and after outlier removal to describe the central tendency and dispersion of Se soil content [30]. The average Se soil content in the study area was compared with the national background values to evaluate the enrichment and distribution characteristics of Se soil in the region.

The Se soil content in the study area ranged from 0.105 to 5.653 mg/kg, with an average of 0.347 mg/kg, which was higher than the national background value. This indicates an enrichment of Se soil. Since the maximum Se content far exceeded the average, the coefficient of variation was used to describe the degree of dispersion in Se soil content and to analyze the distribution of high Se values. The CV for Se in the raw data was 0.72, which decreased to 0.24 after outlier removal, resulting in a CV ratio of 3. This substantial change in CV indicates a high degree of dispersion in Se soil content, with a notable number of high-value areas and regions exhibiting elevated Se soil levels [31].

A Se soil content distribution map was created using kriging interpolation based on the processed data (Figure 1a). Accordingly, as shown in Figure 1b, Se soil content was categorized into five levels [2,13]: Se-deficient (<0.125 mg/kg), Se-marginal (0.125–0.175 mg/kg), Se-sufficient (0.175–0.4 mg/kg), Se-rich (0.4–3.0 mg/kg), and Se-excessive (>3.0 mg/kg). The distribution map shows that Se soil content follows a clear zonal pattern, aligning closely with the distribution of geological strata and parent materials. The Permian is identified as the primary Se-rich strata, contributing to a large area of Se-rich soils in the study region.

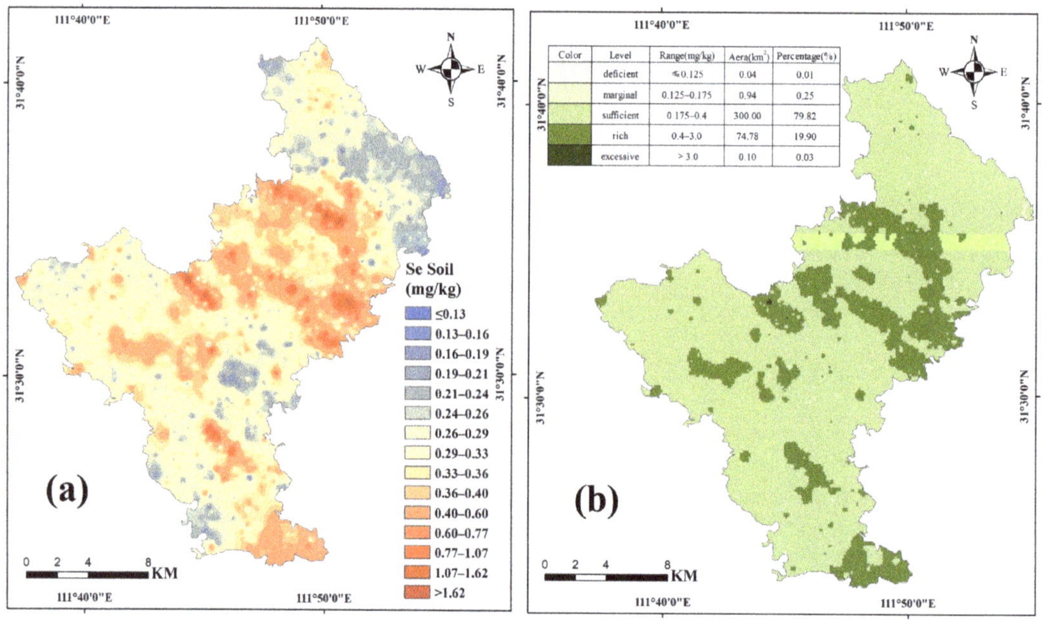

Figure 1. Xiaoyan Town, Nanzhang County: (**a**) distribution map of Se soil content and (**b**) classified map of Se soil content.

2.4. Se Crop Content

The collected crop samples included wheat, rice, rapeseed, tea leaves, tobacco leaves, potatoes, and corn. Table A1 summarizes the average Se content in different crop types. Rice grain, wheat, rapeseed, and tea leaves exhibited Se content within a reasonable range. Crop samples with extremely low (e.g., potato: 0.004 mg/kg; corn: below detection limit) or unusually high Se content (e.g., rice root: 9.94 mg/kg) were excluded to avoid skewing the modeling process. The remaining crop samples were utilized to predict the Se content distribution in the study area and to delineate optimal planting zones for Se-rich crops.

3. Methods

3.1. Fuzzy Weights-of-Evidence (FWofE)

Cheng and Agterberg (1999) improved upon the conventional weights-of-evidence model by introducing the FWofE model, which is a logarithmic, linear model under the Bayesian probability framework [32]. It updates prior probabilities by calculating posterior probabilities under certain conditions, enabling the superposition and analysis of predictor variables for a target variable to generate posterior probability distribution maps. Compared to the traditional weights-of-evidence model, the FWofE model addresses the loss of information caused by discretized layers, effectively characterizing the spatial relationships between evidence layers and targets such as mineral deposits or crops. This results in improved prediction accuracy [13,33]. In recent years, the FWofE model has been widely applied in fields such as mineral prospecting and crop planting planning, achieving promising results (e.g., Zhang et al., 2016; Xie et al., 2023) [13,34].

The FWofE model defines evidence layers as fuzzy sets through a multivalued fuzzy membership function (MSF) and quantifies their fuzziness by calculating membership degrees. This approach enables the processing of multi-class evidence layers. Using MSF to fit training samples, the evidence weights are computed. The computational workflow of the FWofE model involves the following steps: preliminary screening of evidence layers, construction of the FWofE model, calculation of weights for evidence factors, analysis of evidence factor weights and contrasts to determine which evidence layers to retain, plotting the ratio of evidence factor contrast C to its standard deviation S(C), calculation of multi-value membership functions for evidence factors, derivation of fuzzy weights and posterior probabilities for evidence factors [33,35].

3.2. Machine Learning (ML)

Unlike the FWofE model, which identifies Se-rich crop zones based on probabilistic inference, the ML methods predict Se crop content to delineate optimal planting zones.

In recent years, ML methods have been widely applied to solve various geoscientific problems, such as mineral prospecting, crop planting planning, and groundwater quality assessment, yielding promising results [36–38]. Among these, BP-NN and SVR have demonstrated excellent performance in addressing nonlinear problems [24]. In this study, both BP-NN and SVR were used as models to predict and delineate optimal planting zones for Se-rich crops.

3.2.1. Backpropagation Neural Network (BP-NN)

The BP-NN model, based on the error backpropagation algorithm, is one of the most successful and widely used artificial neural networks. The BP-NN is a multilayer network composed of an input layer, one or more hidden layers, and an output layer connected by nodes. Training a BP-NN involves two main processes: forward propagation of input data and backward propagation of errors. Through iterative learning, the BP-NN continuously adjusts the connection weights and thresholds between input and hidden layer nodes, as well as between hidden and output layer nodes. Using the gradient descent method, the network minimizes the mean square error (MSE) between the actual output and the desired output. Hidden layer weights and biases are updated via backpropagation, enabling the multilayer feedforward network to establish appropriate linear or nonlinear relationships between input and output layers [25,39,40].

To enhance the BP-NN's capability in handling nonlinear problems, the ReLU activation function was selected for neuron activation due to its ability to mitigate gradient vanishing and maximize neuron efficiency. The ReLU function is expressed as:

$$\text{ReLu} = \max(0, X), \tag{1}$$

$$\text{Output} = \text{ReLu}(0, W^T + b), \tag{2}$$

where W and b are the trainable parameter matrix and vector composed of neuron weights and biases.

3.2.2. Support Vector Regression (SVR)

SVR, an extension of the support vector machine for regression tasks, employs an ϵ-insensitive loss function, making it well-suited for nonlinear regression analysis problems. The core idea of SVR is to fit the data as closely as possible without violating the boundary conditions. The algorithm aims to find a hyperplane from the given data points, determine the closest relationship between support vectors and the hyperplane, and describe them using a function.

Finding the optimal fitting curve often requires tuning hyperparameters such as ϵ, γ, and C, as well as selecting a kernel function. The radial basis function (RBF) kernel was chosen for this study due to its superior performance in handling nonlinear problems, low model complexity, and fewer required parameters. The RBF kernel is expressed as:

$$K(X, X_m) = \exp\left(-\frac{\|X - X_m\|^2}{2\sigma_m^2}\right), \tag{3}$$

where X_m is the center of the m-th kernel function, and σ_m is the width of the m-th kernel function.

The SVR workflow involves mapping input vectors into a feature space, calculating the dot product of input vectors using the kernel function, assigning weights to the results, and summing them to obtain the output [41,42].

3.2.3. Evaluation of the Prediction Models

In this study, both BP-NN and SVR were implemented as nonlinear regression models to predict optimal planting zones for Se-rich crops. Therefore, the coefficient of determination (R^2), mean absolute error (MAE), and root mean square error (RMSE) were selected as evaluation metrics to assess the prediction performance of the BP-NN and SVR models. The closer R^2 is to 1, the closer the predicted values are to the true values, indicating higher correlation and better interpretability of the prediction results. The formulas for R^2, MAE, and RMSE are, respectively, as follows [16,43]:

$$R^2 = 1 - \frac{\sum_{i=1}^{N} (\widehat{y}_i - y_i)^2}{\sum_{i=1}^{N} (\overline{y} - y_i)^2}, \tag{4}$$

$$MAE = \frac{1}{\text{N}} \sum_{i=1}^{N} |\widehat{y}_i - y_i|, \tag{5}$$

$$RMSE = \sqrt{\frac{1}{N} \sum_{i=1}^{N} (\widehat{y}_i - y_i)^2}, \tag{6}$$

where N is the total number of samples, $\hat{y}_i$ is the predicted Se content of the i-th crop sample, $\bar{y}$ is the mean Se content of the crop samples, and y_i is the true Se content of the i-th crop sample.

4. Results and Discussion

4.1. Factors Affecting Se Crop Content

Studies have shown that total Se soil content, Se speciation and bioavailability, soil physicochemical properties (e.g., pH, Eh, SOM), and the synergistic or antagonistic interactions between Se and other soil elements are key factors that affect Se uptake by crops [43]. Because this study primarily used data on crop and root soil element contents and pH, the following discussion is focused on the synergistic and antagonistic effects of Se with other elements, as well as the impact of soil pH on Se crop content.

Research indicates that elements such as Mn, Mo, and Cu exhibit synergistic effects with Se, while P, S show antagonistic effects [44,45]. For instance, SO_4^{2-} and PO_4^{3-}, as competitive ions, inhibit Se absorption by crops. Therefore, the presence of P, S reduces Se crop uptake [8,46].

In soils, the bioavailability of different Se species is generally in the following order [47]: selenate > organic selenium > selenite > elemental selenium > selenides. Soil pH and Eh significantly influence the distribution of Se species. Selenate is typically found in alkaline surface soils, whereas selenite is more common in acidic subsoils. Therefore, Se crop content generally increases with higher soil pH [8]. Additionally, total Se soil content remains a critical factor affecting Se crop content.

To analyze the correlation between Se crop content and root soil indicators in the study area, we calculated Pearson correlation coefficients for 130 paired samples of Se crop content and root soil indicators, and a heatmap was generated (Figure 2).

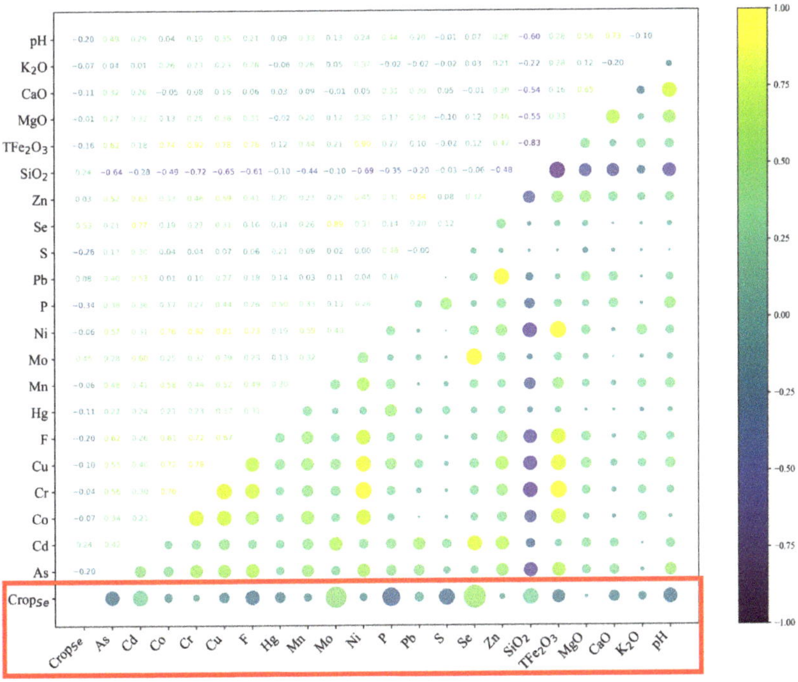

Figure 2. Heat map of correlation between Se crop and root soil elements.

The analysis revealed significant positive correlations between Se crop content and root soil Cd, Mo, Se, and SiO_2 ($p < 0.01$), as well as significant negative correlations with P and S ($p < 0.01$). Among these, the findings for Se, P, and S align well with previous research conclusions [44,46]. The high correlation between root soil Se and Mo (r = 0.89, $p < 0.01$) and Cd (r = 0.77, $p < 0.01$) indicates that Mo and Cd are commonly enriched in Se-rich soils in the study area, suggesting that Se crop content is positively correlated with root soil Mo and Cd. Therefore, in subsequent model construction, data on these elements were prioritized as evidence layers or training data.

4.2. *Prediction of Optimal Planting Zones for Se-Rich Crops*

4.2.1. Fuzzy Weights-of-Evidence

A common challenge in applying the FWofE model for prediction lies in the selection of evidence layers. In this study, the selection of evidence layers for the FWofE method and training data layers for ML models was guided by the correlation analysis results and previous research findings. Because numerous root soil indicators are related to Se crop content, factor analysis was first performed to reduce dimensionality. To avoid the closure effect of the root soil data, the data should be opened by log-ratio transformation before performing factor analysis. Because the centered log-ratio (CLR) transformation preserves the correlation structure between components and is computationally straightforward, it was applied to process the root-zone soil data [48,49]. The parameters and diagnostic metrics of the factor analysis are presented in Table A2. The resulting factor loadings are shown in Table 1, where Factor3 has high negative loadings on Mo and Se, while Factor4 has high negative loadings on P and S. According to the correlation analysis results, Factor3 contains information on root soil elements that are positively correlated with Se crop content, while Factor4 contains information on root soil elements that are negative correlated with Se crop content. Therefore, Factor3 and Factor4 were identified as significantly correlated with Se crop content.

Table 1. Factor loadings of root soil elements and pH.

Elements	Factor1	Factor2	Factor3	Factor4	Factor5
As	0.054	−0.733	0.025	0.089	−0.284
Cd	−0.648	−0.200	−0.353	−0.112	0.038
Co	0.715	0.390	−0.109	0.153	−0.087
Cr	0.750	0.529	−0.021	0.098	0.053
Cu	0.690	−0.308	−0.088	0.004	0.306
F	0.680	−0.104	−0.040	0.120	−0.130
Hg	−0.116	−0.240	−0.197	−0.717	−0.061
Mn	0.090	−0.260	−0.032	0.269	−0.611
Mo	−0.073	−0.425	−0.623	0.285	−0.185
Ni	0.870	−0.027	−0.100	0.169	−0.085
P	−0.281	−0.161	0.320	−0.739	−0.042
Pb	−0.225	−0.025	−0.056	0.164	0.816
S	−0.246	0.271	−0.045	−0.723	−0.027
Se	−0.377	0.284	−0.736	0.197	0.054
Zn	0.250	0.131	0.021	0.207	0.840
SiO_2	0.062	0.913	0.080	0.095	0.016
TFe_2O_3	0.851	0.259	0.153	0.219	0.046
MgO	0.082	0.254	0.588	0.459	0.224
CaO	−0.402	−0.449	0.707	−0.046	−0.084
K_2O	0.367	0.739	0.118	0.217	0.055
pH	−0.151	0.226	0.810	0.248	−0.135

Based on the actual conditions of the study area and FWofE experiments, the final evidence layers included the factor score distribution maps of Factor3 and Factor4, as well as the 500-m nine-ring buffers of the study area's main geological strata (Silurian, Permian, Triassic, Cretaceous, and Paleogene) (Figure 3).

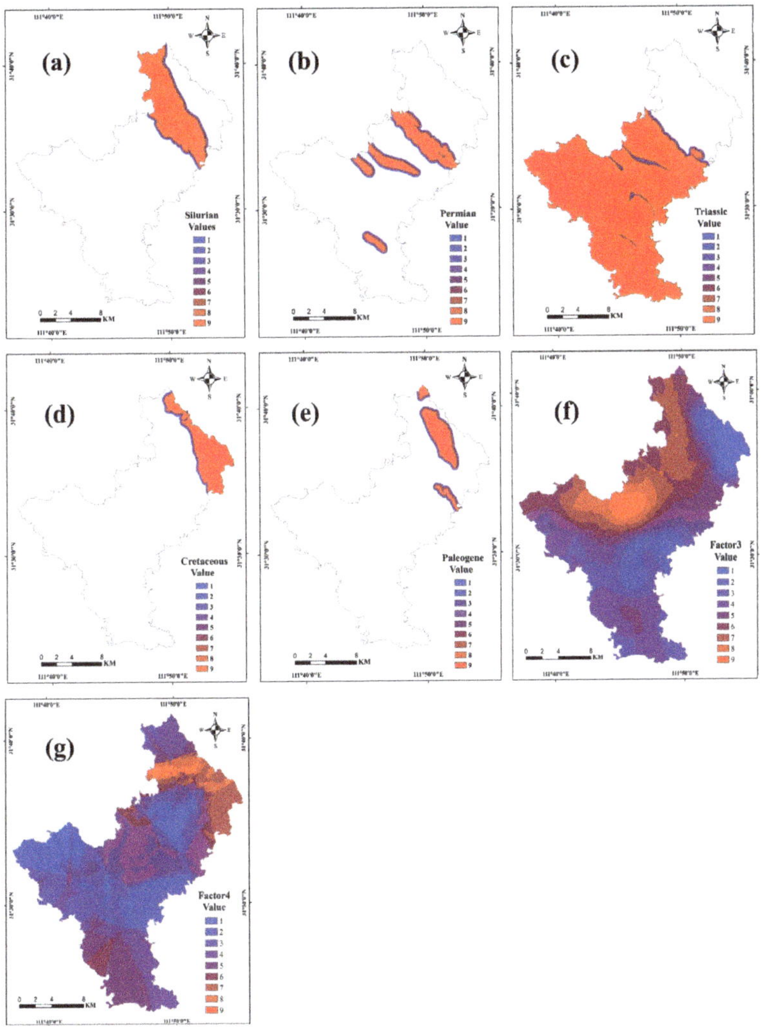

Figure 3. Evidence layers used in the FWofE methods ((**a**): 500-m nine-ring buffers of Silurian; (**b**): 500-m nine-ring buffers of Permian; (**c**): 500-m nine-ring buffers of Triassic; (**d**): 500-m nine-ring buffers of Cretaceous; (**e**): 500-m nine-ring buffers of Paleogene; (**f**): distribution map of Factor3 scores; (**g**): distribution map of Factor4 scores).

Before building the FWofE model, it was necessary to identify Se-rich crop samples. This required setting a threshold to distinguish Se-rich crops, making threshold determination a key step in the modeling process. Due to the availability of standardized Se thresholds only for Se-rich rice according to national regulations and due to the absence of national standards for other crops (which follow regional standards, such as DBS42-002-2022, issued by Hubei Health Commission, China), this study faced challenges in defining consistent criteria for Se-rich crops [50]. Moreover, the regional thresholds for Se-rich crops were set too high, resulting in a scarcity of Se-rich crop samples in the study area. Considering these factors, the Se threshold for Se-rich rice defined by the "Selenium-Enriched Rice" (GB/T 22499-2008) standard was adopted as the basis for training data in the FWofE model. Crops with Se content ranging from 0.040 to 0.300 mg/kg were considered Se-rich for the purpose of this study [51].

Using the FWofE model, eigenvalues and Student's t-distribution values for each evidence layer were calculated, and a schematic diagram illustrating the selection of fuzzy weights was generated. The multivalued fuzzy membership function (MSF) was assigned for each evidence layer, after which the fuzzy weights of the evidence layers were calculated accordingly. The fuzzy weights of the evidence layers are shown in Table A3. By integrating multiple evidence layers, the posterior probabilities of Se-rich crops were calculated, and a posterior probability map of Se-rich crops in the study area was produced. The workflow of the constructed FWofE model is shown in Figure 4a.

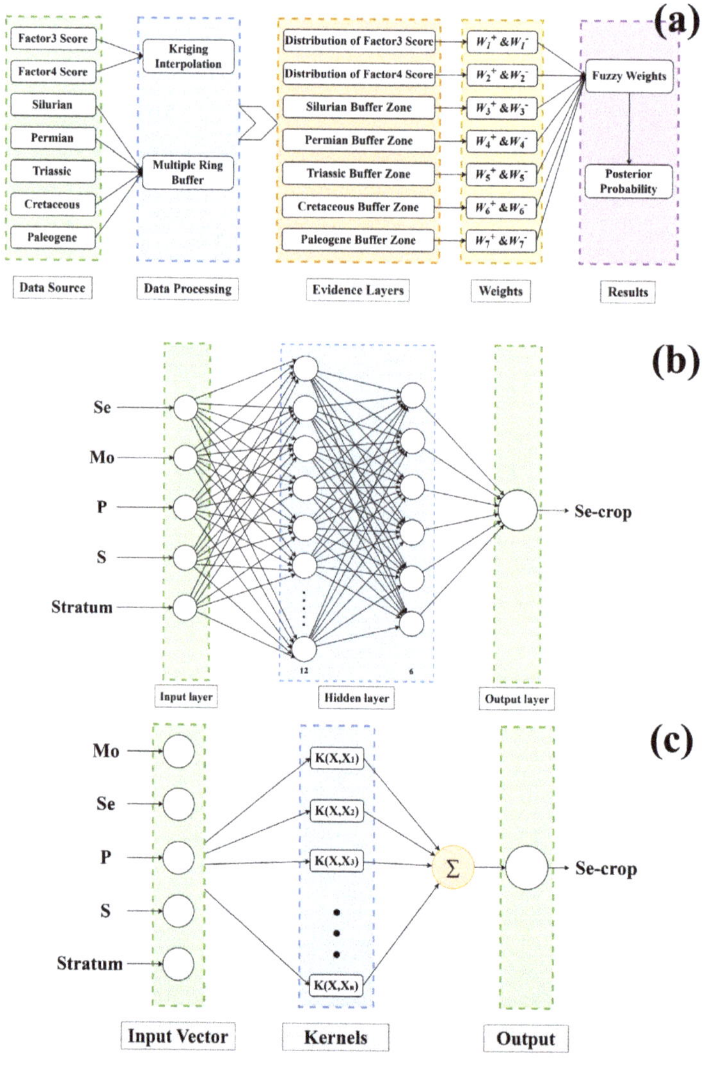

Figure 4. Schematic diagram of the FWofE (a), BP-NN (b), and SVR (c) model.

Using the geometric interval method [13], the posterior probability map of Se-rich crops was divided into three zones: Se-low crop planting zones, second-class Se-rich crop planting zones, and first-class Se-rich crop planting zones (Figure 5a). This division delineated the optimized planting zones for Se-rich crops in the study area. According to Figure 5a, the areas covered by first-class Se-rich crop planting zones, second-class Se-rich crop planting zones, and Se-low crop planting zones were 27.97 km^2, 188.37 km^2, and 159.49 km^2, accounting for 7.44%, 50.12%, and 42.44% of the study area, respectively.

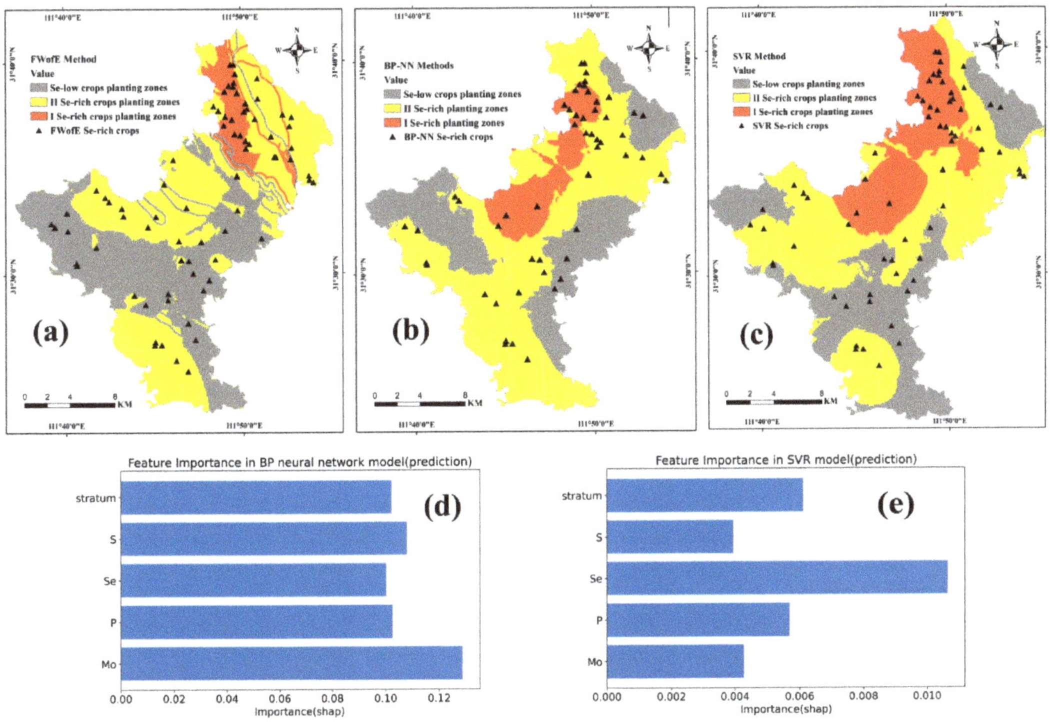

Figure 5. Graded maps of the prediction results and bar chart of the importance of indicators to the prediction results ((**a**): graded map of the FWofE model prediction results; (**b**): graded map of the BP-NN model prediction results; (**c**): graded map of the SVR model prediction results; (**d**): barchart of the importance of input features to the prediction results in the BP-NN model; (**e**): bar chart of the importance of input features to the prediction results in the SVR model).

The prediction results of the FWofE model showed significant spatial correlations with the distributions of geological strata and soil types. First-class Se-rich crop planting zones were primarily concentrated in the northeastern part of the study area, coinciding with the distribution of Silurian sandstone and yellow-brown earth. Conversely, Se crop content was relatively lower in calcareous soil areas. This conclusion aligns with the findings of Liu et al. (2024) [52].

4.2.2. Machine Learning

Similar to the FWofE model, one limitation of ML lies not only in the opacity of its intermediate processes but also in the significant impact of feature selection on prediction accuracy. Before constructing models to predict Se crop content using ML algorithms, preprocessing training data and carefully selecting root soil indicators are essential steps. The crop data were normalized using the max-min normalization method as training data.

Root soil indicators used in BP-NN and SVR models were selected based on correlation analysis results and model training performance.

From the correlation analysis and geological background, the BP-NN and the SVR model used Mo, Se, P, S, and crop geological strata (the geological strata at the location where the crop samples were collected) as input features. The distributions of the root soil content for Se, Mo, P, and S are shown in Figure A1.

In the BP-NN modeling, Mo, Se, P, S, and crop geological strata were used as input layer features, with Se crop content as the output variable. The hyperparameters of the BP-NN model were determined through an empirical approach based on the error curves of the training and testing sets, with the aim of avoiding overfitting. The model included two hidden layers, with 12 neurons in the first layer and 6 neurons in the second, and the complete hyperparameters are provided in Table A4. A training-validation error curve method was applied to monitor model performance and avoid overfitting. The schematic diagram of the BP-NN is shown in Figure 4b. The BP-NN model was used to predict Se crop content, and interpolation was applied to generate a distribution map of Se content in the study area. The prediction showed Se crop content ranged from 0.034 to 0.116 mg/kg. The geometric interval method was used to classify Se crop contents into three levels [13]: Se-low crop planting zones, second-class Se-rich crop planting zones, and first-class Se-rich crop planting zones (Figure 5b). Based on the geometric interval classification of the predicted Se content, the boundary between Se-low crop planting zones and second-class Se-rich crop zones was used as the threshold (0.045 mg/kg) to identify Se-rich crops. First-class Se-rich crop planting zones, second-class Se-rich crop planting zones, and Se-low crop planting zones covered 44.72 km^2, 214.18 km^2, and 116.97 km^2, accounting for 11.90%, 56.98%, and 31.12% of the study area, respectively.

First-class Se-rich crop planting zones were concentrated in the north and northeast of the study area, displaying a significant positive spatial correlation with root soil Se and Mo content and a significant negative spatial correlation with P and S content. These results were consistent with the results of Pearson correlation analysis. To enhance the interpretability of the model, SHAP values were used to assess the importance of each input variable in the BP-NN model [53,54]. In addition, Partial Dependence Plots (PDPs) were introduced to further illustrate the effect of individual variables on the prediction outcomes, thus complementing SHAP by providing insights into the effect direction [55]. The SHAP value bar chart (Figure 5d) revealed that Mo contributed most to the prediction (SHAP = 0.1286), while other features had slightly lower contributions. The PDPs revealed (Figure 6) that Se shows a clear positive effect on the predictions, while Mo, P, S, and stratum exhibit overall negative trends. The PDP trend of Mo appears opposite to its actual effect in predictions, possibly due to its uneven spatial distribution, which may limit the PDP's ability to reflect its true impact.

Similarly, the SVR model used the same set of input vectors to predict Se crop content. The hyperparameters of the SVR model were determined by using the grid search approach. A radial basis function (RBF) kernel was selected as the kernel function, and the complete hyperparameters are provided in Table A4. A learning curve method was used to monitor model performance and avoid overfitting. The schematic diagram of the SVR model is shown in Figure 4c. Using the SVR model, Se crop content was predicted, and interpolation was applied to generate a Se crop content distribution map for the study area. The prediction indicated Se content ranged from 0.036 to 0.112 mg/kg. The geometric interval method was used to classify Se content into three levels [13]; the boundary between Se-low crop planting zones and second-class Se-rich crop zones was used as the threshold (0.044 mg/kg) to identify Se-rich crops. First-class Se-rich crop planting zones, second-class Se-rich crop planting zones, and Se-low crop planting zones covered 77.60 km^2, 176.80 km^2, and 121.46 km^2, accounting for 20.65%, 47.04%, and 32.32% of the study area, respectively (Figure 5c).

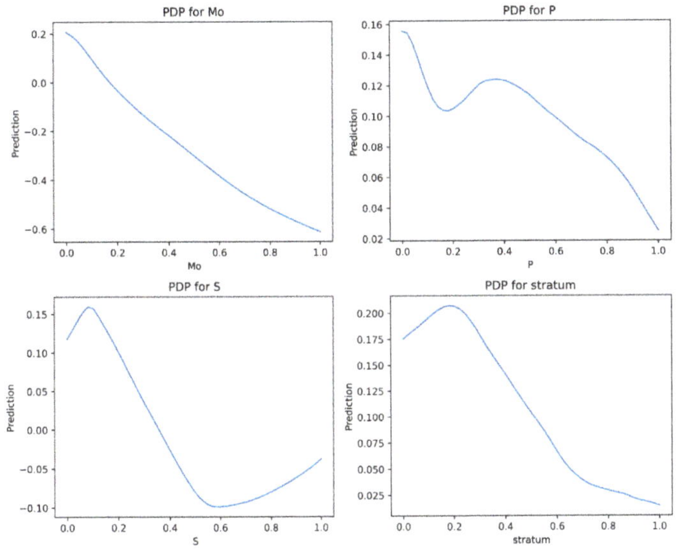

Figure 6. The PDP presentation of input features the BP-NN model.

The first-class Se-rich crop planting zones predicted by SVR were generally consistent with those predicted by the BP-NN, with the main difference being the larger area predicted by SVR. SHAP values were also calculated to assess feature importance in the SVR model. The SHAP value bar chart (Figure 5e) showed that root soil Se content had the highest contribution to prediction (SHAP = 0.0086). The PDPs revealed (Figure 7) that Se has a clear positive impact on the predictions, while Mo, P, S, and stratum exhibit overall negative trends. Compared to the BP-NN model, the SVR model has a potentially limited ability to capture complex nonlinear interactions.

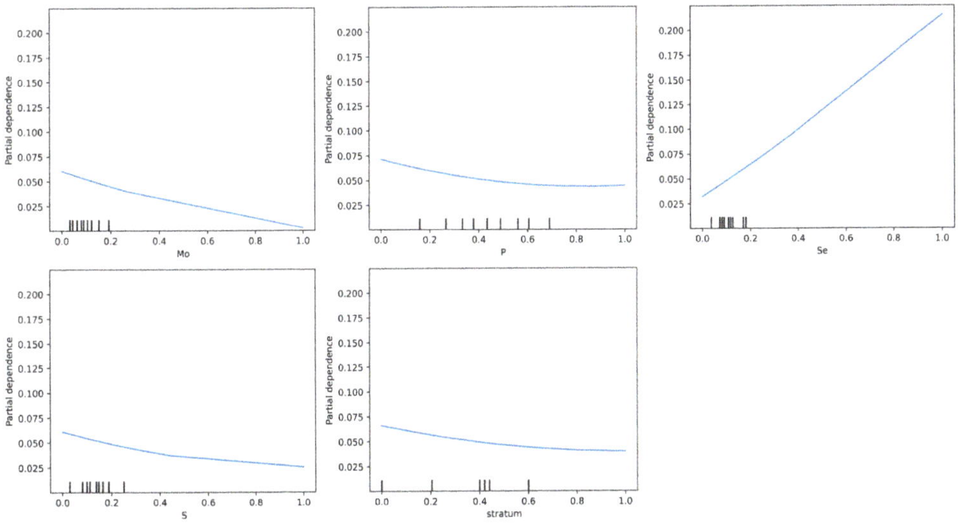

Figure 7. The PDP presentation of input features of the SVR model.

The performance of the two ML models was evaluated using the coefficient of determination (R^2), mean absolute error (MAE) and root mean square error (RMSE). The BP-NN achieved an R^2 of 0.77, an MAE of 0.0148, and an RMSE of 0.0191, while the SVR model achieved an R^2 of 0.71, an MAE of 0.0157, and an RMSE of 0.0218. Both models demonstrated comparable parameters, low prediction errors, and high fitting accuracy, indicating acceptable reliability of the predictions.

Additionally, we compared the performance of our models with those reported in previous studies. For example, Ma et al. (2022) applied both an artificial neural network (ANN) and a multiple linear regression (MLR) model for similar prediction tasks, reporting R^2 values of 0.77 for ANN and 0.67 for MLR. In comparison, our BP-NN and SVR models achieved R^2 values of 0.77 and 0.71, respectively, indicating not only comparable performance with existing ANN models but also notable improvements over traditional regression methods such as MLR [20].

4.2.3. Model Comparison and Discussion

In the process of predicting and delineating optimal planting zones for Se-rich crops in the study area, three models were employed: FWofE, BP-NN, and SVR. The Se-rich crop thresholds used by these models were 0.040 mg/kg, 0.045 mg/kg, and 0.044 mg/kg, respectively. A statistical comparison of the study areas classified as Se-low crop planting zones, second-class Se-rich crop planting zones, and first-class Se-rich crop planting zones, along with the number of Se-rich crop samples within each zone, is presented in Table 2.

Table 2. Statistical summary of prediction results of the FWofE, BP-NN, and SVR models.

		FWofE	**BP-NN**	**SVR**
Se-low crop planting zones	Number of Se-rich crops	25	8	20
	Proportion of Se-rich crops (%)	25.25	13.33	25.32
	Area (km^2)	159.49	116.97	121.46
	Area Proportion (%)	42.44	31.12	32.32
II Se-rich crop planting zones	Number of Se-rich crops	50	40	36
	Proportion of Se-rich crops (%)	50.51	66.67	45.57
	Area (km^2)	188.37	214.18	176.80
	Area Proportion (%)	50.12	56.98	47.04
I Se-rich crop planting zones	Number of Se-rich crops	24	12	23
	Proportion of Se-rich crops (%)	24.24	20.00	29.11
	Area (km^2)	27.97	44.72	77.60
	Area Proportion (%)	7.44	11.90	20.65

As shown in Table 2, the percentage of Se-rich crop samples within the Se-rich crop planting zones delineated by the FWofE, BP-NN, and SVR models were 74.75%, 86.67%, and 74.68%, respectively. The BP-NN exhibited the highest spatial concordance between the predicted Se-rich crop planting zones and the distribution of Se-rich crop samples. The FWofE model predicted smaller Se-rich crop planting zones, whereas the SVR model predicted larger zones. Based on the statistical and comparative analysis, the BP-NN was determined to be the most effective model for predicting and delineating optimal planting zones for Se-rich crops in the study area.

In Se-deficient regions, applying Se fertilizers has proven effective in increasing Se crop content [5,56]. For areas where Se crop content is low, applying Se fertilizers could be a viable solution. The graded results indicate that certain areas in the study region have low Se crop content. By applying Se fertilizers in Se-low crop planting zones and optimizing the use of Se-rich crop planting zones, the yield and quality of local Se-rich crops can be enhanced, thereby promoting the development of Se-rich agricultural products in the region.

4.3. Prediction of Optimal Planting Zones for Se-Rich Wheat, Rice, Rapeseed, and Tea

The above prediction of Se-rich crop planting zones across multiple crops does not provide specific planting recommendations for individual Se-rich crops. Therefore, this section focuses on predicting optimal planting zones for four widely cultivated crops in the study area: wheat, rice, rapeseed, and tea. These crops are grown in drylands, paddy fields, tea gardens, and forested areas, with varying land-use practices such as crop rotation and land-use function changes. As such, the areas covered by drylands, paddy fields, tea gardens, and forested regions were selected as the prediction zones for Se-rich crops.

In the previous prediction of Se-rich crop planting zones, the BP-NN model demonstrated the best performance, with Se-rich crop samples accounting for 86.67% of the predicted Se-rich planting zones. This percentage was higher than that of the FWofE and SVR models, highlighting the BP-NN's ability to effectively handle small-sample datasets and nonlinear relationships. Consequently, the BP-NN model was adopted to predict Se-rich planting zones for wheat, rice, rapeseed, and tea separately. Correlation analysis was conducted for Se content and corresponding root soil indicators for each crop, and based on the results (Figure 8), root soil Se, As, Pb, Mg, Hg, and geological strata were selected as input features for wheat; Hg, Mn, Si, pH, and geological strata for rice; Cd, Mo, Se, Ni, and geological strata for rapeseed; and P, S, Si, Ca, Cu, Fe, and geological strata for tea. The hyperparameters of the BP-NN model were also determined through an empirical approach based on the error curves of the training and testing sets, with the aim of avoiding overfitting, and the complete hyperparameters are provided in Table A4.

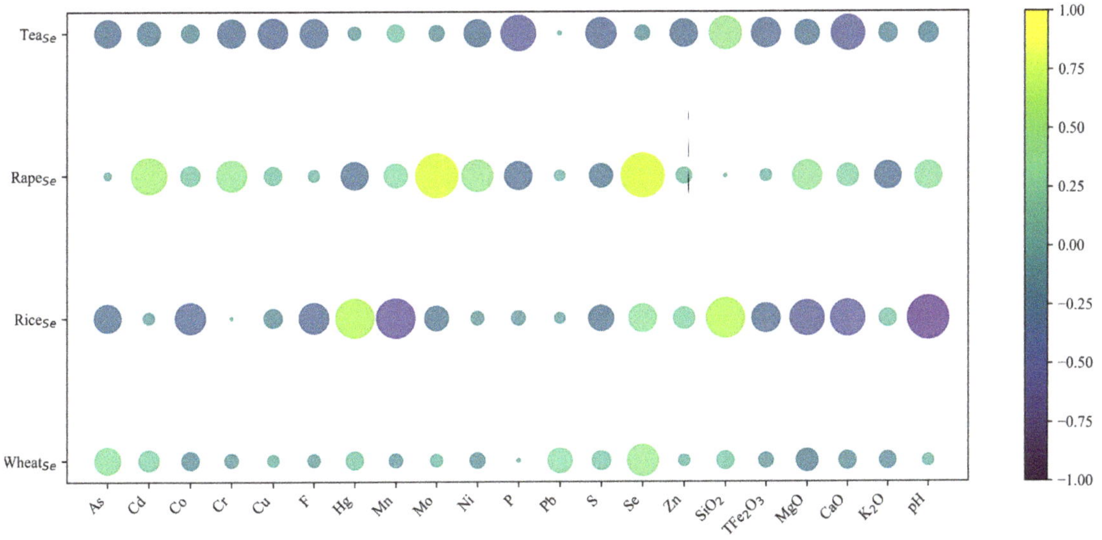

Figure 8. Heat map of correlation between Se of the four kinds of crops (wheat, rice, rapeseed, and tea) and root soil elements.

Using the previously constructed BP-NN model, Se content was predicted for wheat, rice, rapeseed, and tea, and distribution maps for each crop's Se content were generated. The prediction results showed Se content ranges of 0.031–0.096 mg/kg for wheat, 0.025–0.071 mg/kg for rice, 0.016–0.286 mg/kg for rapeseed, and 0.033–0.127 mg/kg for tea. The prediction results were graded into Se-low planting zones, second-class Se-rich planting zones and first-class Se-rich planting zones using the Jenks Natural Breaks (Figure 9) [13]. The threshold for identifying Se-rich crops was defined as the boundary between Se-low and second-class Se-rich planting zones, as determined by this classification. Accordingly, Se-rich thresholds for wheat, rice, rapeseed, and tea were set at 0.047 mg/kg, 0.040 mg/kg, 0.068 mg/kg, and 0.071 mg/kg, respectively. The areas of Se-rich planting zones were 241.48 km^2 (68.30%) for wheat, 213.35 km^2 (60.35%) for rice, 245.59 km^2 (69.46%) for rapeseed, and 179.47 km^2 (50.76%) for tea.

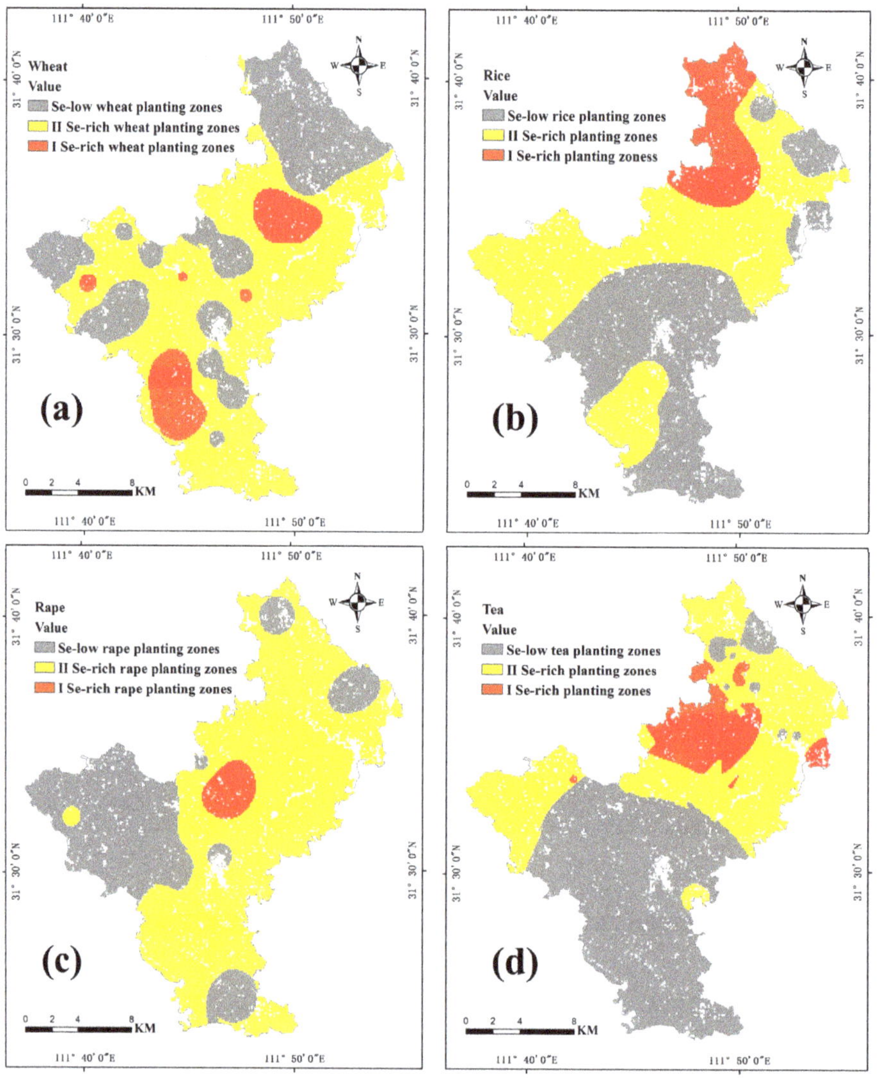

Figure 9. Graded map of the prediction results of wheat (**a**), rice (**b**), rapeseed (**c**), and tea (**d**).

The BP-NN model's evaluation metrics were calculated for each crop to assess the interpretability of the predicted Se-rich planting zones. For wheat, the MAE was 0.0096, the RMSE was 0.0129, and R^2 was 0.64; for rice, the MAE was 0.0035, the RMSE was 0.0037, and R^2 was 0.95; for rapeseed, the MAE was 0.0055, the RMSE was 0.0067, and R^2 was 0.99; and for tea, the MAE was 0.0110, the RMSE was 0.0147, and R^2 was 0.71.

These metrics indicate that the prediction errors for rice and rapeseed were minimal, while the interpretability of the results (based on R^2) was very high for these two crops. The prediction performance for tea was acceptable, whereas the prediction for wheat was less accurate. The superior prediction performance for rice and rapeseed may be attributed to the correlation analysis results: the selected indicators for these two crops had higher correlation coefficients ($|r| > 0.5$) and were more distinct from other indicators, whereas the correlation coefficients for wheat were relatively low ($|r| < 0.4$).

The Se-rich planting zones delineated for rice, rapeseed, and tea using the BP-NN model provide planting recommendations for these crops in Xiaoyan Town, Nanzhang County. Se-rich rapeseed optimized planting zones were associated with soils containing high levels of Mo, Se, Mn, and Cu. Se-rich rice and tea-optimized planting zones were primarily distributed in areas underlain by Silurian strata, characterized by acidic soils. According to Wang et al. (2024), Se-rich tea thrives in acidic soils because tea plants have a higher affinity for selenite. Se-rich rice, on the other hand, is suited to slightly acidic soils, likely due to the concentration of paddy fields in these regions [5].

Additionally, the mobility of Se species differs by soil type: selenate in alkaline soils is prone to leaching due to its high mobility, whereas selenite in acidic soils tends to accumulate due to its lower mobility [57]. Based on these findings, rapeseed is best planted in soils with high Mo, Se, and Mn content, while rice and tea are better suited for acidic soils. Appropriate fertilization strategies should also be implemented to reduce heavy metal content in crops and improve the yield and quality of Se-rich products.

5. Conclusions

This study analyzed Se crop content and corresponding root soil data in Xiaoyan Town, Nanzhang County, employing FWofE, BP-NN, and SVR methods to delineate regions with different Se content levels and identify optimized planting zones for Se-rich crops. These findings provide scientific evidence and planting guidelines for the rational cultivation and sustainable development of Se-rich crops in the area, promoting the development of the local Se-rich industry.

The study's main objective was to develop a reliable model for predicting Se-rich planting zones. Among the FWofE, BP-NN, and SVR models, the BP-NN model demonstrated the best predictive performance. The RMSE and R^2 of the BP-NN model verified by the test data were 0.0191 and 0.77, respectively. Additionally, the correct classification rate of the BP-NN model reached 86.67%. Specific predictions for Se-rich planting zones for wheat, rice, rapeseed, and tea were also made, with the model's R^2 values indicating that the predictions for rice (0.95) and rapeseed (0.99) were the most reliable.

The correlation analysis and spatial analyses revealed certain patterns consistent with previous research on factors influencing Se uptake by crops. Elements that were significantly positively correlated with Se showed overlapping spatial distributions with Se-rich crop planting zones, reflecting their synergistic effects on Se uptake. Conversely, elements with significant negative correlations exhibited an opposite spatial distribution, indicating their antagonistic effects.

Based on these findings, targeted cultivation recommendations are proposed. For Se-rich crop planting zones, appropriate chemical treatments or adjustments to soil physicochemical properties should be implemented to avoid the accumulation of elements that antagonize Se, ensuring the long-term sustainability of Se-rich crop cultivation. For Se-low crop planting zones, applying exogenous Se fertilizers, adjusting soil pH, and modifying redox conditions can enhance Se bioavailability in the soil and increase Se content in crops.

Importantly, the prediction results can assist farmers in selecting optimal areas for Se-rich crop cultivation, improving yield quality and income. Local planners may also use these results for agricultural zoning, soil management, and policy-making aimed at sustainable land use and Se-rich food production.

However, a limitation of this study is that the model does not incorporate Se bioavailability data, which could provide a more comprehensive understanding of Se uptake and improve predictions. Future work could integrate more data and extend the analysis to more crops and regions to validate model applicability.

Author Contributions: Conceptualization, J.L. and S.X.; methodology, J.L. and S.X.; software, J.L., W.Z., W.W. and H.S.; validation, W.Y., W.W. and H.S.; formal analysis, J.L.; investigation, W.Y.; data curation, J.L.; writing—original draft preparation, J.L. and W.Y.; writing—review and editing, S.X. and E.J.M.C.; visualization, J.L. and W.Z.; supervision, J.L. and S.X. All authors have read and agreed to the published version of the manuscript.

Funding: This research was financially supported by the New geochemical exploration technology for shallow-covered landscapes in desert Gobi and alpine steppes (No. 2024ZD1002400) under the subproject "Geochemical intelligent prospecting prediction and software development for copper-nickel and gold deposits" (No. 2024ZD1002406) and the Geochemical evaluation of land quality in Nanzhang County, Hubei Province (No. HBJTD20160101).

Institutional Review Board Statement: Not applicable.

Informed Consent Statement: Not applicable.

Data Availability Statement: The original contributions presented in this study are included in the article. Further inquiries can be directed to the corresponding author.

Acknowledgments: We are very thankful for all the editors and reviewers who have helped us improve the manuscript.

Conflicts of Interest: The authors declare no conflicts of interest.

Appendix A

Table A1. The average Se content in different crop types.

Crop	Wheat	Rice Grain	Rice Root	Rapeseed	Tea Leaf	Tobacco Leaf	Potato	Corn
Se content (mg/kg)	0.048	0.059	9.94	0.073	0.08	0.04	0.004	below detection limit

Table A2. Factor analysis parameters and diagnostic metrics.

Item	Description
KMO value	0.676
Bartlett's test	$p = 0.000$
Factor extraction criterion	Eigenvalues > 1
Rotation method	Varimax

Table A3. Fuzzy weights of the evidence layers.

Factor3	class	2	3	4	5	6	7	8	9	
	weight	−0.2	−0.2	−0.2	−0.2	−0.2	0.39	0.39	0.39	
Factor4	class	1	2	3	4	5	6	7	8	9
	weight	−0.11	−0.11	−0.11	−0.11	0.18	0.18	0.18	0.18	0.18
Silurian	class	3	4	5	6	7	8	9		
	weight	−1.21	−1.21	−1.21	0.79	0.86	0.86	0.86		
Permian	class	7	8	9						
	weight	−2.26	−2.26	−0.15						

Table A3. *Cont.*

Triassic	class	1	2	3	4	5	6	7	8	9
	weight	0.48	0.48	0.48	0.48	−0.34	−0.35	−0.35	−0.35	−0.35
Cretaceous	class	1	2	3	4	5	6	7	9	
	weight	0.54	0.54	0.54	0.54	0.34	0.34	0.34	0.34	
Paleogene	class	1	2	3	4	5	6	7	8	9
	weight	0.97	0.97	0.97	0.97	0.1	−0.26	−0.26	−0.26	−0.26

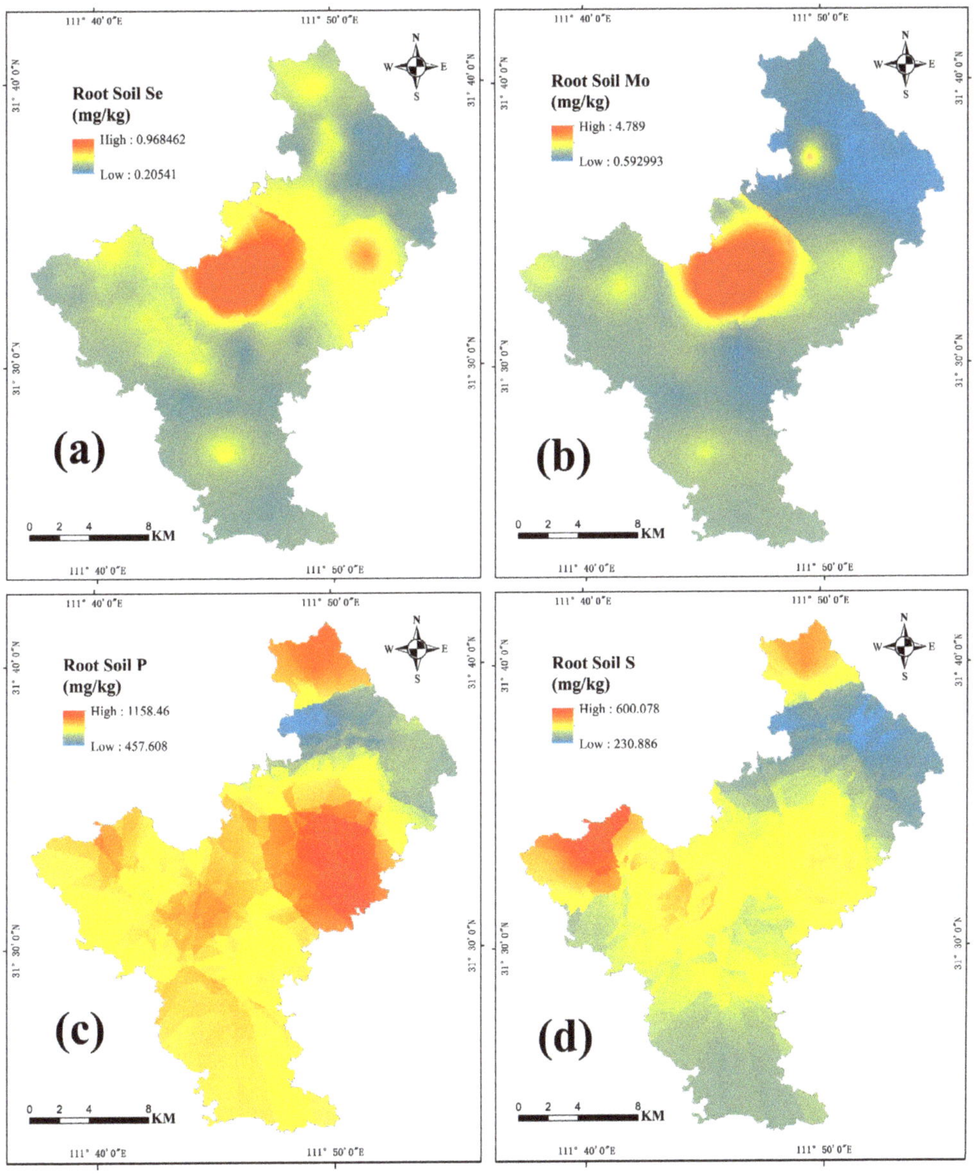

Figure A1. The distributions of root soil elements ((**a**): Se, (**b**): Mo, (**c**): P, and (**d**): S).

Table A4. The hyperparameters of the models.

	BP-NN					SVR	
Hyperparameters	**Crops**	**Wheat**	**Rice**	**Rapeseed**	**Tea**	**Hyperparameters**	**Crops**
epoch	3637	1815	1096	1115	589	C	100
learning rate	0.001	0.001	0.1	0.003	0.004	epsilon	0.000001
hidden layer size			12, 6			gamma	0.01
activation			ReLU			kernel	rbf
loss function			MSELoss				

References

1. Mao, J.; Pop, V.J.; Bath, S.C.; Vader, H.L.; Redman, C.W.; Rayman, M.P. Effect of low-dose selenium on thyroid autoimmunity and thyroid function in UK pregnant women with mild-to-moderate iodine deficiency. *Eur. J. Nutr.* **2016**, *55*, 55–61. [CrossRef] [PubMed]
2. Dinh, Q.T.; Cui, Z.; Huang, J.; Tran, T.A.T.; Wang, D.; Yang, W.; Zhou, F.; Wang, M.; Yu, D.; Liang, D. Selenium distribution in the Chinese environment and its relationship with human health: A review. *Environ. Int.* **2018**, *112*, 294–309. [CrossRef] [PubMed]
3. Pecoraro, B.M.; Leal, D.F.; Frias-De-Diego, A.; Browning, M.; Odle, J.; Crisci, E. The health benefits of selenium in food animals: A review. *J. Anim. Sci. Biotechnol.* **2022**, *13*, 58. [CrossRef] [PubMed]
4. Doaei, S.; Mardi, A.; Zare, M. Role of micronutrients in the modulation of immune system and platelet activating factor in patients with COVID-19; a narrative review. *Front. Nutr.* **2023**, *10*, 1207237. [CrossRef]
5. Wang, K.; Fang, Q.; He, P.; Tu, Y.; Liu, Z.; Li, B. Unveiling the potential of selenium-enriched tea: Compositional profiles, physiological activities, and health benefits. *Trends Food Sci. Technol.* **2024**, *145*, 104356. [CrossRef]
6. Genchi, G.; Lauria, G.; Catalano, A.; Sinicropi, M.S.; Carocci, A. Biological activity of selenium and its impact on human health. *Int. J. Mol. Sci.* **2023**, *24*, 2633. [CrossRef]
7. Werkneh, A.A.; Gebretsadik, G.G.; Gebru, S.B. Review on environmental selenium: Occurrence, public health implications and biological treatment strategies. *Environ. Chall.* **2023**, *11*, 100698. [CrossRef]
8. Natasha; Shahid, M.; Niazi, N.K.; Khalid, S.; Murtaza, B.; Bibi, I.; Rashid, M.I. A critical review of selenium biogeochemical behavior in soil-plant system with an inference to human health. *Environ. Pollut.* **2018**, *234*, 915–934. [CrossRef]
9. Ren, X.; Wang, Y.; Sun, J.; Liang, K.; Zhu, H.; Li, Y.; Gao, J.; Zhang, Y.; Huang, S.; Zhu, D. Legal Standards for Selenium Enriched Foods and Agricultural Products: Domestic and International Perspectives. *Nutrients* **2024**, *16*, 3659. [CrossRef]
10. Navarro-Alarcon, M.; Cabrera-Vique, C. Selenium in food and the human body: A review. *Sci. Total Environ.* **2008**, *400*, 115–141. [CrossRef]
11. Huang, Y.; Wang, Q.; Gao, J.; Lin, Z.; Banuelos, G.S.; Yuan, L.; Yin, X. Daily dietary selenium intake in a high selenium area of Enshi, China. *Nutrients* **2013**, *5*, 700–710. [CrossRef]
12. Zhang, L.; Song, H.; Guo, Y.; Fan, B.; Huang, Y.; Mao, X.; Liang, K.; Hu, Z.; Sun, X.; Fang, Y.; et al. Benefit-risk assessment of dietary selenium and its associated metals intake in China (2017–2019): Is current selenium-rich agro-food safe enough? *J. Hazard. Mater.* **2020**, *398*, 123224. [CrossRef] [PubMed]
13. Xie, S.; Wan, X.; Dong, J.; Wan, N.; Jiang, X.; Carranza, E.J.M.; Wang, X.; Chang, L.; Tian, Y. Quantitative prediction of potential areas likely to yield Se-rich and Cd-low rice using fuzzy weights-of-evidence method. *Sci. Total Environ.* **2023**, *889*, 164015. [CrossRef] [PubMed]
14. Bai, J. Research on Problems and Countermeasures of Organic Agriculture Development in Nanzhang County. Master's Thesis, Central China Normal University, Wuhan, China, 2020. (In Chinese with English Abstract)
15. Yan, J.; Tang, J. Study on the development pattern of geologic hazards and influencing factors in Nanzhang County. *China Constr.* **2023**, *8*, 118–120. (In Chinese)
16. Liu, X.; Ma, Q.; Song, Z.; Ye, Z.; Zhai, X.; Zhang, M.; Zhang, L.; Wang, Q. Research on the spatial pattern distribution of soil selenium using machine learning methods integrating geographic proximity in complex terrain. *J. Soils Sediments* **2024**, *24*, 2776–2790. [CrossRef]
17. Zhang, L.; Gao, R.; Zhang, H.; Luo, X.; Song, Z. Study on the Influence of Consumers' Purchase Intention of Selenium-Rich Agricultural Products. *Appl. Sci.* **2023**, *13*, 1859. [CrossRef]
18. Dinh, Q.T.; Wang, M.; Tran, T.A.T.; Zhou, F.; Wang, D.; Zhai, H.; Peng, Q.; Xue, M.; Du, Z.; Bañuelos, G.S. Bioavailability of selenium in soil-plant system and a regulatory approach. *Crit. Rev. Environ. Sci. Technol.* **2019**, *49*, 443–517. [CrossRef]
19. Mora, M.; Durán, P.; Acuña, J.; Cartes, P.; Demanet, R.; Gianfreda, L. Improving selenium status in plant nutrition and quality. *J. Soil Sci. Plant Nutr.* **2015**, *15*, 486–503. [CrossRef]

20. Ma, X.; Yang, Z.; Yu, T.; Guan, D.X. Probability of cultivating Se-rich maize in Se-poor farmland based on intensive field sampling and artificial neural network modelling. *Chemosphere* **2022**, *309*, 136690. [CrossRef]
21. Zhang, L.; Yu, T.; Zheng, G.; Tang, Q.; Peng, M.; Li, C.; Hou, Q.; Yang, Z. Using machine learning to predict selenium content in crops: Implications for soil health and agricultural land utilization in longevity regions. *Sci. Total Environ.* **2025**, *964*, 178520. [CrossRef]
22. Falegnami, A.; Tomassi, A.; Gunella, C.; Amalfitano, S.; Corbelli, G.; Armonaite, K.; Fornaro, C.; Giorgi, L.; Pollini, A.; Caforio, A.; et al. Defining conceptual artefacts to manage and design simplicities in complex adaptive systems. *Heliyon* **2024**, *10*. [CrossRef] [PubMed]
23. Guo, R.; Ren, R.; Wang, L.; Zhi, Q.; Yu, T.; Hou, Q.; Yang, Z. Using machine learning to predict selenium and cadmium contents in rice grains from black shale-distributed farmland area. *Sci. Total Environ.* **2024**, *912*, 168802. [CrossRef]
24. Hagenauer, J.; Helbich, M. A geographically weighted artificial neural network. *Int. J. Geogr. Inf. Sci.* **2022**, *36*, 215–235. [CrossRef]
25. Huang, B.Y.; Lu, Q.X.; Tang, Z.X.; Tang, Z.; Chen, H.P.; Yang, X.P.; Zhao, F.J.; Wang, P. Machine learning methods to predict cadmium (Cd) concentration in rice grain and support soil management at a regional scale. *Fundam. Res.* **2024**, *4*, 1196–1205. [CrossRef] [PubMed]
26. Zheng, J.; Yang, J.; Liu, X.; Chen, X.; Zhou, W.; Cao, G.; Wang, M. Study on Ecological Restoration Countermeasures for Typical Karst Rocky Desertification Areas in the Middle Reach of Yangtze River—Study on the Development Laws and Influencing Factors of Geological Hazards in Nanzhang County, Xiangyang City, Hubei Province, as an Example. *J. Huanggang Norm. Univ.* **2021**, *41*, 27–30. (In Chinese)
27. *DZ/T 0145-2017*; Code of Practice for Soil Geochemical Survey. Ministry of Land and Resources of the People's Republic of China: Beijing, China, 2017; p. 42.
28. *DZ/T 0295-2016*; Specification of Land Quality Geochemical Assessment. Ministry of Land and Resources of the People's Republic of China: Beijing, China, 2016; p. 57.
29. Zhao, W.; Cui, S.; Zou, X.; Tang, C.; Xia, J.; Jin, S. Geochemical characteristics of stream sediments and metallogenic prognosis of Heku area, Hunan Province. *Geol. China* **2014**, *41*, 638–647. (In Chinese with English Abstract)
30. Moharana, P.C.; Jena, R.K.; Kumar, N.; Singh, R.S.; Rao, S.S. Assessment of soil organic and inorganic carbon stock at different soil depths after conversion of desert into arable land in the hot arid regions of India. *Carbon Manag.* **2021**, *12*, 153–165. [CrossRef]
31. Bai, Y.; Chen, K.X.; Chen, C.; Li, F.L.; Zhang, J.C.; Wei, L.X.; Si, K.F.; Zheng, X.W.; Hu, Y.F.; Wu, Y.; et al. Stream sediment geochemistry and prospecting targets in Harper, Liberia. *Geophys. Geochem. Explor.* **2024**, *48*, 382–392. (In Chinese with English Abstract) [CrossRef]
32. Cheng, Q.; Agterberg, F. Fuzzy weights of evidence method and its application in mineral potential mapping. *Nat. Resour. Res.* **1999**, *8*, 27–35. [CrossRef]
33. Deng, J.; Zhan, M.G.; Zhou, W.J.; Wu, S.; Huang, N.; Zhang, R.; Xie, S. Quantitative prediction of mineral resources in typical gold deposits in Guangxi, China using a fuzzy weights of evidence method. *J. Geomech.* **2021**, *27*, 374–390. (In Chinese with English Abstract) [CrossRef]
34. Zhang, Z.; Zuo, R.; Xiong, Y. A comparative study of fuzzy weights of evidence and random forests for mapping mineral prospectivity for skarn-type Fe deposits in the southwestern Fujian metallogenic belt, China. *Sci. China Earth Sci.* **2016**, *59*, 556–572. [CrossRef]
35. Cheng, Q.; Chen, Z.; Ali, K. Application of Fuzzy Weights of Evidence Method in Mineral Resource Assessment for Gold in Zhenyuan District, Yunnan Province, China. *Earth Sci.-J. China Univ. Geosci.* **2007**, *32*, 175–184. (In Chinese with English Abstract) [CrossRef]
36. Xie, S.; Huang, N.; Deng, J.; Wu, S.; Zhan, M.; Carranza, E.J.M.; Zhang, Y.; Meng, F. Quantitative prediction of prospectivity for Pb–Zn deposits in Guangxi (China) by back-propagation neural network and fuzzy weights-of-evidence modelling. *Geochem. Explor. Environ. Anal.* **2022**, *22*, geochem2021-085. [CrossRef]
37. Adeniyi, O.D.; Brenning, A.; Maerker, M. Spatial prediction of soil organic carbon: Combining machine learning with residual kriging in an agricultural lowland area (Lombardy region, Italy). *Geoderma* **2024**, *448*, 116953. [CrossRef]
38. Singha, S.; Pasupuleti, S.; Singha, S.S.; Singh, R.; Kumar, S. Prediction of groundwater quality using efficient machine learning technique. *Chemosphere* **2021**, *276*, 130265. [CrossRef] [PubMed]
39. Shao, W.; Guan, Q.; Tan, Z.; Luo, H.; Li, H.; Sun, Y.; Ma, Y. Application of BP-ANN model in evaluation of soil quality in the arid area, northwest China. *Soil Tillage Res.* **2021**, *208*, 104907. [CrossRef]
40. Wang, Y.; Yang, Z.; Chen, G.; Zhan, L.; Zhang, M.; Zhou, M.; Sheng, W. Influencing factors of selenium transformation in a soil-rice system and prediction of selenium content in rice seeds: A case study in Ninghua County, Fujian Province. *Environ. Sci. Pollut. Res. Int.* **2024**, *31*, 995–1006. [CrossRef]
41. Abdollahpour, S.; Kosari-Moghaddam, A.; Bannayan, M. Prediction of wheat moisture content at harvest time through ANN and SVR modeling techniques. *Inf. Process. Agric.* **2020**, *7*, 500–510. [CrossRef]

42. Paidipati, K.K.; Chesneau, C.; Nayana, B.; Kumar, K.R.; Polisetty, K.; Kurangi, C. Prediction of rice cultivation in India—Support vector regression approach with various kernels for non-linear patterns. *Agriengineering* **2021**, *3*, 182–198. [CrossRef]
43. Wang, Y.; Yu, T.; Yang, Z.; Bo, H.; Lin, Y.; Yang, Q.; Liu, X.; Zhang, Q.; Zhuo, X.; Wu, T. Zinc concentration prediction in rice grain using back-propagation neural network based on soil properties and safe utilization of paddy soil: A large-scale field study in Guangxi, China. *Sci. Total Environ.* **2021**, *798*, 149270. [CrossRef]
44. Yang, H.; Yang, X.; Ning, Z.; Kwon, S.Y.; Li, M.L.; Tack, F.M.G.; Kwon, E.E.; Rinklebe, J.; Yin, R. The beneficial and hazardous effects of selenium on the health of the soil-plant-human system: An overview. *J. Hazard. Mater.* **2022**, *422*, 126876. [CrossRef] [PubMed]
45. Chauhan, R.; Awasthi, S.; Tripathi, P.; Mishra, S.; Dwivedi, S.; Niranjan, A.; Mallick, S.; Tripathi, P.; Pande, V.; Tripathi, R.D. Selenite modulates the level of phenolics and nutrient element to alleviate the toxicity of arsenite in rice (*Oryza sativa* L.). *Ecotoxicol. Environ. Saf.* **2017**, *138*, 47–55. [CrossRef]
46. Zhou, X.; Yang, J.; Kronzucker, H.J.; Shi, W. Selenium Biofortification and Interaction With Other Elements in Plants: A Review. *Front. Plant Sci.* **2020**, *11*, 586421. [CrossRef] [PubMed]
47. Fordyce, F.M. Selenium deficiency and toxicity in the environment. In *Essentials of Medical Geology: Revised Edition*; Springer: Berlin/Heidelberg, Germany, 2012; pp. 375–416.
48. Darabi-Golestan, F.; Hezarkhani, A. Evaluation of elemental mineralization rank using fractal and multivariate techniques and improving the performance by log-ratio transformation. *J. Geochem. Explor.* **2018**, *189*, 11–24. [CrossRef]
49. Mi, T.; Hao, L.; Zhao, X.; Lu, J.; Zhao, Y. The Study of Stream Sediment Geochemical Data Processing by Using k-Means Algorithm and Centered Logratio Transformation—An Example of a District in Hunan, China. *Geochem. Int.* **2018**, *56*, 1233–1244. [CrossRef]
50. *DBS42/002-2022*; Selenium Content Requirement for Organic Selenium Enriched Foods. Health Commission of Hubei Province: Wuhan, China, 2022.
51. *GB/T 22499-2008*; Rich Selenium Paddy. State General Administration of the People's Republic of China for Quality Supervision and Inspection and Quarantine, Standardization Administration of the People's Republic of China: Beijing, China, 2008; p. 8.
52. Liu, R.; Zhou, W.; Yu, D.; Shang, G.; Zhou, Y.; Xu, Z.; Tang, D.; Cui, H.; Liao, C. Spatial distribution and enrichment characteristics of selenium in paddy soil and rice around the Dongting Lake. *Environ. Pollut.* **2024**, *359*, 124552. [CrossRef]
53. Lundberg, S.M.; Lee, S.I. A unified approach to interpreting model predictions. *Adv. Neural Inf. Process. Syst.* **2017**, 30. [CrossRef]
54. Antonini, A.S.; Tanzola, J.; Asiain, L.; Ferracutti, G.R.; Castro, S.M.; Bjerg, E.A.; Ganuza, M.L. Machine Learning model interpretability using SHAP values: Application to Igneous Rock Classification task. *Appl. Comput. Geosci.* **2024**, *23*, 100178. [CrossRef]
55. Kazemi, F.; Asgarkhani, N.; Jankowski, R. Optimization-based stacked machine-learning method for seismic probability and risk assessment of reinforced concrete shear walls. *Expert Syst. Appl.* **2024**, *255*, 124897. [CrossRef]
56. Ros, G.; Van Rotterdam, A.; Bussink, D.; Bindraban, P.S. Selenium fertilization strategies for bio-fortification of food: An agro-ecosystem approach. *Plant Soil* **2016**, *404*, 99–112. [CrossRef]
57. De Feudis, M.; D'Amato, R.; Businelli, D.; Guiducci, M. Fate of selenium in soil: A case study in a maize (*Zea mays* L.) field under two irrigation regimes and fertilized with sodium selenite. *Sci. Total Environ.* **2019**, *659*, 131–139. [CrossRef] [PubMed]

applied sciences

MDPI

Article

Enhancing Prospecting Target Prediction Precision: A Multi-Source Data Mining Approach in Gansu's Beishan Area

Lihui Zhu [1,2,3], Runsheng Han [1,*], Yan Zhang [1,*], Hao Fu [4], Jianmin Luo [2,3] and Yunzhi Luo [2,3]

1 Faculty of Land Resources Engineering, Kunmig University of Science and Technology, Kunming 650093, China
2 Gansu Geological Big Data Engineering Research Center, Lanzhou 730000, China
3 Geological Survey of Gansu Province, Lanzhou 730000, China
4 149th Team, Gansu Coal Geology Bureau, Lanzhou 730020, China
* Correspondence: 11301124@kust.edu.cn (R.H.); 20190021@kust.edu.cn (Y.Z.)

Abstract: The success of geological prospecting depends on the accuracy of target area prediction. Traditional qualitative research methods rooted in theoretical frameworks have shown significant limitations, especially in their inability to fully exploit the latent value of existing geological information. Applying big data concepts and methodologies to geological information mining has emerged as an effective way to improve the accuracy of prospecting target prediction. This study is founded on the core principle of geoscience big data: to "uncover correlations within data to address geological issues". Taking geochemical prospecting and aeromagnetic data from the Beishan area in Gansu Province as a case in point, this study emphasizes the significance of meticulous data processing in averting potential errors. A suite of prospecting models was developed through multi-source data mining to identify potential gold deposits. Notably, aeromagnetic data were innovatively employed for the first time to predict the occurrence of non-magnetic minerals, which are primarily structurally altered rock-type and quartz vein-type gold deposits. The developed prospecting model was used to predict metallogenesis in the Beishan area of Gansu Province. The prospecting target area was delineated, accounting for 3.67% of the study area. Verification using field sampling data revealed that the actual mineralization rate in the level-I target area reached 52.6%. The research results suggest that this approach can substantially enhance the accuracy of prospecting target area prediction.

Keywords: big data; multi-source data information mining; aeromagnetic; geochemical information restoration; quantitative prediction model for prospecting target areas; gold mine

1. Introduction

In recent years, big data have been successfully utilized in finance, the internet, medicine, logistics, and media [1–3], but their practical application is relatively weak [4–6]. The essence of big data is a new way of thinking: deeply mining data to study correlations between data, extract valuable information, solve scientific and technological problems, and approach the event itself with the highest probability [7]. Big data are triggering a profound revolution in the field of geoscience [8,9]. Academicians Zhao Pengda and Guo Huadong have repeatedly emphasized the important viewpoint that "The concept of 'big data' represents a novel scientific research idea and method grounded in theoretical science". The difference between the prediction of mineral prospecting target areas based on ore genesis theory and the in-depth mining of geological information (data) lies in the former studying the causal relationship of the ore-formation process, whereas the latter focuses on the correlation between mineral prospecting information (data) [1,10–12].

The accuracy of predicting the target area for prospecting determines the effectiveness of the prospecting efforts [13–16]. As geological prospecting advances, the quantity of mineral deposits uncovered using traditional predictive prospecting concepts (theory-driven) and methodologies (examining cause-and-effect relationships) is diminishing [17–21]. The application of big data thinking (data-driven) and methods (correlation analysis) has resulted in the construction of quantitative prediction models. There is now a consensus within the industry that existing geological and mineral deposit information (data) can be harnessed to improve the accuracy of prospecting target area predictions [22–32]. The element Au occurs naturally in the form of particles, with a low and uneven distribution. This leads to some gold deposits (points) not displaying corresponding abnormal occurrences of the element Au. This increases the difficulty of geological prospecting and metallogenic prediction research through elemental anomalies. The geochemical sampling and testing process inherently constrains the capacity to accurately reflect information on Au abundance due to representativeness and other factors. This diminishes the applicability of the data. Consequently, some Au deposits (points) lack corresponding geochemical anomalies for implausible phenomena [33–35]. In response to the aforementioned issues, this paper uses the Beishan area of Gansu Province as a case study. Guided by the concept of "data-driven" geological big data, the approach emphasizes the reduction of noise in raw data and the restoration of geochemical information through the in-depth mining of geochemical and magnetic data in the area. The concept of using geophysical and geochemical data to develop predictive models has been employed to quantitatively identify prospective areas for gold deposits. This approach has shown notable success in enhancing the precision of target area prediction, offering promising insights into the integration of geophysical and geochemical data in gold prospecting.

2. Geology and Data

2.1. Geological Setting

The Beishan area in Gansu Province is a significant metallogenic zone in Northwest China. It is situated at the convergence of the three ancient continental systems of Kazakhstan, Siberia, and Tarim. It exhibits a complex geological structure, frequent magmatic activity, and strong metamorphism (Figure 1). The area is of particular interest regarding the formation of gold, copper, iron, lead–zinc, tungsten, vanadium, uranium, nickel, and other ores in China [36–40]. The stratigraphy in the Beishan area is fully developed, and strata from the Archean to the Cenozoic are all exposed in the area. The exposed strata mainly include the Neoarchean–Paleoproterozoic Dunhuang Group, the Changchengian Qianluzigou Group, and the Gudongjing Group [41]. There are also the Jixianian Pingtoushan Formation; the Qingbaikouan Dahuoluoshan Formation; the Nanhua–Sinian Xichangjing Group; the Cambrian Shuangyingshan Formation and Xishuangyingshan Formation; the Ordovician Luoyachuanshan Formation; Huaniushan Group; Xilinkebo Formation; Baiyunshan Formation; the Silurian Heijianshan Formation and Gongpoquan Group; the Devonian Sangejing Formation and Dundunshan Group; the Carboniferous Hongliuyuan Formation; Jijitai Formation; Baishan Formation and Saozishan Formation; the Permian Shuangbaotang Formation; Fangshankou Formation; Jinta Formation and Hongyanjing Formation; the Triassic Erduanjing Formation; the Neogene Kuquan Formation; and the Quaternary System [42]. As exemplified in the Beishan metallogenic belt of Gansu Province, the gold mineralization systems are dominantly classified into quartz vein-type, structural altered rock-type, and magmatic–hydrothermal-type deposits. These gold deposits are typically controlled by ductile shear zones and fault systems, with mineralization closely associated with Paleozoic magmatic activities and regional metamorphic–hydrothermal processes [43–45]. Copper deposits in this region primarily manifest as porphyry-type, magmatic–hydrothermal-type, and sedimentary–metamorphic-type mineralization. Notably, the porphyry copper deposits are genetically linked to the Late Paleozoic (Hercynian) intermediate–acidic intrusive complexes, while sedimentary–metamorphic copper ores are hosted in Proterozoic to Carboniferous marine sedimentary sequences subjected to subsequent tectonic reworking [46,47]. This metallogenic belt is characterized by its wealth of mineral species, the variety of deposit types, and the prevalence of large-scale deposits. It possesses considerable economic value and holds significant potential for scientific research. The stratigraphic succession in the Beishan area has undergone extensive geological evolution, encompassing distinct evolutionary phases that span two primary geotectonic units: the southern margin of the Tianshan–Xingmeng Orogenic System (I) and the eastern sector of the Tarim Block (III) [48–50]. The region showcases a diverse paleogeographic environment and a perpetually changing land–sea distribution pattern. It also exhibits a high degree of intrusive rock development, with an exposed area of approximately 20,480 km^2, which constitutes 19% of the total North Mountain region area. These rocks have been formed over an extensive timespan, from the Proterozoic to the Jurassic, and encompass a diverse range of rock types, from ultramafic to granitoid. A majority of these rocks are granitoids, originating from the Silurian to Permian periods [51–53]. The Beishan region in Gansu Province boasts favorable metallogenic conditions, abundant mineral resources, a diverse array of minerals, and a relatively comprehensive range of mineral types [43–45]. However, the composition is complex, the sizes vary, and the overall level of exploration remains relatively low. By the end of 2023, a total of 62 distinct solid mineral species had been identified within the Beishan region of Gansu Province, with a total of 622 locations—including mineralization points—having been documented. Among these, 115 gold deposits (points) have been discovered, with the primary formation type being hydrothermal deposits from the late Paleozoic, Mesoproterozoic, Archean–Proterozoic, and early Paleozoic eras, followed by structural altered rock-type, volcanic rock-type, and sedimentary–metamorphic-type deposits [54–56].

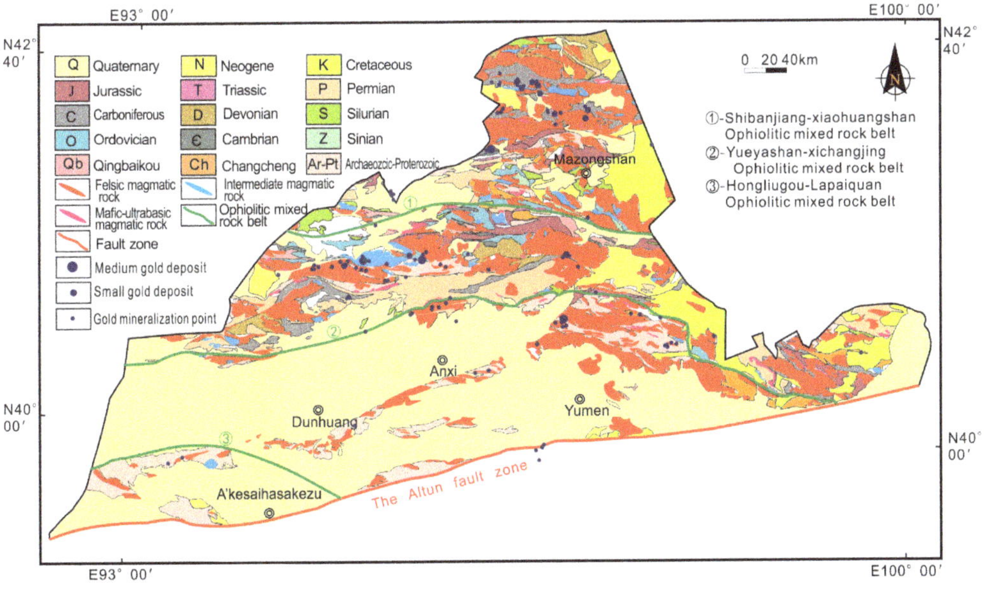

Figure 1. Geological sketch map of the Beishan area in Gansu (distribution of gold deposits).

2.2. Data Sources

The datasets for this study are detailed as follows: Regional geological data consist of Geological and mineral resources map of Gansu Province (1:1,000,000), tectonic phase map of Gansu Province (1:1,000,000), and a total of 32 sets of regional geological maps and descriptions at a scale of 1:200,000 covering the entire Beishan area. Mineral geological data consist of a total of 32 sets of 1:200,000 mineral geological maps and descriptions covering the entire Beishan area of Gansu Province, 10 sheets of 1:50,000 regional geological data, 15 sheets of 1:50,000 mineral prospecting survey data, approximately 30 detailed geological and mineral resources investigation documents, the mineral database of Gansu Province, and mineral data from the geological records of Gansu Province. Geochemical exploration data consist of a 1:200,000 regional geochemical database for the Beishan area of Gansu Province, containing stream sediment samples' locations and test results for 39 elements, including Au, Cu, U, Th, Ag, Mo, Sn, W, As, Sb, Bi, and Hg, for a total of 29,783 data points. Aeromagnetic data consist of 2,512,247 aerial magnetic measurement points in the Beishan area of Gansu Province at scales of 1:50,000 and 1:100,000.

3. Methodology

3.1. Research Ideas and Research Procedures

3.1.1. Research Ideas

A quantitative optimization of Au mineralized targets in the Beishan area of Gansu Province was conducted, based on the fundamental premise of the geoscience big data "theory of change-driven model to data-driven model". A database of mineral deposit and mineralization point information is constructed based on the characteristics of mineral distribution in the study area. A quantitative study of the correlation between results and variables (information and conditions) is conducted using a proven mathematical model (multivariate statistical analysis) and an inference model (artificial intelligence and machine learning), which are combined with a 1:200,000 stream sediment survey and aeromagnetic data in the study area. This method can accurately determine the degree of similarity between "prediction units" (areas where mineral deposits have not yet been discovered) and "ore-bearing units" (areas where mineral deposits have been found). This enables the correlation between multiple parameters and geological elements to be revealed, the potential value of data information to be harnessed, the efficiency and accuracy of target area selection to be enhanced, and comprehensive geological surveys and quantitative target area selection for regional mineral prospecting to be conducted more effectively.

3.1.2. Research Procedures

By fully utilizing the computing, searching, and processing capabilities of computers, it is possible to achieve a rough approximation of the quantitative optimization of regional mineral prospecting target areas, based on big data concepts and methods. This can be accomplished through the following steps (Figure 2):

1. The objective is to compile and categorize data related to regional and mineral surveys and mineral resources within the specified study area, at varying scales, including 1:50,000 and 1:200,000. These data must be organized and classified according to a standardized geological background information system. Subsequently, a database of information on the principal gold deposits (points) within the Beishan area should be established, based on statistical analysis.
2. The geochemical data at a scale of 1:200,000 and aeromagnetic data of various scales were processed individually, leading to the creation of separate geochemical and aeromagnetic databases. These databases were then combined in readiness for mathematical modeling.

3. A comprehensive investigation into the relationship between regional geophysical and geochemical exploration data and the Au deposits (points) previously identified in the study area is necessary. This will involve developing a conditional correlation model, associating predicted deposits (points) with geophysical and geochemical exploration units, and creating a database of optimal models that incorporate known ore-bearing units.
4. SPSS Statistics Trial Version (20), which is the software provided by IBM China Headquarters located in Shanghai, China, was utilized to construct the quantitative preferred series model for the Au mine regional mineral exploration target area, employing specification units, sample units, and anomaly units, respectively.
5. Utilizing the data from the optimal results, a map of optimal prospecting targets is constructed, providing comprehensive information on predicted mineral species within the study area. The relevant parameters undergo statistical analysis, and the target area is subsequently verified in the field to evaluate the effectiveness and optimization of the optimal model.

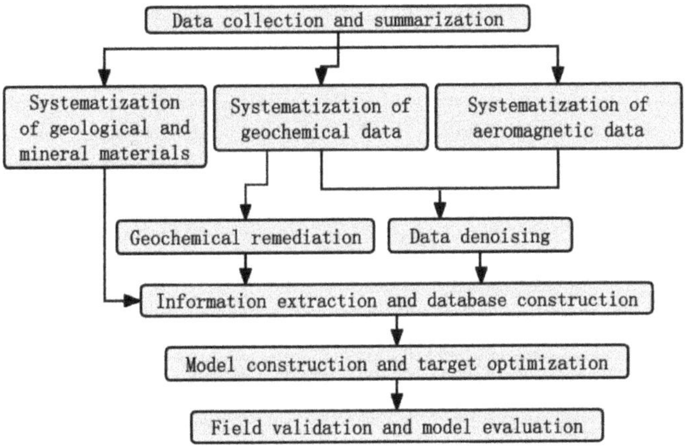

Figure 2. Research flow diagram.

3.2. Raw Data Processing

In big data research, the quality of the raw data directly determines the accuracy and validity of the results [1,57,58]. Due to the various errors that may exist in the raw data, these errors can directly affect the accuracy of the results of big data research if not properly processed. Therefore, noise reduction processing of raw data is particularly important.

3.2.1. Raw Data Noise Reduction Processing

A total of 29,783 geochemical samples were collected from 29 map sheets of 1:200,000 hydrogeological sediment surveys in the Beishan area of Gansu Province. Each sample underwent analysis for 39 geochemical elements. Given that these data were collected by various units over different years, significant systematic errors exist for the same element across different map sheets. To enhance the accuracy of the preferred target area, it is essential to conduct noise reduction, including determining the lower limit of anomalies and frame-by-frame adjustment on the original data.

To ascertain the lower threshold of anomalies within the raw data, the research team employed the concept of "machine vision" and independently devised the "linear approximation" method and algorithmic model software. This enabled them to accurately compute the lower limit of positive anomalies and the upper limit of negative anomalies for each element within each map frame of the study area simultaneously, significantly enhancing the precision and efficiency of calculating critical parameters. The "geochemical data sub-frame adjustment software" was utilized to simultaneously adjust the 39 elements measured in the 1:200,000-scale water system sediment survey in the study area to regional background reference. Additionally, the "geophysical data sub-frame adjustment software" was employed to conduct sub-frame adjustment processing on aeromagnetic data from various periods, survey areas, and scales across the entire region, ensuring the aeromagnetic data were standardized within the scope of the research area.

3.2.2. Information Extraction

The process of extracting mineral information is as follows: A total of 115 Au deposits (points) were identified in the selected area based on the aforementioned mineral geological data. The gold mineralization systems are dominantly classified into quartz vein-type, structural altered rock-type, and magmatic–hydrothermal-type deposits. These gold deposits are typically controlled by ductile shear zones and fault systems, with mineralization closely associated with Paleozoic magmatic activities and regional metamorphic–hydrothermal processes. Of these, 109 Au deposits (points) are located within the 1:200,000 stream sediment survey area. This study extracts 20-dimensional information on these deposits, including their location, stratum, structure, magmatic rock characteristics, ore body characteristics, ore-bearing structure (rock), resource quantity, and data source for each deposit (point). This allows basic geology to be utilized by a computer. The deposits, mineralization locations, stratigraphy, structural features, magmatic rocks, ore body characteristics, ore-bearing structures (rocks), resources (reserves), data sources, and other 20-dimensional information were extracted and subjected to mathematical statistical analysis using a computer.

The extraction of geochemical information is as follows: The geochemical data from the 1:200,000 water system sediment survey in the Beishan area underwent processing, resulting in the formation of 24,715 units (samples) through gridding at a 2 km $\times$ 2 km resolution. The background upper limit, lower limit, background value, median (mean) value, background difference rate, standard deviation, variation coefficient, skewness, kurtosis, upper limit difference rate, lower limit rate of change, goodness of fit, number of iterations, number of samples, and overall maximum and minimum values were extracted. The values, arithmetic mean, standard deviation, coefficient of variation, number of samples, and other statistical parameters were calculated for each geochemical parameter. A total of 33 geochemical parameters were considered. A stepwise regression simulation model was constructed for the Au element and other related elements. The fitted theoretical value Auh of Au was calculated to enhance the potential value of geochemical prospecting information in the quantitative and optimal research of regional prospecting target areas.

Extraction of aeromagnetic information: The aeromagnetic survey data, collected during various periods, were leveled and processed. Data points that intersected and overlapped with different datasets were cropped, adhering to the principle of prioritizing the dataset with the largest scale. This resulted in 2,512,247 effective physical measurement points.

Integration of geophysical and geochemical exploration data: To facilitate the simultaneous quantitative processing of geophysical and geochemical exploration data within the study area, the maximum (ΔTd) and minimum (ΔTx) values of the aeromagnetic data, after adjustment, were selected as the new aeromagnetic parameters. These parameters were positioned at the center of the coordinate points of the chemical exploration data and extended 2 km on each side. This selection takes into account both positive and negative aeromagnetic anomalies, as well as the characteristics of the aeromagnetic gradient zone. After the sorting process, a "Basic database of Au mineral exploration information (24,715 × 42)" was created for the study area, consisting of 24,715 samples and 42 variables (39 measured values of chemical elements, 1 fitted value of Au element Auh, and 2 aeromagnetic parameters).

3.2.3. Restoration of Au Geochemical Information

The geochemical map of the Au element, following the adjustment of the research area, is divided into five categories: extremely high anomaly, high anomaly, medium anomaly, low anomaly, and background. A statistical analysis of the distribution frequency of the 109 Au deposits (points) within the research area across different areas and the area percentage of different partitions is presented in Figure 3. As illustrated in Figure 3, (1) 42.6% of the Au deposits (points) are distributed in the Au anomaly area, but they are not closely related to the medium and high anomalies; (2) 95.5% of the Au deposits (points) are distributed in the low-anomaly area and background area, which collectively occupy 81.6% of the total area of the study area. A mere 4.5% of the Au deposits (points) are situated within the medium- and high-Au-element anomaly areas. It is evident that the use of Au element anomaly information alone is insufficient for the effective reduction of the prospecting target area. (3) If the prospecting target area is only studied based on Au element anomalies, nearly 57.4% of the information will be lost. The above conclusion demonstrates that it is challenging to effectively conduct research on optimizing Au prospecting target areas in the study area by solely examining Au and related element anomalies. Consequently, this study has rectified the Au geochemical information.

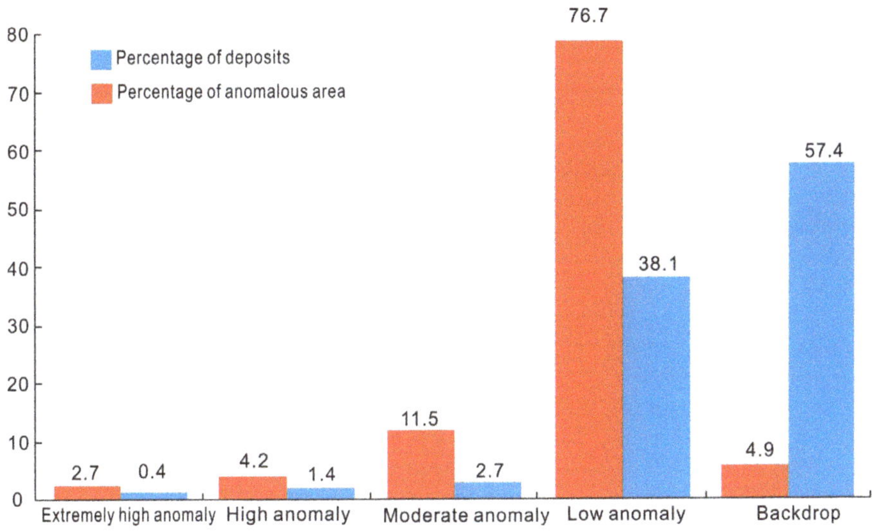

Figure 3. Comparison of the percentage of gold deposits in the Beishan Area of Gansu Province and the percentage of area with different levels of abnormality.

Considering the absence of natural chemical reactions that can decompose elemental Au particles into an ionic state, the Au element in dispersed flow predominantly exists in an extremely uneven metallic state and is present in granular form. Considering the representativeness of field sampling and assay sampling, along with the geographical location (water system location) of the gold deposit, the surface landscape, weathering conditions, and other comprehensive factors, it has not been observed that 57.4% of the Au deposits (points) in the study area lack corresponding Au element anomalies. According to relevant geochemical theories, all metallogenesis involves the enrichment of a series of related elements. Elements associated with Au metallogenesis, including Ag, As, Sb, Hg, Cu, and Bi, are typically evenly distributed in the water system sediments in ionic states. Consequently, these elements must share a specific relationship with the Au element. The authors utilized samples with anomalous Au values, along with the entire sample set from the region, to develop a stepwise regression simulation model (Auh) for Au and other related elements (Table 1). The numbers in the model represent the weights of each element or variable within the model. The elements listed in Table 1 were selected using the correlation analysis method in statistics. As shown in Table 1, elements such as Bi, Pb, Cu, Zn, Sb, SiO_2, and Y show a positive correlation with Auh, while elements such as Zn, K_2O, La, and Be exhibit a negative correlation with Auh.

$$\begin{aligned} \text{Auh} = {} & 2.76 + \text{Bi} \times 0.76 + \text{Pb} \times 0.02 + \text{Cu} \times 0.02 - \text{Zn} \times 0.01 + \text{Sb} \times 0.07 + \text{SiO}_2 \times 0.01 + \text{Y} \times 0.02 - \text{K}_2\text{O} \times 0.10 + \text{B} \times 0.004 \\ & - \text{La} \times 0.01 + \text{Th} \times 0.02 - \text{Be} \times 0.05 - \text{Na}_2\text{O} \times 0.251 - \text{Li} \times 0.004 - \text{MgO} \times 0.01 + \text{CaO} \times 0.01 + \text{Au} \times 0.002 - \text{Nb} \times 0.004 \\ & - \text{U} \times 0.01 - \text{Fe}_2\text{O}_3 \times 0.01 + \text{Al}_2\text{O}_3 \times 0.005 - \text{Mo} \times 0.01 \end{aligned} \quad (1)$$

Table 1. Stepwise regression fitting model of elemental Au for 1:200,000 water system sediment measurements.

Variant	Constant	Bi	Pb	Cu	Zn	Sb	SiO_2	Y	K_2O	B	La	Th
Factor	2.79	0.76	0.02	0.02	−0.01	0.07	0.01	0.02	−0.10	0.004	−0.01	0.02
Variant	Be	Na_2O	Li	MgO	CaO	Au	Nb	U	Fe_2O_3	Al_2O_3	Mo	
Factor	−0.05	−0.02	−0.004	−0.01	0.01	0.002	−0.004	−0.01	−0.01	0.005	−0.01	

The theoretical value of the Au element (Auh) is calculated using the entire sample in the research area to compensate for the loss of information due to sampling representativeness and other accidental factors, thereby increasing the correspondence between Au element anomalies and existing Au deposits (points), and enhancing the potential value of geochemical prospecting information in the quantitative optimization of prospecting target areas within Au-mining regions. Geochemical maps for Au and Auh fitting values were constructed separately, dividing the areas into five regions: extremely high anomaly, high anomaly, medium anomaly, low anomaly, and background. The frequency of occurrence of the 109 Au deposits (points) within the region was determined, as was the percentage of the region's area occupied by each of the different anomaly regions. It is noteworthy that out of the identified Au mineralization points in the region, only 2 were not located within the geochemical anomaly area of the Au element or the Auh fitting value. In other words, 98.2% of the Au occurrences (points) in the entire region fall at least within the anomalous area of Au or Auh element fitting value. Therefore, it is believed that the Auh fitting value effectively compensates for the impact of the extremely uneven distribution of Au element in the water system sediment on the accuracy of the quantitative optimization of the target area for regional mineral exploration. This result has been well-validated in the subsequent stages of the metallogenic optimization model.

3.2.4. Specification of Unit Division

Centering on the 109 known Au deposits (points) in the region, a grid is divided, with each deposit (point) having a side length of 2 km. It is established that anomalous samples exhibiting Au elements or Auh fitting values within the grid have been identified as containing mineral samples. The mean value of each variable (element) within the grid is taken as the parameter (variable) of the known ore-bearing unit. The "Basic Database of Au Mineralization Exploration Information (24,715 × 42)" is combined with the "Modeling Database of Au Mineralization Exploration Information (24,824 × 42)" to form a comprehensive repository of information on Au mineralization exploration. This database encompasses 109 known ore-bearing units, representing 0.44% of the total.

4. Results and Discussion

4.1. Construction of a Quantitative Optimization Model for Prospecting Target Areas

Taking the "Au mineralized geophysical information modelling database (24,824 × 42)" as the research object, the relevant statistical analysis modules in SPSS Statistics Trial Version (20) are organically combined with corresponding algorithms to construct a series of "Quantitative optimization models for Au mineralized areas and prospecting target areas in the Beishan area of Gansu Province based on geophysical and geochemical information specifications". The set of models consists of three models.

1. Model I

A random selection of the 4943 ore-free units (units approximately 20% of the whole units) and 109 known ore-bearing units were processed using a stepwise discriminant model in SPSS software to construct the "Quantitative Optimization Model for Regional Prospecting Targets for Au Mineralization in the Beishan Area of Gansu Province I". The model and its parameters are presented in Table 2.

$$R1 = SiO_2 \times 0.208 + \Delta Tx \times 0.004 + \Delta Td \times 0.003 + Al_2O_3 \times 0.267 + Na2O \times 0.251 + Ti \times 0.0006 + Ba \times 0.0008 - Pb \times 0.03 \\ -V \times 0.014 - Au \times 0.04 + CaO \times 0.06 - 19.44 \quad (2)$$

R1: Value of the specification unit identification function (model I).

Table 2. Quantitative optimization series model of Au mine regional mineral search target area based on physical and chemical exploration information.

Model No.	Variables and Parameters							Validity Check	Typical Deposits
Model I	Variable	constant	SiO_2	ΔTx	ΔTd	Al_2O_3	Na_2O	$R_0 = -1.98$ $F_p = 16.26$ Mineral positive conviction rate, 82.2% Mineral-free positive conviction rate, 98.3% Ore-bearing units: 60	The Xiaoxi Gong medium-sized gold deposit in Subei County and the Laojin Chang medium-sized gold deposit in Guazhou County
	Parameter	−19.44	0.208	0.004	−0.003	0.267	0.251		
	Contribution		32.645	22.784	16.283	7.904	4.466		
	Variable		Ti	Ba	Pb	V	Au		
	Parameter		0.001	0.001	−0.034	−0.014	−0.043		
	Contribution		2.652	2.594	2.473	1.869	1.838		
Model II	Variable	constant	SiO_2	Au	Zr	CaO	Sr	$R_0 = -0.89$ $F_p = 6.45$ Mineral positive conviction rate, 100% Mineral-free positive conviction rate, 88.1% Ore-bearing units: 78	The Xiaocaohu small gold deposit in Anxi County and the Xinjinchang small gold deposit in Xihu Township, Anxi County
	Parameter	−20.08	0.216	−0.119	0.012	0.064	0.002		
	Contribution		25.092	18.628	16.981	7.465	6.778		
	Variable		MgO	Mn	ΔTd				
	Parameter		0.153	0.001	−0.002				
	Contribution		6.685	5.696	5.302				
Model III	Variable	constant	Auh	Y	Al_2O_3	ΔTx	V	$R_0 = -1.41$ $F_p = 0.47$ Mineral positive conviction rate, 85.7% Mineral-free positive conviction rate, 94.6% Ore-bearing units: 21	Jinchanggou gold deposit in Subei County and the Jingoujing gold deposit in Anxi County
	Parameter	0.18	−1.821	−0.126	0.341	0.002	0.026		
	Contribution		15.260	13.269	10.611	6.122	5.327		
	Variable		B	CaO	Pb	Sr	Cu		
	Parameter		0.022	0.105	−0.029	−0.002	0.042		
	Contribution		4.389	4.338	4.105	3.172	3.013		

Description: $F_{0.01} = 3$, there are a total of 102 ore-bearing units in the serial model (93.6% of the total number of occupied ore-bearing units).

$$R0 = \frac{NA \times \overline{RA} + NB \times \overline{RB}}{NA + NB} \quad (3)$$

where R0 is the discrimination threshold; NA and NB are the number of units in A (population with minerals) and B (population to be judged), respectively; $\overline{RA}$ and $\overline{RB}$ are the mean value of the discrimination function values for units A and B [59].

If $\overline{RA} > \overline{RB}$, when R1 > R0, the research unit is judged to be population A (preferably with ore), and when R1 < R0, the research unit is judged to be population B (preferably without ore). If $\overline{RA} < \overline{RB}$, the result is reversed [60,61].

The results of model I's optimization are as follows: The aforementioned discriminant function (model I) is used to calculate the discriminant value (R1) for each research unit within the "Au Mineralization Prospecting Information Specification Unit Modeling Database (24,824 × 42)" in the Beishan region. The discrimination threshold is set at R0 = −1.98. Units with R1 < R0 are deemed to be ore-bearing units of preference. Model I identifies 467 preferred units as ore-bearing units of preference, of which 60 are known ore-bearing units, representing 12.85% of the total number of ore-bearing units of preference. This figure is 29.2 times higher than the proportion of units with minerals (ratio of known ore-bearing units to total projected units) (0.44%).

2. Model II

We identified 467 preferred units in model I and removed them from the units. Then, 4943 ore-free units without minerals and 109 known ore-bearing units were randomly selected. Using a stepwise discriminant model in SPSS software, we constructed the "Quantitative Optimization Model for Regional Prospecting Targets for Au Mineralization in the Beishan Area of Gansu Province II". The parameters of the model are shown in Table 2.

$$R2 = SiO_2 \times 0.216 - Au \times 0.119 + Zr \times 0.012 + CaO \times 0.064 + Sr \times 0.002 + MgO \times 0.153 + Mn \times 0.001 - \Delta Td \times 0.002 - 20.08 \quad (4)$$

The results of model II's optimization are as follows: The discriminant function (model II) presented in Table 2 is employed to calculate the discriminant value (R2) of each research unit within the "Au mineralization exploration information modeling database" (24,824 × 42), which encompasses the Beishan area. The discriminant threshold value is R0, which is −0.89. Units with an R2 value less than this are deemed to be ore-bearing units. Model II identifies 665 research units as preferred ore-bearing units, of which 78 are known ore-bearing units, representing 11.73% of the total number of preferred ore-bearing units. This figure is 26.66 times higher than the proportion of units with minerals (0.44%).

3. Model III

Following the removal of the ore-bearing units of preference identified by models I and II, 4943 ore-free units and 109 known ore-bearing units were randomly selected and processed using the stepwise discriminant model in SPSS software. This formed the basis of the "Quantitative Optimization Model for Prospecting Target Areas for Au Mineralization Information in the Beishan Area of Gansu Province III". The model parameters are presented in Table 2.

$$R3 = -Auh \times 1.821 - Y \times 0.126 + Al_2O_3 \times 0.341 + \Delta Tx \times 0.002 + V \times 0.026 + B \times 0.022 + CaO \times 0.105 - Pb \times 0.029 - Sr \times 0.002 + Cu \times 0.042 + 0.18 \quad (5)$$

The results of model III's optimization are as follows: The discriminant function (model III) in Table 2 is used to calculate the discriminant value (R3) of each research unit in the "Au mineral exploration information specification unit modeling database (24,824 × 42)" in the Beishan area. R0 = −1. 41 is the critical value of the discriminant, and R3 < R0 determines that the research unit is a preferred ore-bearing unit, and model III identified 550 research units as preferred ore-bearing units, of which 21 are known ore-bearing units, accounting for 3.81% of the total number of preferred ore-bearing units, which is 8.68% higher than the proportion of units with minerals (0.44%).

4.2. Discussion of the Model's Validity

1. The effectiveness of optimal model I:

Mathematical validity analysis:

The F-test is represented by the following equation [62–64]:

$$\mathrm{Fp} = \frac{NA + NB - P - 1}{(NA + NB - 2)P} \times \frac{NA \times NB}{NA + NB} D2 \tag{6}$$

where F_P is the F-distribution value calculated by the discriminant function (model); $F_{0.01}$ is the theoretical value of the F-distribution table (obtained by checking the F-test table) at the significance level of 0.01 (1% error); A and B are the two discriminant classification populations (A—discriminant ore-bearing unit population; B—discriminant non-ore-bearing unit population); NA and NB are the sample sizes of the two populations, respectively; $D2 = \overline{R}A - \overline{R}B$ (Marth distance); and P is the first degree of freedom (number of variables). When $F_P > F_{0.01}$, the two total samples are significantly different, and the constructed discriminant function (model) is effective in discriminating between unknown samples. In Table 2, $F_P = 16.26$, which is much larger than $F_{0.01} = 3$, indicating that the constructed discriminant model is highly effective. Preferred model I in Table 2 has a positive prediction rate of 82.2% for ore-bearing units and 98.3% for non-ore-bearing units. This model identifies 82.2% of the known ore-bearing units as preferred ore-bearing units. Therefore, the authors believe that this model can be used to reliably identify the mineralization of all research units [65–67].

Analysis of the effectiveness of the target area optimization results: Optimal model I identified 467 exploration units as preferred ore-bearing units, of which 60 were known ore-bearing units, with a mineralized unit ratio of up to 12.85%, which is 29.2 times higher than the proportion of units with minerals (0.44%). Utilizing preferred model I to optimize each exploration unit can result in 55.05% of known ore-bearing units being distributed within 1.88% of the exploration units, significantly reducing the size of the preferred target area for regional prospecting.

The researcher ranks each preferred mineral unit in ascending order of R1 value and divides it into three levels using the "golden section" method [68]. We provide a summary of the preferred statistical parameters at all levels, as presented in Table 3. A total of 89 preferred ore-bearing units have been identified in level I, of which 24 are known ore-bearing units, representing 26.97% of all units. This figure is 61.29 times higher than the proportion of units with minerals (0.44%). A total of 144 level-II preferred ore-bearing units have been identified, of which 28 are known ore-bearing units. This represents 19.44% of the total, which is 44.18 times higher than the proportion of units with minerals (0.44%). A total of 233 preferred ore-bearing units have been identified in level III, of which 8 known ore-bearing units have been confirmed. This represents a known ore-bearing unit proportion of 3.43%, which is 7.8 times higher than the proportion of units with minerals (0.44%) (Table 3). Consequently, the model constructed in this study is highly effective in quantitatively optimizing the prospecting target area for the Au deposit in the Beishan region of Gansu Province.

2. The effectiveness of optimal model II

Mathematical validity analysis: The results of the "F-test" for preferred model II are shown in Table 2. $F_P = 6.45$, which is much greater than $F_{0.01} = 3$, indicating that the constructed discriminant model is highly effective. Table 2 shows that optimal model II has a positive prediction rate of 100% for the ore-bearing units and 88.1% for the non-ore-bearing units. This model identifies 100% of the known ore-bearing units as preferred ore-bearing units, and it is considered that the model has a high reliability for identifying the ore-bearing characteristics of all the exploration units.

Table 3. Statistics of quantitative optimization results of model I.

Class of Preferred Units	Number of Preferred Units	Ratio of Preferred Units to Total Projected Units	Number of Known Ore-Bearing Units	Ratio of Known Ore-Bearing Units to Preferred Units	Increase Multiplier for Proportion of Units with Minerals
Level I	89	0.36%	24	26.97%	61.29
Level II	144	0.58%	28	19.44%	44.18
Level III	233	0.94%	8	3.43%	7.80
Total	467	1.88%	60	12.85%	29.20

Description: Total number of projected units: 24,824. The number of known ore-bearing units: 109. Proportion of units with minerals (ratio of known ore-bearing units to total projected units: 0.44%.

Analysis of the validity of the preferred results: Preferred model II identified 665 exploration units as preferred ore-bearing units, of which 78 were known to contain minerals, with a proportion of ore-bearing units as high as 11.73%, which is 26.66 times higher than the proportion of units with minerals (0.44%). Using preferred model II to optimize each exploration unit, 71.6% of the known ore-bearing units can be distributed into 2.68% of the exploration units, significantly reducing the size of the preferred target area for regional prospecting.

The preferred ore-bearing units are sorted in ascending order of R2 value from small to large and divided into three levels using the "golden section" method. The statistical parameters of the preferred units at each level are summarized (Table 4).

Table 4. Statistics of quantitative optimization results of model II.

Class of Preferred Units	Number of Preferred Units	Ratio of Preferred Units to Total Projected Units	Number of Known Ore-Bearing Units	Ratio of Known Ore-Bearing Units to Preferred Units	Increase Multiplier for Proportion of Units with Minerals
Level I	127	0.51%	20	15.75%	35.79
Level II	205	0.83%	35	17.07%	38.71
Level III	332	1.34%	23	6.92%	15.72
Total	665	2.68%	78	11.73%	26.66

Description: Total number of projected units: 24,824. The number of known ore-bearing units: 109. Proportion of units with minerals (ratio of known ore-bearing units to total projected units): 0.44%.

There are 127 level-I preferred ore-bearing units, of which 20 are known ore-bearing units, accounting for 15.74% of the total. This is 35.79 times higher than the proportion of units with minerals (0.44%) and means that 15.74% of the known ore-bearing units are spread over only 0.51% of the exploration units. There are 205 preferred ore-bearing units in level II, of which 35 are known ore-bearing units, accounting for 17.07% of the total. This is 38.71 times higher than the proportion of units with minerals (0.44%). There are 332 preferred ore-bearing units in level III, of which 23 are known ore-bearing units, and the Ratio of known ore-bearing units to preferred units is 6.92%, which is 15.72 times higher than Proportion of units with minerals (0.44%) (Table 4). Therefore, the model constructed in level II is highly effective in quantitatively selecting target areas for prospecting Au deposits in the Beishan area of Gansu Province.

3. The effectiveness of optimal model III

Mathematical validity analysis: The results of the F-test for optimal model III are shown in Table 2. $F_P = 0.47$, which is much smaller than $F_{0.01} = 3$, The constructed discriminant model is mathematically ineffective (optimal model I for F-test calculation and parameters). In Table 2, preferred model III has a positive prediction rate of 85.7% for the ore-bearing units and a positive prediction rate of 94.6% for the non-ore-bearing units. This model identifies 85.7% of the known ore-bearing units as preferred ore-bearing units, and it is considered that the model has a high reliability in regard to identifying the mineralization of each exploration unit. However, due to its poor F-test result, its preferred results should be used with caution.

Analysis of the validity of the preferred results: Optimal model III identified 550 research units as preferred ore-bearing units, including 21 known ore-bearing units, with a ratio of known ore-bearing units of 3.81%, which is 8.68 times higher than the proportion of units with minerals (0.44%). Using preferred model III to optimize each research unit can result in 19.3% of the known ore-bearing units being distributed in 2.2% of the research units, which to some extent narrows the scope of the preferred target area for regional prospecting.

The preferred ore-bearing units are arranged in ascending order of R3 value, from smallest to largest, and divided into three categories using the "golden section" method. The summary of the preferred statistical parameters at each level is presented in Table 5.

Table 5. Statistics of quantitative optimization results of model III.

Class of Preferred Units	Number of Preferred Units	Ratio of Preferred Units to Total Projected Units	Number of Known Ore-Bearing Units	Ratio of Known Ore-Bearing Units to Preferred Units	Increase Multiplier for Proportion of Units with Minerals
Level I units	105	0.42%	6	5.71%	12.98
Level II units	170	0.68%	4	2.35%	5.34
Level III units	275	1.11%	11	4%	9.09
Total	550	2.22%	21	3.81%	8.68

Description: Total number of projected units: 24,824. The number of known ore-bearing units: 109. Proportion of units with minerals (ratio of known ore-bearing units to total projected units): 0.44%.

There are 105 preferred ore-bearing units at level I, of which 6 are known ore-bearing units, representing 5.71% of the total. This is 12.98 times higher than the proportion of units with minerals (0.44%), with 5.71% of known ore-bearing units spread over only 0.42% of the exploration units. There are 170 preferred ore-bearing units in level II, of which 4 are known ore-bearing units, and the proportion of known ore-bearing units is 2.35%, which is 5.34 times higher than the proportion of units with minerals (0.44%). There are 275 preferred ore-bearing units in level III, of which 11 are known ore-bearing units, and the proportion of known ore-bearing units is 4%, which is 9.09 times higher than the proportion of units with minerals (0.44%) (Table 5). Therefore, the constructed model III has a certain effect on the quantitative optimization of the prospecting target area for Au deposits in the Beishan area of Gansu Province.

4. Discussion of a series of models for the optimization of prospecting target areas in the Au mining area

In models I and II, $F_p > F_{0.01}$, which means that models I and II are highly effective in a statistical sense. Although $F_p < F_{0.01}$ in model III, each of the three models has a very high positive prediction rate for both sample types. Furthermore, the proportion of ore-bearing units in each model and classification is greater than the proportion of units with minerals (0.44%). Therefore, all three models are considered to have a very good optimization effect. The results of the validity test, the optimization outcomes, and the analysis of the ore-bearing rate in each classification of each model indicate that model I displays the most optimal results, model II is the second most optimal, and model III exhibits relatively poor optimization effects. The application of the optimization results of each model should be treated differently.

Optimal models I, II, and III are all based on the theoretical foundation of mathematical statistics. The determination of each model is based on the correlation between different variables and the existing Au deposits (points) in the area. This approach eliminates the influence of subjective human factors and the uncertainties and differences inherent in geological and metallogenic theories. Of the 109 known ore-bearing units in the entire region, 102 known ore-bearing units (representing 93.6% of the total) are distributed across the three preferred models. It is postulated that 93.6% of the ore-bearing units in the entire region can be clearly delineated into three categories within the 42-dimensional space (comprising 39 geochemical elements, 2 aeromagnetic variables, and 1 fitted value of the Au element).

Preferred model I: A total of 60 known ore-bearing units are highly separated from other non-ore-bearing units in 10 dimensional spaces, such as SiO_2, ΔTx, ΔTd, Al_2O_3, and Au. Typical mineral deposits include the Xiaoxi Gong medium-sized gold deposit in Subei County and the Laojin Chang medium-sized gold deposit in Guazhou County (Table 2). Further study is required to determine whether this type of Au deposit has a different genesis than other Au deposits in the study area. However, it is evident that there are notable differences in regional prospecting research based on regional geophysical and geochemical information, which serves as a valuable reference for more in-depth research on the genesis of regional Au deposits and regional prospecting.

Moreover, the individual contributions of parameter variables such as SiO_2, Al_2O_3, Na_2O, ΔTx, and ΔTd in preferred model I are markedly greater than the contribution of the Au element. This may indicate that Au deposits (points) in the study area are predominantly formed in the fractured granite belt, and that a majority of Au element anomalies do not result in the output of Au deposits (points). The coefficient of the Au variable in preferred model I is negative, and its individual contribution to the model is 1.84%. A preliminary investigation indicates that there is no evident correlation between the elevated geochemical anomaly of the Au element and the delineation of an Au ore target area. However, this observation merely reflects the fact that 69% of Au deposits (points) are situated in low-anomaly regions, and 9.6% of Au deposits (points) are even distributed in negative anomaly areas. Therefore, this paper argues that the anomalous Au element and associated combined anomalies alone cannot be used to optimally study regional Au metallogenesis. This conclusion requires attention and further in-depth research in future metallogenic optimization studies.

Preferred model II: A total of 78 known ore-bearing units are distributed within this category of preferred ore-bearing units. Typical mineral deposits include the Xiaocaohu small gold deposit in Anxi County (carlin-type) and the Xinjinchang small gold deposit in Xihu Township (structural altered rock-type), Anxi County (Table 2). The individual contribution of Au to the model in this model is as high as 18.6%, and its coefficient is negative. Combined with its $R0 = -0.89$ (which is negative), it is believed that the ore-bearing units identified by this model are more likely to be associated with higher Au element anomalies.

Preferred model III: Only six known ore-bearing units are distributed among the preferred level-I ore-bearing units. Typical mineral deposits include the Jinchanggou gold deposit in Subei County and the Jingoujing gold deposit in Anxi County (Table 2). The Auh variable in this model alone contributes 15.3% to the model, and its coefficient is negative. Combined with its $R0 = -1.41$ (negative), it is believed that the adjustment and restoration of the Au element of the regional geochemical information have solved the problem of some Au deposits (points) not having corresponding Au element anomalies. At the same time, it confirms the irreplaceable and important role of the Auh parameter in the quantitative optimization of prospecting target areas, providing new ideas and methods for future research in the processing of geochemical information and optimization of metallogenesis.

5. Discussion on the contribution of aeromagnetic data in the models

In preferred mode I, the variable coefficient of ΔTx is positive, while the variable coefficient of ΔTd is negative. The individual contributions of the two exceed 10%, and their cumulative height is 39.1%. This highlights the crucial importance of regional aeromagnetic data in identifying promising regions for gold mineral exploration, and the significant discrepancy is beneficial for prospecting. It can be inferred from the aeromagnetic geophysical characteristics that the output of Au deposits (points) in the region is closely related to the aeromagnetic gradient zone. This conclusion is consistent with the established fact that regional Au deposits are predominantly located in fault zones and contact zones between different geological bodies. In model II, although the contribution of the ΔTd variable is not as significant as Au, it also exceeds 5% and holds a certain weight in the model; in model III, the contribution of the ΔTx variable is much higher than that of Au, also exceeding 5%, and it holds a certain weight in the model.

In model I, the individual contribution of the ΔTx and ΔTd aeromagnetic data variables to the optimal model is significantly greater than that of the Au element. This outcome provides definitive evidence that when aeromagnetic data are transformed into a format suitable for use with geochemical data to construct an optimal model for regional Au prospecting target areas, aeromagnetic data exhibit excellent discriminant ability. This result not only calls into question the traditional geological research conclusion that aeromagnetic results are ineffective in determining Au mineral prospecting targets, but also corroborates the assertion that geophysical information is not influenced by sampling representativeness and is a more reliable and accurate measurement than geochemical information. In conclusion, this evidence demonstrates the potential for significant insights to be gained from extensive datasets. From this, it becomes clearer that there is infinite potential in massive data information, which can only be fully explored by changing the way we think, based on the concept of big data and the method of quantitative research.

4.3. Results of Quantitative Target Selection Based on Geophysical and Geochemical Prospecting Information for Au Mining Region

Preferred models (discriminant functions) I, II, and III, as presented in Table 2, were utilized to process the "Au mineralization prospecting information specification unit modeling database (24,824 × 42)" within the designated study area. Subsequently, the discriminant values (R1, R2, and R3) for each model were calculated for each research unit. Identify the inverse value of each discriminant value, calculate the sum of the discriminant values for each ore-bearing unit, and determine the RZ value (RZ represents the comprehensive discriminant value derived from a series of preferred models for each ore-bearing unit, and its magnitude directly reflects the reliability of prospecting for ore-bearing units). The delineation of the preferred ore-bearing regions is retained, while the low and isolated areas are excluded. The largest comprehensive discriminant value in each concentration area and its location are selected to represent the area. The maximum RZ values in each concentration area are sorted in descending order, and the top 300 preferred ore-bearing unit concentration areas are selected as the preferred ore target areas for the Au ore area in the Beishan area based on the geophysical and geochemical information.

The 300 prospecting target areas are classified into three categories according to the "golden section rule". The first category, level I, comprises 57 areas, including 11 known ore-bearing target areas, with a mineralization rate of 19.29% in level-I target areas. The 93 level-II Au prospecting target areas, including the 11 mineralized target areas, exhibit a mineralization rate of 11.87% in target areas. A total of 150 level-III Au prospecting target areas were identified, of which 5 mineralized target areas were confirmed to contain ore, yielding a target area ore-finding rate of 3.33%. The ore-finding rate of each level of the target area is considerably higher than the average ore-finding rate of the anomaly unit. It is therefore postulated that the quantitative optimal model for prospecting Au in the anomaly unit is highly accurate for determining and optimizing the Au prospecting target areas in this region (Table 6).

Table 6. Statistics of quantitative optimization results for the target area of the Au mine.

Class of Target Areas	Number of Preferred Target Areas	Ratio of Preferred Target Areas to Total Projected Target Areas	Number of Known Ore-Bearing Target Areas	Ratio of Known Ore-Bearing Target Areas to Preferred Target Areas	Increase Multiplier for Proportion of Target Areas with Minerals
Level I	57	0.23%	11	19.29%	43.85
Level II	93	0.37%	11	11.82%	26.88
Level III	150	0.60%	5	3.33%	7.57
Total	300	1.21%	27	9.00%	20.45

Description: Total number of projected target areas: 24,824. The number of known ore-bearing target areas: 109. Proportion of target areas with minerals (ratio of known ore-bearing target areas to total projected target areas): 0.44%.

The proportion of the area of each level of preferred target area is calculated. The aggregate area of the 57 level-I prospecting target areas is 838.75 km^2, with the smallest area measuring 4.05 km^2 and the largest area measuring 29.37 km^2. The cumulative area of the 93 level-II prospecting target areas is 1022.27 km^2, with the smallest area measuring 0.5 km^2 and the largest area measuring 33.77 km^2. The cumulative area of the preferred I- and II-level target areas is 1861.02 km^2, representing a mere 3.67% of the total area of the study region (49,430 km^2).

A total of 150 I- and II-level target areas have been identified for prospecting Au mineralization regions, with the selection process based on quantitative optimization using geophysical and geochemical prospecting information. Of these, 22 target areas are known ore-bearing target areas, representing a ratio of 33.3 times higher than proportion of target areas with minerals. The quantitatively optimized I- and II-level target areas are distributed across a mere 3.67% of the total area of the study area. It is believed that the quantitative target selection model for Au mine searching based on geophysical and geochemical exploration information is highly accurate for determining and selecting the target area for Au mine searching in the Beishan area of Gansu Province. In comparison with the traditional selection method, it is hypothesized that the precision of the target area is significantly enhanced and the area of the target area is considerably reduced.

4.4. Field Inspection to Verify the Situation

The Gansu Provincial Bureau of Geology and Mineral Resources Development and its subordinate units conducted field inspections of 46 blank target areas (target areas where no mineral deposits had been found) in the level-I target area. There were 19 target areas with Au grades greater than 0.5 g/t in the pick-up block samples in Table 7 (including 13 target areas with Au grades greater than 1 g/t). The mineralization grade of the blank target areas was 41.3%. Combining the 11 known ore-bearing target areas, the actual ore-discovering rate in the level-I target area can reach 52.6%. The authors believe that the method of quantitatively selecting Au ore target areas based on geophysical and geochemical information has significant results in the Beishan area of Gansu Province (Figure 4).

Table 7. List of blank target area-checking results.

Number of Target Area	Au (10^{-6})	Number of Target Area	Au (10^{-6})	Data Supplier
I-2-6	1.96–2.02	I-2-9	0.63	The Fourth Geological Survey Institute of Gansu Provincial Geology and Mining Bureau
I-2-10	0.57	I-2-11	3.65	
I-2-12	2.76–3.43	I-2-13	0.97	
I-2-14	0.51	-	-	
I-2-7	1.42–1.71	I-2-4	1.42	Geological Survey of Gansu Province
I-2-8	0.86–0.93	I-2-1	0.59	
I-2-5	1.63–1.84	I-2-2	2.34–3.78	
I-2-3	1.16	I-2-4	2.76–2.95	
I-2-14	0.66	I-2-15	1.74	The Third Geological Survey Institute of Gansu Provincial Geology and Mining Bureau
I-2-16	1.52–2.13	I-2-17	1.06–4.42	
I-2-18	1.62–1.74	I-2-19	2.85–5.66	

Description: The Third Survey Institute, the Fourth Survey Institute of Gansu Province Geology and Mining Bureau, and the Geological Survey of Gansu Province completed the field validation work, and the Comprehensive Research Office of Gansu Province Geology and Mining Bureau summarized the validation results.

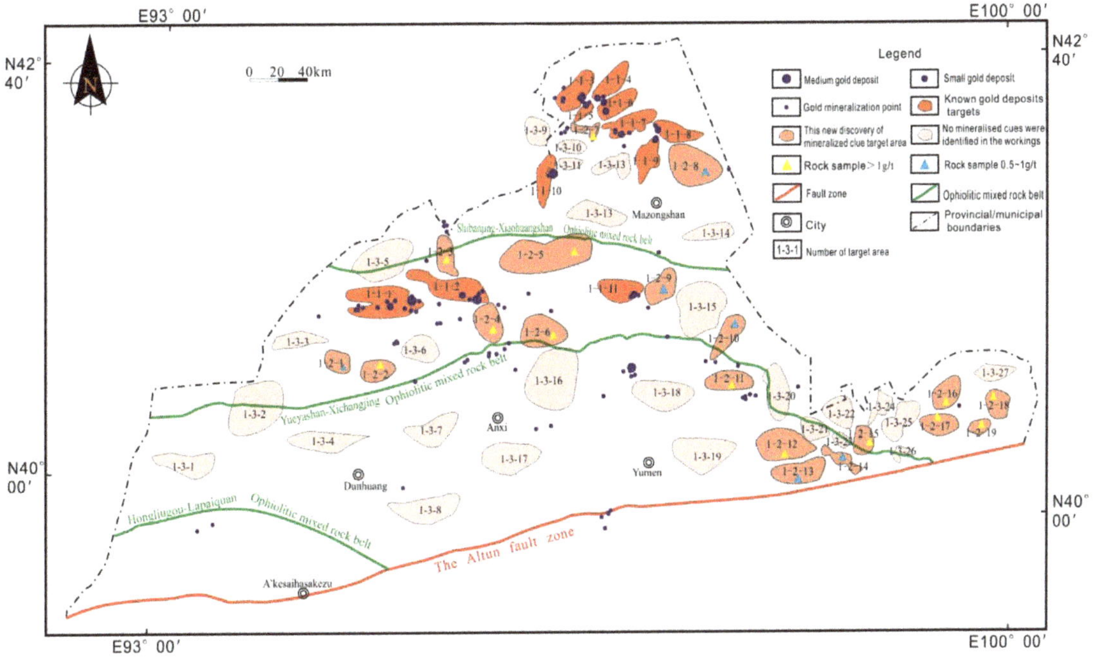

Figure 4. Schematic representation of the results of the inspection of the prospecting target area for level-I Au deposits based on geophysical and geochemical information in the Beishan area of Gansu Province.

5. Conclusions

The utilization of big data, artificial intelligence, and other methodologies and technologies enables a digitally driven approach to the exploitation of geological data. This approach involves studying the correlation of geophysical and geochemical exploration information (data) to achieve a quantitative prediction of mineral prospecting targets and increase the probability of discovering mineral deposits. Compared to the conventional "theory-driven" and "causality analysis" qualitative prediction research methods, the constraints imposed by pertinent geological experience (theory) and the impact of subjective variables are effectively negated.

The self-developed software for determining the prediction lower limit of geochemical exploration data and processing data leveling by amplitude and area effectively eliminates the original data's systematic errors, realizes the leveling of different-scale geochemical exploration data, and addresses the long-standing challenges of data processing and the "loss of low and slow anomalies". This lays a solid foundation for enhancing the effectiveness of mineral prediction and discovery.

The essential information of the classification of mineral deposits was restored, the potential value of the geochemical exploration data was effectively exploited, and the correlation between the mineralization elements and the predicted mineral species was significantly improved. The predicted target area was reduced to 3.67% of the total study area. Within this, the actual observation rate of minerals in the level-I predicted target area is as high as 52.6%, thereby markedly enhancing the accuracy of the predictions for the target area.

The innovative application of aeromagnetic data in predicting the mineralization of non-magnetic mineral species, such as Au, has altered the conclusion that aeromagnetic results are ineffective for identifying target areas for gold mining, as previously suggested in traditional geological research.

Author Contributions: L.Z., conceptualization, methodology, software, and writing—original draft preparation; R.H., writing—review and editing, funding acquisition; Y.Z., resources, supervision, and writing—review; H.F., visualization; J.L., methodology and conceptualization; Y.L., investigation and data curation. All authors have read and agreed to the published version of the manuscript.

Funding: This research was financially supported by the Science and Technology Plan of Gansu Province "Key Research and Development Program of Gansu Province" (No. 21YF5NA040); the National Natural Science Foundation Project of China (No. 41572060, No. 42172086, and No. U1133602).

Institutional Review Board Statement: Not applicable.

Informed Consent Statement: Not applicable.

Data Availability Statement: All data used in this study are available in the article and the published literature.

Acknowledgments: We thank the Comprehensive Research Office of the Gansu Provincial Bureau of Geology and Mineral Resources for providing data on field verification of the target area.

Conflicts of Interest: The authors declare no conflict of interest.

References

1. Luo, J.M.; Wang, X.W.; Song, B.T.; Yang, Z.M.; Zhang, Q.; Zhao, Y.Q. Discussion on the method for quantitative classification of magmatic rocks: Taking it's application in West Qinling of Gansu Province for example. *Acta Petrol. Sin.* **2018**, *34*, 326–332.
2. Wang, C.; Ma, X.; Chen, J.; Chen, J. Information extraction and knowledge graph construction from geoscience literature. *Comput. Geosci.* **2018**, *112*, 112–120. [CrossRef]
3. Enkhsaikhan, M.; Holden, E.J.; Duuring, P.; Liu, W. Understanding ore-forming conditions using machine reading of text. *Ore Geol. Rev.* **2021**, *135*, 104200. [CrossRef]
4. Austin, J.R.; Blenkinsop, T.G. Local to regional scale structural controls on mineralization and the importance of a major lineament in the eastern Mount Isa Inlier. Australia: Review and analysis with autocorrelation and weights of evidence. *Ore Geol. Rev.* **2009**, *35*, 298–316. [CrossRef]
5. Liu, C.; Ji, X.; Dong, Y.; He, M.; Yang, M.; Wang, Y. Chinese mineral question and answering system based on knowledge graph. *Expert Syst. Appl.* **2023**, *231*, 120841. [CrossRef]
6. Zhuang, C.; Liu, C.; Zhu, H.; Ma, Y.; Shi, G.; Liu, Z.; Liu, B. Constraint information extraction for 3D geological modelling using a span-based joint entity and relation extraction model. *Earth Sci. Inf.* **2024**, *17*, 985–998. [CrossRef]
7. Sobhana, N.V.; Mitra, P.; Ghosh, S.K. Conditional random field based named entity recognition in geological text. *Int. J. Comput. Appl.* **2010**, *1*, 143–147. [CrossRef]

8. Qiu, Q.; Tian, M.; Huang, Z.; Xie, Z.; Ma, K.; Tao, L.; Xu, D. Chinese engineering geological named entity recognition by fusing multi-features and data enhancement using deep learning. *Expert Syst. Appl.* **2024**, *238*, 121925. [CrossRef]
9. Zhou, Y.Z.; Chen, S.; Zhang, Q.; Xiao, F.; Wang, S.G.; Liu, Y.P.; Jiao, S.T. Advances and prospects of big data and mathematical geoscience. *Acta Petrol. Sin.* **2018**, *34*, 255–263.
10. Zhou, Y.Z.; Xiao, F. Overview: A glimpse of the latest advances in artificial intelligence and big data geoscience research. *Earth Sci. Front.* **2024**, *31*, 1–6.
11. Abbaszadeh, M.; Hezarkhani, A.; Soltani-Mohammadi, S. An SVM-based machine learning method for the separation of alteration zones in Sungun porphyry copper deposit. *Geochemistry* **2013**, *73*, 545–554. [CrossRef]
12. Zuo, R.G. Identifying geochemical anomalies associated with Cu and Pb-Zn skarn mineralization using principal component analysis and spectrum-area fractal modeling in the Gangdese Belt, Tibet (China). *J. Geochem. Explor.* **2011**, *111*, 13–22. [CrossRef]
13. Filzmoser, P.; Hron, K.; Reimann, C. Robust factor analysis for compositional data. *Comput. Geosci.* **2009**, *35*, 1854–1861. [CrossRef]
14. Maepa, F.; Smith, R.S.; Tessema, A. Support vector machine and artificial neural network modelling of orogenic gold prospectivity mapping in the Swayze greenstone belt, Ontario, Canada. *Ore Geol. Rev.* **2021**, *130*, 103968. [CrossRef]
15. Chen, Z.; Wu, Q.; Han, S.; Zhang, J.; Yang, P.; Liu, X.; Lang, M. The metallogenic tectonic implication of the volcanic rocks of the Dahalajunshan Formation in the Early Carboniferous in the West Tianshan based on big data analytics. *Arab. J. Geosci.* **2022**, *15*, 1658. [CrossRef]
16. Zhang, S.; Xiao, K.Y.; Carranza, E.J.M.; Yang, F. Maximum entropy and random forest modeling of mineral potential: Analysis of gold prospectivity in the Hezuo—Meiwu district west Qinling Orogen, China. *Nat. Resour. Res.* **2019**, *28*, 645–664. [CrossRef]
17. Shabani, A.; Ziaii, M.; Monfared, M.S.; Shirazy, A.; Shirazi, A. Multi-Dimensional Data Fusion for Mineral Prospectivity Mapping (MPM) Using Fuzzy-AHP Decision-Making Method, Kodegan-Basiran Region. East Iran. *Minerals* **2022**, *12*, 1629. [CrossRef]
18. Nti, I.K.; Quarcoo, J.A.; Aning, J.; Fosu, G.K. A mini-review of machine learning in big data analytics: Applications, challenges, and prospects. *Big Data Min. Anal.* **2022**, *5*, 81–97. [CrossRef]
19. Zhang, S.; Carranza, E.J.M.; Xiao, K.Y.; Wei, H.T.; Yang, F. Mineral prospectivity mapping based on isolation forest and random forest: Implication for the existence of spatial signature of mineralization in Outliers. *Nat. Resour. Res.* **2023**, *31*, 1981–1999. [CrossRef]
20. Qiu, Q.; Ma, K.; Lv, H.; Tao, L.; Xie, Z. Construction and application of a knowledge graph for iron deposits using text mining analytics and a deep learning algorithm. *Math. Geosci.* **2021**, *55*, 423–456. [CrossRef]
21. Jean, G.E.; Bancroft, M. An XPS and SEM study of gold deposition at low temperatures on sulphide mineral surfaces: Concentration of gold by adsorption /reduction. *Geochim. Cosmochim. Acta* **1985**, *49*, 979–987. [CrossRef]
22. Sadeghi, B.; Khalajmasoumi, M.; Afzal, P.; Moarefvand, P.; Yasrebi, A.B.; Wetherelt, A.; Foster, P.; Ziazarifi, A. Using ETM+ and ASTER sensors to identify iron occurrences in the Esfordi 1:100,000 mapping sheet of Central Iran. *J. Afr. Earth Sci.* **2013**, *85*, 103–114. [CrossRef]
23. Hu, H.; Wen, Y.; Chua, T.S.; Li, X. Toward scalable systems for big data analytics: A technology tutorial. *IEEE Access* **2014**, *2*, 652–687. [CrossRef]
24. Rodriguer-galiano, V.F.; Chica, O.M.; Chica, R.M. Predictive modelling of gold potential with the integration of multisource information based on random forest: Acase study on the Rodalquilar area Southern Spain. *Int. J. Geogr. Inf. Sci.* **2014**, *28*, 1336–1354. [CrossRef]
25. Carranza, E.J.M.; Laborte, A.G. Random forest predictive modeling of mineral prospectivity with small number of prospects and data with missing values in Abra (Philippines). *Comput. Geosci.* **2015**, *74*, 60–70. [CrossRef]
26. Rodgiguez, G.; Sanchez, C.; Chica, O.M.; Chica, R.M. Machine learning predictive models for mineral prospectivity: An evaluation of neural networks random forest regression trees and supportvector machines. *Ore Geol. Rev.* **2015**, *71*, 804–818.
27. Baumann, P.; Mazzetti, P.; Ungar, J.; Barbera, R.; Barboni, D.; Beccati, A.; Bigagli, L.; Boldrini, E.; Bruno, R.; Calanducci, A.; et al. Big data analytics for earth sciences: The Earth Server approach. *Int. J. Digit. Earth* **2016**, *9*, 3–29. [CrossRef]
28. Li, T.F.; Xia, Q.L.; Zhao, M.Y.; Gui, Z.; Leng, S. Prospectivity mapping for tungsten polymetallic mineral resources Nanling Metallogenic Belt South China: Use of Random Forest Algorithm from a Perspective of Data Imbalance. *Nat. Resour. Res.* **2019**, *29*, 203–227. [CrossRef]
29. Mohamed, A.; Najafabadi, M.K.; Wah, Y.B.; Zaman, E.A.K.; Maskat, R. The state of the art and taxonomy of big data analytics: View from new big data framework. *Artif. Intell. Rev.* **2020**, *53*, 989–1037. [CrossRef]
30. Chen, G.X.; Huang, N.; Wu, G.P.; Luo, L.; Wang, D.T.; Cheng, Q.M. Mineral prospectivity mapping based on wavelet neural network and Monte Carlo simulations in the Nanling W-Sn metallogenic province. *Ore Geol. Rev.* **2022**, *143*, 104765. [CrossRef]
31. Li, Y.S.; Peng, C.; Ran, X.J.; Xue, L.F.; Chai, S.L. Soil geochemical prospecting prediction method based on deep convolutional neural networks-Taking Daqiao Gold Deposit in Gansu Province, China as an example. *China Geol.* **2022**, *5*, 71–83.
32. Ren, W.X.; Luo, J.M.; Sun, B.N.; Wang, H.T.; Wang, Y.X. Application of geochemical data in gold prospecting and target selecting: Taking the Yushishan area in Gansu Province as a case. *Acta Petrol. Sin.* **2018**, *34*, 3225–3234.

33. Safari, S.; Ziaii, M.; Ghoorchi, M.; Sadeghi, M. Application of concentration gradient coefficients in mining geochemistry: A comparison of copper mineralization in Iran and Canada. *J. Min. Environ.* **2018**, *9*, 277–292.
34. Zuo, R.; Xiong, Y. Big data analytics of identifying geochemical anomalies supported by machine learning methods. *Nat. Resour. Res.* **2018**, *27*, 5–13. [CrossRef]
35. Ziaii, M.; Safari, S.; Timkin, T.; Voroshilov, V.; Yakich, T. Identification of geochemical anomalies of the porphyry-Cu deposits using concentration gradient modelling: Acase study Jebal-Barez area Iran. *J. Geochem. Explor.* **2019**, *199*, 16–30. [CrossRef]
36. Huston, D.L.; Sie, S.H.; Suter, G.F.; Cooke, D.R. Both RATrace elements in sulfide minerals from Eastern Australian volcanic-hosted massive sulfide deposits: Part, I. Proton Microprobe analyses of pyrite, chalcopyrite, and sphalerite, and Part II. Selenium Levels in Pyrite: Comparison with δ34S values and implications for the source of sulfur in volcanogenic hydrothermal systems. *Econ. Geol.* **1995**, *90*, 1167–1196.
37. Han, B.F.; Guo, Z.J.; Zhang, Z.C.; Zhang, L.; Chen, J.F.; Song, B. Age, geochemistry, and tectonic implications of a late Paleozoic stitching pluton in the North Tian Shan suture zone, Western China. *Geol. Soc. Am. Bull.* **2010**, *122*, 627–640. [CrossRef]
38. Kwok, S.W.; Carter, C. Multiple decision trees. In *Machine Intelligence and Pattern Recognition*; Elsevier: Amsterdam, The Netherlands, 1990; Volume 9, pp. 327–335.
39. Kempe, U.; Seltmann, R.; Graupner, T.; Rodionov, N.; Sergeev, S.A.; Matukov, D.; Kremenetsky, A.A. Concordant U-Pb SHRIMP ages of U-rich zircon in granitoids from the Muruntau gold district (Uzbekistan): Timing of intrusion, alteration ages, or meaningless numbers. *Ore Geol. Rev.* **2015**, *65*, 308–326. [CrossRef]
40. Feng, W.Y.; Zheng, J.H.; Shen, P. Petrology, mineralogy, and geochemistry of the Carboniferous Katbasu Au-Cu deposit, western Tianshan, Northwest China: Implications for petrogenesis, ore genesis, and tectonic setting. *Ore Geol. Rev.* **2023**, *161*, 105659. [CrossRef]
41. Teng, C.; Dong, M.; Yang, X.; Xiao, D.; Shao, J.; Cao, J.; Su, Y.; Lu, W. Zircon U-Pb Geochronology and Geochemical Constraints of Tiancang Granites, Southern Beishan Orogenic Belt: Implications for Early Permian Magmatism and Tectonic Evolution. *Minerals* **2025**, *15*, 426. [CrossRef]
42. Li, R.; Su, S.; Sun, H.; Liu, R.; Xia, Y. Petrogenesis and Tectonic Significance of Early Permian Intermediate–Felsic Rocks in the Southern Beishan Orogen, Northwest China: Geochronological and Geochemical Constraints. *Minerals* **2024**, *14*, 114. [CrossRef]
43. Chen, B.; Shen, Y.; Liu, T. Ore-controlling factors and metallogenic models of gold deposits in the Beishan orogenic belt, NW China. *Minerals* **2021**, *11*, 345.
44. Zhang, F.; Li, G. Geological characteristics and prospecting indicators of quartz vein-type gold deposits in the Beishan area. *J. Geochem. Explor.* **2022**, *235*, 106952.
45. Lu, X. Metallogenic regularity and prospecting indicators of porphyry copper deposits in the Beishan area: A case study of the Gongpoquan deposit. *Appl. Sci.* **2023**, *13*, 5890.
46. Shen, Z.; Zeng, Q. Tectonic evolution and its constraints on gold-copper mineralization in the Beishan orogenic belt. *Ore Geol. Rev.* **2020**, *124*, 103678.
47. He, Z.; Zhan, G. Geochemical characteristics and metallogenic potential of sedimentary rock-hosted gold deposits in the Beishan area. *J. Asian Earth Sci.* **2021**, *215*, 104803.
48. Laurent-Charvet, S.; Charvet, J.; Shu, L.S.; Ma, R.S.; Lu, H.F. Palaeozoic late collisional strike-slip deformations in Tianshan and Altay, Eastern Xinjiang, NW China. *Terra Nova* **2002**, *14*, 249–256. [CrossRef]
49. Keith, M.; Smith, D.J.; Jenkin, G.R.T.; Holwell, D.A.; Dye, M.D. A review of Te and Se systematics in hydrothermal pyrite from precious metal deposits: Insights into ore-forming processes. *Ore Geol. Rev.* **2018**, *96*, 269–282. [CrossRef]
50. Large, R.R.; Mukherjee, I.; Gregory, D.D.; Steadman, J.A.; Maslennikov, V.V.; Meffre, S. Ocean and atmosphere geochemical proxies derived from trace elements in marine pyrite: Implications for ore genesis in sedimentary basins. *Econ. Geol.* **2017**, *112*, 423–450. [CrossRef]
51. Mao, Q.G.; Xiao, W.J.; Wang, H.; Ao, S.J.; Windley, B.; Song, D.F.; Sang, M.; Tan, Z.; Li, R.; Wang, M. Prolonged Late Mesoproterozoic to Late Triassic Tectonic Evolution of the Major Paleo-Asian Ocean in the Beishan Orogen (NW China) in the Southern Altaids. *Front. Earth Sci.* **2022**, *9*, 825852. [CrossRef]
52. Kempe, U.; Graupner, T.; Seltmann, R.; Boorder, H.D.; Dolgopolova, A.; Zeylmans Van Emmichoven, M. The Muruntau gold deposit (Uzbekistan)- A unique ancient hydrothermal system in the southern Tien Shan. *Geosci. Front.* **2016**, *7*, 495–528. [CrossRef]
53. Dong, L.L.; Wan, B.; Yang, W.Z.; Deng, C.; Chen, Z.; Yang, L.; Cai, K.D.; Xiao, W.J. Rb-Sr geochronology of single gold-bearing pyrite grains from the Katbasu gold deposit in the South Tianshan, China and its geological significance. *Ore Geol. Rev.* **2018**, *100*, 99–110. [CrossRef]
54. Groves, D.I.; Foster, R.P. Archaean lode gold deposits. In *Gold Metallogeny and Exploration*; Foster, R.P., Ed.; Springer: Boston, MA, USA, 1991; pp. 63–103.
55. Xu, X.W.; Ma, T.L.; Sun, L.Q. Characteristics and dynamic origin of the large−scale Jiaoluotage ductile compressional zone in the eastern Tianshan Mountains, China. *J. Struct. Geol.* **2003**, *25*, 1901–1915. [CrossRef]

56. Yousefi, M.; Kreuzer, O.P.; Nykänen, V.; Hronsky, J.M.A. Exploration information systems-a proposal for the future use of GIS in mineral exploration targeting. *Ore Geol. Rev.* **2019**, *111*, 103005. [CrossRef]
57. Xi, Y.Z.; Li, Y.B.; Liu, J.J.; Wu, S.; Lu, N.; Liao, G.X.; Wang, Q.L. Application of Analytic Hierarchy Process in Mineral Prospecting Prediction Based on an Integrated Geology-Aerogeophysics-Geochemistry Model. *Minerals* **2023**, *13*, 978. [CrossRef]
58. Agterberg, F. New applications of the model of de Wijs in regional geochemistry. *Math. Geol.* **2007**, *39*, 1. [CrossRef]
59. Parsa, M.; Maghsoudi, A. Assessing the effects of mineral systems-derived exploration targeting criteria for random Forests-based predictive mapping of mineral prospectivity in Ahar-Arasbaran area, Iran. *Ore Geol. Rev.* **2021**, *138*, 104399. [CrossRef]
60. Allard, D.; Comunian, A.; Renard, P. Probability aggregation methods in geoscience. *Math. Geosci.* **2012**, *44*, 545–581. [CrossRef]
61. Chhabra, A.B.; Sreenivasan, K.R. Negative dimensions: Theory, computation, and experiment. *Phys. Rev. A* **1991**, *43*, 1114. [CrossRef]
62. Chen, Y.; Lu, L.; Li, X. Application of continuous restricted Boltzmann machine to identify multivariate geochemical anomaly. *J. Geochem. Explor.* **2014**, *140*, 56–63. [CrossRef]
63. Enkhsaikhan, M.; Liu, W.; Holden, E.J.; Duuring, P. Auto-labelling entities in low-resource text: A geological case study. *Knowl. Inf. Syst.* **2021**, *63*, 695–715. [CrossRef]
64. Abedini, M.; Ziaii, M.; Negahdarzadeh, Y.; Ghiasi-Freez, J. Porosity classification from thin sections using image analysis and neural networks including shallow and deep learning in Jahrum formation. *J. Min. Environ.* **2018**, *9*, 513–525.
65. Parsa, M.; Maghsoudi, A.; Yousefi, M. Spatial analyses of exploration evidence data to model skarn-type copper prospectivity in the Varzaghan district, NW Iran. *Ore Geol. Rev.* **2018**, *92*, 97–112. [CrossRef]
66. Shirazi, A.; Hezarkhani, A.; Beiranvand, P.A.A. Fusion of Lineament Factor (LF) Map Analysis and Multifractal Technique for Massive Sulfide Copper Exploration: The Sahlabad Area, East Iran. *Minerals* **2022**, *12*, 549. [CrossRef]
67. Chorley, R.J.; Haggett, P. Trend-surface mapping in geographical research. *Trans. Inst. Br. Geogr.* **1965**, *37*, 47–67. [CrossRef]
68. Journel, A.; Zhang, T. The necessity of a multiple-point prior model. *Math. Geol.* **2006**, *38*, 591–610. [CrossRef]

applied sciences

MDPI

Article

Research on the Prediction of Concealed Uranium Deposits Using Geo-Electrochemical Integrated Technology in the Guangzitian Area, Northern Guangxi, China

Xiaohan Zhang [1], Meilan Wen [1,2,*], Qiaohua Luo [3,*], Yunxue Ma [1], Yuheng Jiang [1], Yuxiong Jiang [1], Wei Ye [1] and Jiali Zhang [1]

1 College of Earth Sciences, Guilin University of Technology, Guilin 541006, China; 1020230054@glut.edu.cn (X.Z.); 1020220051@glut.edu.cn (Y.M.); q785460307@163.com (Y.J.); yuxiongjiang@glut.edu.cn (Y.J.); vv46967x2@163.com (W.Y.); zjl2017@glut.edu.cn (J.Z.)

2 Applied Nuclear Technology in Geosciences Key Laboratory of Sichuan Province, Chengdu University of Technology, Chengdu 610059, China

3 Nuclear Geological Brigade of Guangxi Zhuang Autonomous Region, 310, Guilin 541199, China

* Correspondence: mlwen@glut.edu.cn (M.W.); we250519@163.com (Q.L.)

Abstract

To achieve a significant breakthrough in the exploration of uranium resources in the Guangzitian area of northern Guangxi, China, an innovative combination of exploration methods was implemented at the peripheral regions of the Guangzitian uranium deposit under the guidance of the following principle: "exploring the edges and identifying the bottom, delving deep and un-covering blind spots". This study introduces geo-electrochemical integrated technology for prospecting research at the peripheral areas of the Guangzitian deposit. By validating the technology's effectiveness on known geological sections, distinct geo-electrochemical extraction anomalies were identified above recognized ore bodies. Simultaneously, soil ionic conductivity and thermally released mercury anomalies were observed, partially indicating the presence of concealed uranium deposits and fault structures. These findings demonstrate that geo-electrochemical integrated technology is effective in detecting buried uranium mineralization in this region. Subsequently, a geological-geoelectrical prospecting model was established through a systematic analysis of anomaly characteristics and metallogenic regularity, and it was subsequently applied to unexplored areas. As a result, one key anomaly verification zone, one Class A comprehensive anomaly zone, two Class B comprehensive anomaly zones, and one Class C comprehensive anomaly zone were identified within the unexplored research area. Drilling engineering validation was conducted in the No. I key anomaly verification zone, resulting in the discovery of an industrial-grade uranium ore body. This achievement not only provides critical technical support but also develops a robust theoretical foundation for future mineral exploration endeavors.

Keywords: geo-electrochemical integrated technology; carbonaceous-siliceous-pelitic uranium deposits; prospecting prediction; element anomaly characteristics; Guangzitian uranium deposit

1. Introduction

Uranium resources are crucial strategic resources and energy minerals, serving as fundamental raw materials for the development of the nuclear industry. The Guangzitian uranium deposit is situated in the north of Guilin City, Guangxi Zhuang Autonomous Region, China, within the Xiang-Gui segment of the Nanling uranium metallogenic belt. As one of the typical representatives of carbonaceous-siliceous-pelitic uranium deposits in China, it plays a significant role in uranium geological research and resource exploitation [1]. The Guangzitian deposit, together with nearby uranium deposits of the same type—such as Tuditang, Dajiangbei, and Kuangshijiao—as well as a series of ore deposits and mineralization points, forms the northern Guangxi uranium ore field [2]. If infill exploration can connect isolated uranium deposits and occurrences in the northern Guangxi uranium ore field, the defined resource inventory will be substantially expanded. However, the periphery of the Guangzitian deposit is predominantly characterized by agricultural land use (rice paddies and orchards). Due to the obscuration caused by burial depths and the overlying cover layer, the geochemical exploration anomalies remain poorly defined, and discerning the mineralization characteristics is challenging [3]. Over the years, relevant geological units have conducted extensive geological exploration operations in this region, yet the results have remained suboptimal. Therefore, by following the principle of "exploring the edges and identifying the bottom, delving deep and uncovering blind spots" for the Guangzitian deposit, an innovative combination of methods has been employed. Geo-electrochemical integrated technology was introduced to extract effective exploration information; summarize the distribution patterns of ore-forming elements; integrate the results of anomaly verification; and enhance the positioning and prediction accuracy of hidden ore bodies within the deep edges and periphery of the deposit [4].

Geo-electrochemical integrated technology comprises three principal methodologies: geo-electrochemical extraction measurement, soil ionic conductivity measurement, and thermally released soil mercury measurement [5]. As a geochemical exploration method integrating disciplines such as geochemistry, geo-electrochemistry, and geophysics, geo-electrochemical integrated technology's origins trace back to the CHIM developed by Soviet scholars in the late 1960s [6]. It was introduced to China in the 1980s by Xu Bangliang and Fei Xiquan. In 1989, Pan Yongfei proposed standardizing the technology. Significant methodological and technological enhancements were achieved in 2003 by Luo Xianrong and Kang Ming, particularly through the development of low-voltage-dipole geoelectrical extraction de-vices, facilitating substantial advancements in China. In 2017, scholars such as Wen Meilan and Sun Binbin drafted the "Technical Regulations for Geo-electrochemical Measurement (Draft for Comment)" jointly. Following over three decades of development, the electrochemical mechanism behind Earth's halo formation has been progressively refined, and associated instrumentation has undergone significant upgrades.

Numerous scientific research and industrial organizations have achieved significant success in exploring concealed mineral deposits across various mineral types and geological coverage areas by employing geo-electrochemical integrated technology. The minerals involved encompass over ten categories, including gold, silver, copper, lead, zinc, tungsten, tin, arsenic, antimony, nickel, and uranium. For instance, in the arid Gobi Desert region of Qinghai, China, this technique demonstrated significantly superior precision in targeting lead-zinc ore bodies at approximately 100 m depth compared to displaced rock geochemical surveys [7]. In the thick loess-covered terrain of Shanxi, soil geochemical surveys yielded only weak responses to gold ore bodies buried at depths of 38–200 m, whereas distinct geo-electrochemical anomalies accurately reflected the location of gold mineralization [7]. On the frozen plateau of Tibet, soil geochemical surveys could only identify shallow mineralization, while geo-electrochemical surveys revealed clear anomalies directly above the ore bodies [8]. Particularly in the eolian sand-shallow cover area of Inner Mongolia, 1:50,000 scale geo-electrochemical surveys successfully delineated multi-element anomalous targets coincident with known lead-zinc-silver mineralization. Eventually, through geophysical IP (Induced Polarization) gradient surveys and drilling programs, a 6 m thick Ag-Cu-rich ore body was discovered at a depth of over 540 m, achieving a breakthrough in mineral exploration. In contrast, the soil geochemical survey only showed sporadic anomalies [9]. These collective results validate the efficacy of geo-electrochemical technology in thick-cover regions. Consequently, they provide the methodological foundation for this study's application in targeting concealed carbonaceous-siliceous-pelitic uranium deposits within the thick Quaternary cover sequences of northern Guangxi. The applied techniques have proven effective for deposits of multiple genetic types, and they include the following: high-temperature, medium-temperature, low-temperature hydrothermal, sedimentary exhalative, stratabound, sediment-hosted and modified, volcanogenic, volcanic-sedimentary metamorphic, skarn, and sandstone-hosted deposits [10–12]. Currently, studies on the application of geo-electrochemical techniques for exploring concealed uranium deposits have covered various regions. These include the Four Mile East and Goulds Dam in Australia; the Dongsheng and Dachengliang areas within the Ordos Basin in Inner Mongolia, China, for sandstone-hosted uranium deposits; the Huxi area, Xiangshan area, and Shengyuan Basin in Jiangxi, China, for volcanic rock-hosted uranium deposits; and the Motianlun area in Guangxi and the Lujing area in Hunan, China, for granite-hosted uranium deposits [13–20]. Considerable prospecting achievements have been attained in these studies. However, in previous research, there were challenges, such as the large and uneven spacing of geo-electrochemical measurement points and lines, making it difficult to predict abnormal planes and causing local anomalies to be overlooked. Another problem was inconsistent supply current, voltage, electrification duration, and extraction liquid selection, making it hard to evaluate the extraction effect. Moreover, there is a lack of systematic research on carbonaceous-siliceous-pelitic uranium deposits. Therefore, this study summarizes previous explorations, selects the best working parameters, and utilizes geo-electrochemical integrated technology to search for concealed carbonaceous-siliceous-pelitic uranium deposits in the thickly covered area outside the Guangzitian deposit. The aim is to develop a set of exploration techniques and methods suitable for finding concealed uranium deposits in this region and to solve the deep and peripheral exploration problems in the Guangzitian area. This is of great significance for increasing mineral reserves in the northern Guangxi uranium ore field and researching understudied topics with respect to the application of geo-electrochemical integrated technology in the search for concealed carbonaceous-siliceous-pelitic uranium deposits.

2. Regional Geological Setting

The study area is situated at the junction of the Yangtze Block, the Yuechengling Fold Belt within the Northern Guangxi Block, the Nanhua Mobile Belt, and the Haiyangshan Uplift in the Central-Northeastern Guangxi Fold System. The regional topography descends from the west to the east, with the western area comprising a mid-low mountainous region at elevations exceeding 1000 m, which is characterized by steep terrain, intense dissection, and the frequent occurrence of sharp peaks and conical summits along ridgelines. To the east lies a hilly area with altitudes ranging from 300 to 500 m, and this area is characterized by gentle slopes, rounded mounds, partially exposed bedrock, and a relative height difference of less than 200 m. The uranium deposit is situated within the transitional zone between the mid-low mountains and hilly domains (Figure 1) [21].

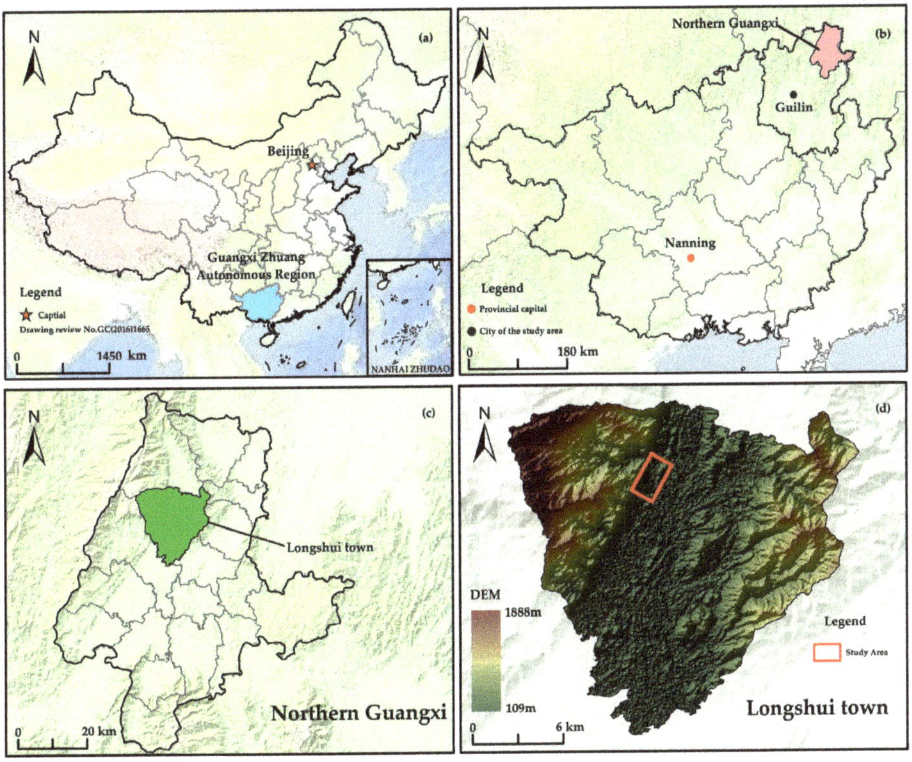

Figure 1. Location diagram of the study area in various administrative divisions of China, Guangxi and the northern of Guangxi (**a**–**d**).

The stratigraphic sequence within the study area exhibits relatively complete exposure, with strata from the Proterozoic to Quaternary systems displaying varying degrees of outcrop, except for the absence of the Silurian and Triassic systems. The lithology is predominantly composed of sandstone, with the Devonian, Carboniferous, and Ordovician systems being the most extensively distributed. Notably, the Middle-Upper Devonian series constitutes the principal ore-bearing strata within the mining field. The tectonic evolution demonstrates multiphase superposition characteristics. The Caledonian Orogeny resulted in the folding and uplift of the Lower Paleozoic strata, resulting in the formation of the Yuechengling Anticline, the Daxijiang Syncline, and the Fenghuangtian Anticline. This process was accompanied by large-scale magmatic intrusions and the development of synkinematic faults, such as the Baishi Major Fault. Subsequently, the Indochinese Orogeny reactivated the uplift and basement faults of the Upper Paleozoic strata, forming several fault zones within the Devonian system on the western wing of the Daxijiang Syncline. These include the F1 and F3 interlayer compression fracture zones, the NE-trending F2 fault zone group (F_2^0, F_2^1, F_2^2, and F_2^3), the F11 cross-cutting fault zones, and the east-west fault zone group. All these structures are closely associated with mineralization processes. The Yanshanian block faulting ultimately shaped the Daxijiang syncline, which comprises platform cover strata displaying arc-shaped bending with an eastward convex apex, which is characterized by a narrow and uplifted northern termination and broad subsided southern portion superimposed upon the Caledonian fold basement rift trough. Under E-W-trending extensional stress during the Himalayan period, the Daxijiang-Longshui fault-depression basin was formed (Figure 2) [22]. The location and morphology of uranium orebodies are controlled by fault-bounding grabens and interlayer fracture zones; the orebodies mainly occur as stratiform, lenticular, vein-type, and irregular shapes [23]. Magmatic rocks are widely distributed in the area, predominantly consisting of Caledonian medium-grained to coarse-grained and medium-grained to fine-grained porphyritic biotite granites, exhibiting an average uranium content of 16.5×10^{-6} and zircon U-Pb isotopic ages of 414 Ma. These are accompanied by Yanshanian fine-grained biotite granites with higher uranium enrichment (26.2×10^{-6}) and isotopic ages ranging from 152 to 201 Ma (zircon U-Pb dating), and they are classified as uranium-enriched granitic bodies. The multiphase and multistage magmatic intrusions ultimately constituted the Yuechengling composite granite batholith, which is presumed to have contributed partial metallogenic materials for uranium mineralization [24].

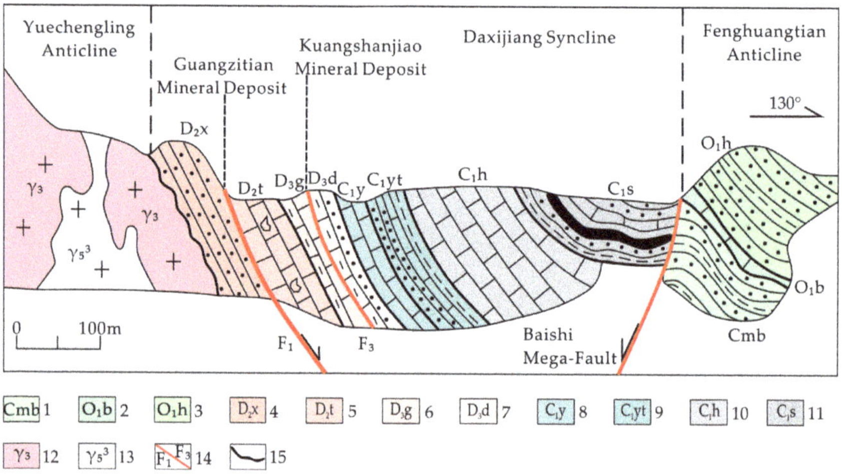

Figure 2. Schematic geological cross-section of the Daxijiang faulted basin, Northern Guangxi [24]. 1—Bianxi Formation: Sandstone interbedded with shale; 2—Baidong Formation: limestone; 3—Huang'ai Formation: sandstone interbedded with shale; 4—Xindu Formation: sandstone with basal conglomerate; 5—Tangjiawan Formation: dolomite and bioclastic limestone; 6—Guilin Formation: limestone and dolomite; 7—Dongcun Formation: argillaceous dolomite and sandstone; 8—Yaoyunling Formation: limestone; 9—Yingtang Formation: sandstone and shale interbedded with limestone; 10—Huangjin Formation: limestone; 11—Simen Formation: sandstone and shale with lenticular limestone, coal-bearing; 12—Caledonian granite; 13—Yanshanian granite; 14—mineralized fault zone and numbers; 15—coal sea.

3. Technical Principles, Methodology, Sampling, and Analysis

3.1. Technical Principles

(1) Geo-electrochemical extraction measurements, also known as the geo-electrochemical extraction method, are carried out under natural conditions. Due to the potential difference between various minerals within deep ore bodies, electrochemical dissolution occurs, generating metal ions, colloids, or complexes. These ions migrate toward the surface through groundwater flow, diffusion, and electrodynamic effects, forming ion halos. Upon the application of an artificial electric field, the pre-existing electrochemical equilibrium is perturbed, initiating the directional migration of cations and anions towards respective electrodes. This process establishes a continuous ionic replenishment chain through successive displacement, facilitating the upward transportation of deep-seated mineralization signals to shallow horizons. A renewed dynamic equilibrium is subsequently achieved following sustained current application. This technique directly detects the surface ion anomalies

corresponding to the composition of deep ore bodies, and the method is characterized by its simplicity and high efficiency. Moreover, practical applications demonstrate that the distribution of these anomalies often aligns with the vertical projection of deep ore bodies, making it especially suitable for the exploration of concealed mineral deposits [11,25,26];

(2) The soil ionic conductivity measurement technique operates on analogous electrochemical dissolution principles driven by oxidation-reduction potential gradients within concealed ore bodies. When ore masses become intersected by fissure water systems or groundwater tables, a natural electrolytic environment is established. Contrasting oxidative conditions between upper and lower ore body sections generate electrochemical potential disparities, thereby inducing electron transfer processes that disrupt electrical neutrality. This disequilibrium triggers the directional migration of cations and anions through electrochemical mobilization, ultimately culminating in the enrichment of ionic species within near-surface soil horizons to form characteristic ionic dispersion halos [25,27]. Electrical conductivity, though a geophysical parameter, serves as an effective geo-chemical indicator by measuring the total conductive ion concentration in soils to reflect the co-enrichment halos of U^{6+}, WO_4^{2-}, SO_4^{2-}, Ca^{2+}, and Fe^{3+} ions. This integrated approach demonstrates superior capability in revealing deep-seated mineralization signals compared to individual element analysis. In particular, in geologically complex terrains where conventional geophysical and geochemical exploration methods fail to achieve breakthroughs, soil ionic conductivity measurements have proven to be an exceptional technique for mineral exploration [28–30];

(3) Thermally released soil mercury measurement represents a mineral exploration technique grounded in mercury's distinctive geochemical properties (liquid state at ambient temperatures, high volatility, and strong thiophilicity and migration capacity). The fundamental principle of this methodology lies in the ubiquitous presence of mercury anomalies surrounding metallic ore deposits, where mercury vapor in soils originates from concealed mineralization bodies [31,32]. Mercury in hydrothermal systems predominantly occurs as native mercury or mercury compounds within sulfide minerals or migrates along fractures to form primary halos. Mercury vapor ascending through groundwater systems or structural discontinuities becomes adsorbed by soil to form geochemical anomalies. However, these anomalies are frequently associated with tectonic activities rather than directly indicating mineralization bodies, and they are susceptible to anthropogenic interference from vegetation cover or agricultural activities. Consequently, the integration of stepwise thermal release protocols becomes imperative for distinguishing genuine anomalies, thereby effectively delineating concealed ore bodies and fault zones. This methodology demonstrates broader applicability beyond mineral exploration, extending to hydrogeological investigations and engineering geological assessments involving structural fracture systems [33,34].

3.2. Sample Collection and Analytical Testing

In accordance with the "from known to unknown" principle, a feasibility test of the method was conducted on a known section of Line L00, which is located approximately 3 km north of the Guangzitian deposit. This section—where the ore body is exposed through drilling engineering and geological conditions that are similar to those of the study area—features 53 measurement points established at 25 m intervals. In the unknown area, covering approximately 2.75 square kilometers outside the Guangzitian deposit (Figure 3), a total of 21 survey lines and 462 measurement points were systematically arranged based on a mineral exploration grid of 100 m × 25 m for predictive research.

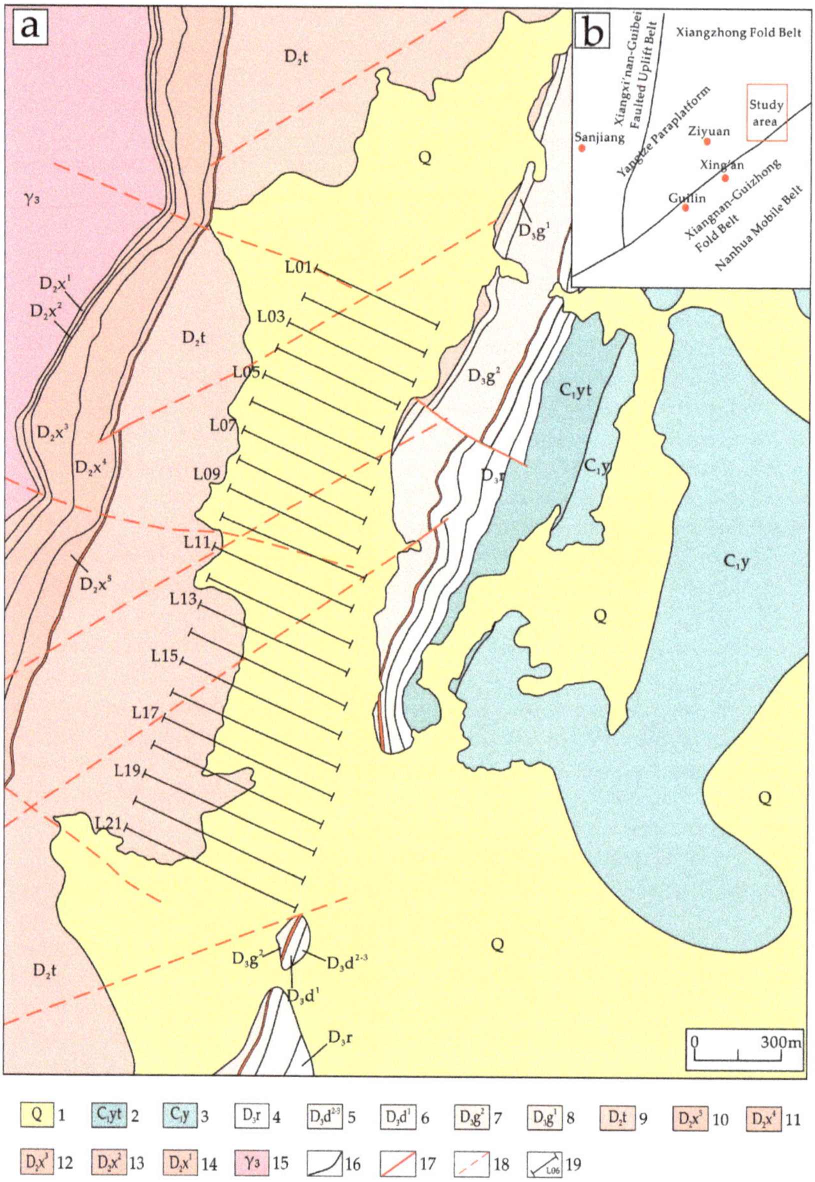

Figure 3. Geological map (**a**) and tectonic location map (**b**) of the study area. Revised based on a specific work report from the Nuclear Geological Brigade of Guangxi Zhuang Autonomous Region, 310. 1—Quaternary; 2—Yintang Formation, Lower Carboniferous; 3—Yaoyunling Formation, Lower Carboniferous; 4—Rongxian Formation, Upper Devonian; 5—Dongcun Formation, Upper Devonian (Members 2—3); 6—Dongcun Formation, Upper Devonian (Member 1); 7—Guilin Formation, Upper Devonian (Member 2); 8—Guilin Formation, Upper Devonian (Member 1); 9—Tangjiawan Formation, Middle Devonian; 10—Xindu Formation, Middle Devonian (Member 5); 11—Xindu Formation, Middle Devonian (Member 4); 12—Xindu Formation, Middle Devonian (Member 3); 13—Xindu Formation, Middle Devonian (Member 2); 14—Xindu Formation, Middle Devonian (Member 1); 15—Caledonian granite; 16—stratigraphic unconformity boundary; 17—fault; 18—inferred fault; 19—exploration line number.

This electrochemical extraction measurement utilizes an independently powered, low-voltage dipole electric extraction device developed by the Hidden Ore Deposit Prediction Research Institute of the Guilin University of Technology. This device comprises two main components: a power supply system and an ion absorber. The power supply employs low-voltage direct currents, while the ion absorber consists of specially treated adsorption materials with specific specifications, in addition to cylindrical graphite carbon rods wrapped in filter paper. A wire is extended from one end of the carbon rod and connected to the power-supply system, thereby forming the ground electricity-extraction device (Figure 4). The key features of this device include its compact size, low operating voltage, self-contained power supply, and ease of operation [35].

The field sampling work was carried out in strict compliance with the technical specifications for geo-electrochemical extraction and measurement. Two sampling pits, each approximately 30 cm in depth, were established at 1 m intervals perpendicular to the measurement line's direction between sampling points (based on the visible residual layer soil). Initially, the soil samples required for measuring soil ionic conductivity and thermally released mercury were collected. Subsequently, ground electro-extraction electrodes, in conjunction with the extraction solution, were parallelly placed within the sampling pits and backfilled sequentially according to the soil-extraction sequence. Following a 48 h electro-extraction period, the ground electro-extraction device was retrieved [36].

Following air-drying at room temperature, the adsorption materials wrapped around the anode and cathode of the ion extraction apparatus were sent to the Testing Center of Guilin Institute of Mineral Geology under the China Nonferrous Metals Industry. The samples were decomposed using nitric-acid-hydrogen peroxide-aqua regia extraction. Subsequently, elements such as U, Th, W, Mo, and As were analyzed and tested via inductively coupled plasma mass spectrometry (ICP-MS). The detection limits of the current determination method are shown in Table 1 [20,37]. The soil samples were dried and sieved through a 100-mesh aperture sieve. Then, the Hidden Ore Deposit Prediction Research Institute of the Guilin University of Technology analyzed and tested the samples using an electrical conductivity meter and a mercury analyzer.

Table 1. Detection limit of inductively coupled plasma mass spectrometry method.

Element	Detection Limit (μg/g)	Element	Detection Limit (μg/g)
U	0.0065	Mo	0.0048
Th	0.1670	As	0.0717
W	0.0498	Pb	0.8620

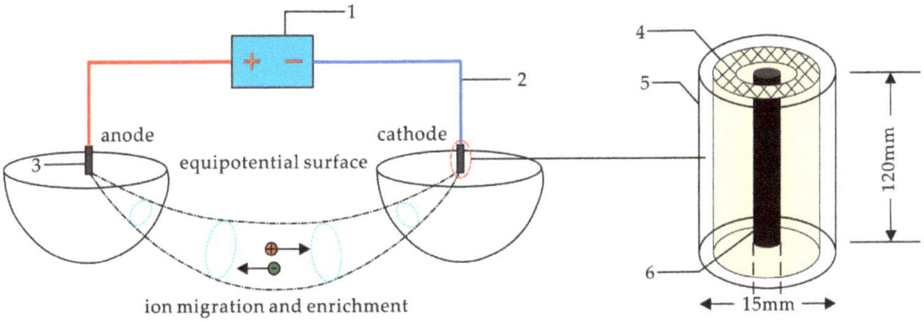

Figure 4. Schematic diagram of an independently powered dipole geoelectric extraction device and its electrode structure. 1—DC power supply; 2—conductive wires; 3—ion absorber; 4—adsorption materials; 5—filter paper; 6—graphite carbon rods.

4. Research on the Effectiveness of Mineral Exploration

To validate the applicability and effectiveness of the geo-electrochemical integration technology for detecting concealed uranium deposits in the study area, a methodological effectiveness study was conducted on the known Line L00's profile located north of the research region. This selected profile exhibits analogous geological conditions and contains ore bodies confirmed through drilling engineering verification [36]. Analyses of element content profile maps revealed that geo-electrochemical extraction anomalies, soil ionic conductivity, and thermal soil mercury release all exhibited varying degrees of geochemical anomalies (Figure 5).

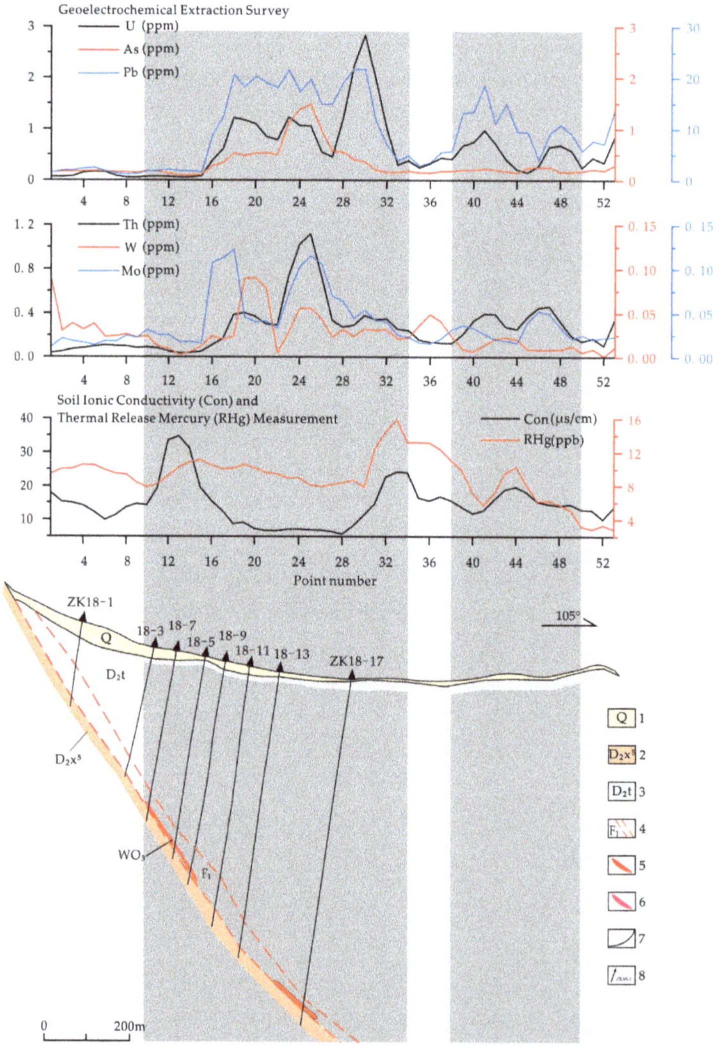

Figure 5. Integrated geological-geo-electrochemical anomaly profile of Line L00. 1—Quaternary; 2—Xindu Formation, Middle Devonian (Member 5); 3—Tangjiawan Formation, Middle Devonian; 4—interlayer fracture zones and serial number; 5—industrial-grade uranium ore body; 6—industrial-grade tungsten ore body; 7—stratigraphic boundary; 8—borehole and serial number.

4.1. Anomaly Characteristics of the Known Profile

(1) The geo-electrochemical extraction anomalies of U, As, Th, and W exhibited a single-peak pattern, while the Pb and Mo anomalies displayed a rabbit-ear-shaped configuration, predominantly distributed between points 15 and 32, with a composite anomaly width of approximately 425 m. The peak concentration of U, the primary ore-forming element, reached 2.84×10^{-6}, representing 6.8 times the background value, which provides a clear indicator for the lower segment of the KT4-11 uranium orebody at depths of −400 to −480 m. The associated element Mo showed an anomaly intensity ranging from 0.03×10^{-6} to 1.22×10^{-6}, corresponding to 1.31–4.69 times the background value. Among all of the elements analyzed via geo-electrochemical extraction, Mo exhibited the strongest spatial correlation with the two underlying mineralized zones of the known orebody;

(2) The soil ionic conductivity anomalies occurred between points 9 and 18, exhibiting a lambda-shaped morphology locally, with a peak value of 34.78 μs/cm and a composite anomaly width of approximately 250 m. These anomalies correlate well with the upper segment of the KT4-11 uranium orebody and the industrial-grade tungsten ore body at depths of 0 to −160 m. While the soil thermally released mercury anomalies did not dis-play distinct peaks above the two uranium orebody segments, the general trend of the anomaly curve indicated elevated values relative to the background, forming an anomaly zone that is spatially consistent with the overlying mineralization;

(3) Between the two orebody segments, the anomalous peaks of various elements were observed. It is inferred that mobile ions extracted through geo-electrochemical methods not only migrated vertically upward from the ore-hosting positions but also ascended along interlayer fractured zones to reach the surface. However, the known profile along Line L00 exhibited a negative topographic depression characterized by elevated peripheries and a depressed central area. Consequently, mineralization-related components may have been transported under the combined effects of rainwater runoff, chemical leaching, and gravitational forces, ultimately accumulating in the topographically low areas to form superimposed strong anomalies through enrichment processes;

(4) At points 32–45, where engineering control was absent, the anomalies of elements such as U, Th, Mo, and Pb were observed, along with anomalies in soil ionic conductivity and thermally released soil mercury. Two possible reasons for these anomalies can be inferred: first, the upward migration of elements from the underground areas to the surface through fracture zones may have resulted in the formation of anomalies; second, there might have been undiscovered uranium ore bodies beneath these points [38].

The results of the feasibility test indicate that the abnormal patterns and characteristics detected by the three methods effectively reflect the location of deep ore bodies. This demonstrates that geo-electrochemical integrated technology is both feasible and efficient for exploring concealed uranium deposits in this region [39].

4.2. Geology-Geoelectric Prospecting Model

4.2.1. Indicators in Geological Exploration

The Middle Devonian Xindu Formation (D2x) and Tangjiawan Formation (D2t) constitute the principal uranium-hosting strata in the study area. The F1 interlayer fracture zones and the secondary fractured zones—developed along their contact interface—serve as critical ore-bearing structures and primary migration pathways for mineralized fluids. Uranium mineralization bodies, strictly constrained by lithology and stratigraphic position, exhibit stratiform and lenticular geometries along these fractured zones. Three essential prospecting indicators for carbonaceous-siliceous-argillaceous uranium deposits in this region have been identified: ① clastic rocks enriched with phosphates, vanadates, and Fe-Mn constituents, intercalated with variegated carbonate and siliceous lithologies; ② interlayer compression fracture zones and cross-cutting fault zones; and ③ wall-rock alterations, including carbonatization, silicification, bleaching, and iron mineralization. These geological features collectively provide diagnostic criteria for carbonaceous-siliceous-pelitic uranium deposits exploration in the study area [40,41].

4.2.2. Geo-Electrochemical Extraction and Measurement Markers of Soil Ionic Conductivity and Thermally Released Mercury

The geo-electrochemical extraction survey revealed distinct elemental anomalies of U, Th, W, Mo, As, and Pb with local overlaps, effectively indicating the presence of deep-seated concealed uranium mineralization. Th, as a decay product of U, shares similar geochemical affinities with its parent element, while Mo demonstrates the most effective geochemical signature for carbonaceous-siliceous-argillaceous uranium mineralization [42]. Therefore, three critical prospecting indicators can be established: ① U-Th-Mo anomalies as direct exploration criteria; ② W-As-Pb anomalies exhibiting strong correlation with U as indirect indicators; and ③ coincident lambda-shaped soil ionic conductivity anomalies and rabbit-ear-shaped thermally released mercury anomalies serving as significant structural-geochemical markers (Figure 6) [43].

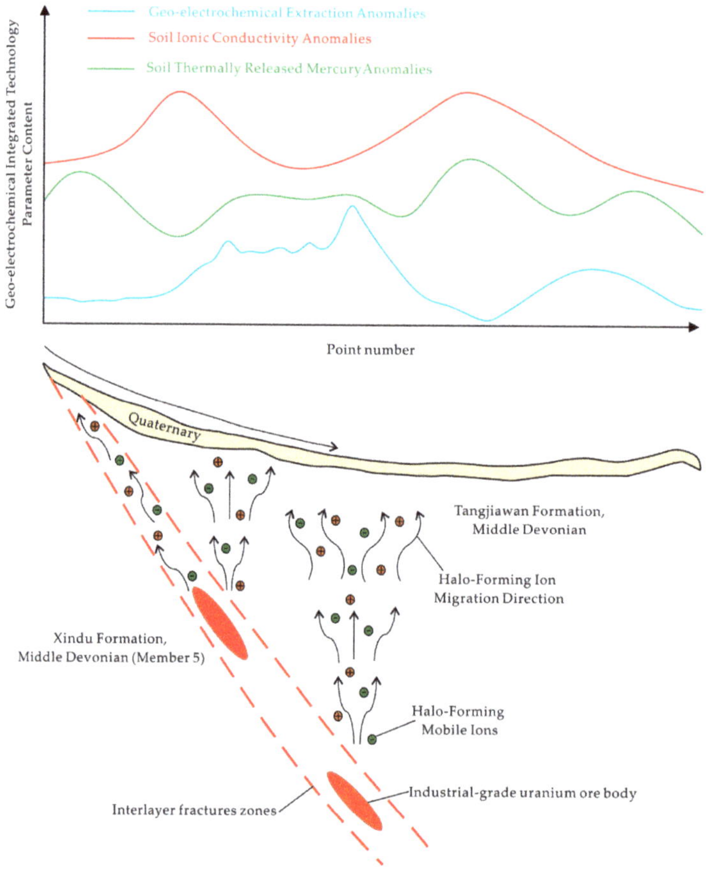

Figure 6. Schematic diagram of the geological-geo-electrochemical prospecting model in the study area.

5. Research on Mineral Prospecting Prediction

5.1. Statistical Analysis of Parameters

To study the effectiveness of the proposed method, geo-electrochemical integrated technology was employed to conduct mineral exploration predictions in the study area. The content data of the six obtained elements were statistically analyzed using GeoChem Studio 2.5.9 and SPSS 27.0 software (Table 2). Through examination of the skewness and kurtosis values of each element, it was evident that none of the elements followed a normal distribution. Consequently, a logarithmic transformation was applied to the original data. As shown in the box plot (Figure 7), although outliers remained present in Th, W, Mo, and As after logarithmic transformation, the mean values of these elements were closer to their medians, and they were centrally located within the box. This indicates the objectivity of the statistical analysis, suggesting that "the data of major elements approximate a normal distribution, while the logarithmic-transformed data of trace elements conform to a normal distribution" [36].

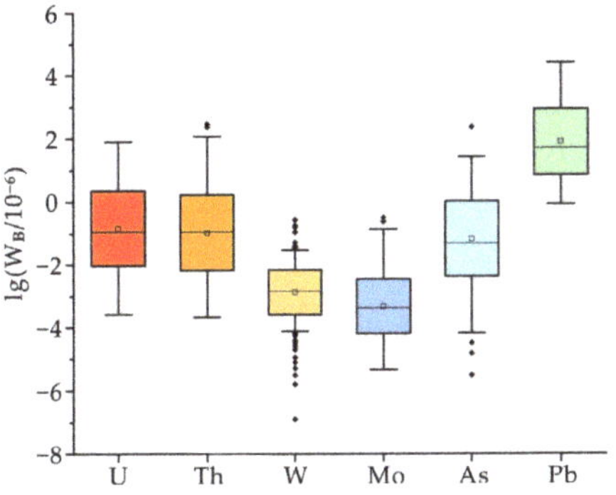

Figure 7. Box diagram of elemental content after logarithmic conversion.

Table 2. Statistical parameters of geo-electric chemistry element contents.

Data Category	Parameters	U	Th	W	Mo	As	Pb
Original data	Sample Size N_1	462	462	462	462	462	462
	Maximum	6.72	11.72	0.56	0.60	10.58	82.47
	Minimum	0.02	0.02	0.01	0.01	0.01	0.91
	Average	0.89	0.76	0.07	0.05	0.60	11.58
	Standard Deviation	1.24	1.21	0.06	0.06	0.83	13.53
	Skewness	2.55	4.74	4.57	4.52	4.89	1.89
	Kurtosis	6.66	30.54	28.99	30.18	46.90	3.40
	Coefficient of Variation CV_1	1.39	1.59	0.85	1.17	1.38	1.17
Eliminate the data with exceptional values	Sample Size N_2	393	405	441	433	378	338
	Average	0.45	0.43	0.06	0.04	0.30	4.65
	Standard Deviation	0.39	0.36	0.03	0.03	0.24	2.82
	Coefficient of Variation CV_2	0.87	0.83	0.47	0.74	0.82	0.61
CV_1/CV_2		1.60	1.92	1.81	1.58	1.68	1.92

Note: The unit for element content is μg/g.

The coefficient of variation (CV_1) of the original data serves as an indicator of element dispersion; however, it may occasionally be susceptible to interference from high background values or non-mineralization factors, such as stratigraphy, weathering, and diagenetic modifications. Such interference can obscure the distinction between mineralization anomalies and uniformly distributed high backgrounds. Therefore, the ratio of the coefficient of variation of the original data (CV_1) to that of the background data (CV_2) provides a quantitative measure of the specific contribution to dispersion caused by mineralization. As shown in Figure 8, the CV_1/CV_2 ratios of the primary mineralization element U, its decay product Th, and associated elements W, Mo, As, and Pb are all greater than 1. This indicates that the high-value anomaly of U in the study area is primarily attributable to mineralization or hydrothermal activity. Moreover, this anomaly not only reflects the intensity of mineralization but also exhibits synergistic behavior with elements such as Th, Mo, As, and Pb during migration and precipitation processes.

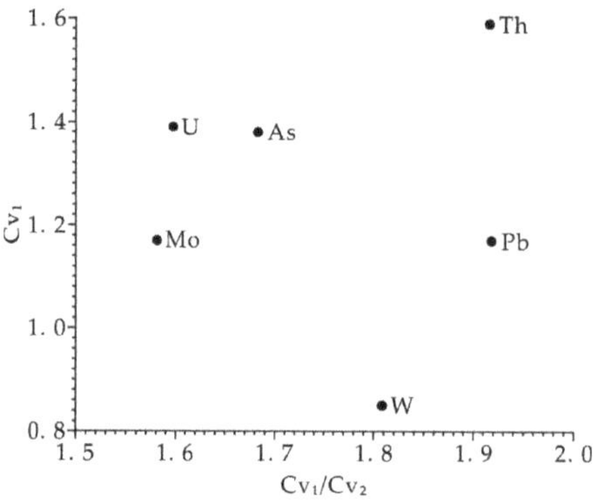

Figure 8. Degree of element dispersion in study area.

5.2. Planar Anomaly Characteristics of Geoelectrochemical Extraction Elements

In total, 462 sampling sites were stratified into subregions according to land-use types (mountainous areas, orchards, and paddy fields) within the study area. The boxplot methodology was systematically applied to iteratively eliminate single-element outliers within each subregion until normal distribution compliance was achieved, thereby establishing representative mean "$\bar{x}$" values. Subregion-processed datasets were generated by calculating the ratio of raw-element concentrations to their respective subregional means. These normalized values were then ranked in ascending order, with cumulative frequency thresholds of 85%, 92%, and 98%, designated as the lower limits of outer, intermediate, and inner geochemical anomaly zones, respectively (Table 3).

Table 3. Elemental anomaly zoning values in the study area.

Element	Original Data						Data After Partition Processing					
	Maximum	Minimum	Average	85%	92%	98%	Maximum	Minimum	Average	85%	92%	98%
U	6.72	0.03	0.89	1.59	2.16	3.12	31.74	0.10	2.96	5.43	7.42	10.13
Th	11.72	0.03	0.76	1.42	1.99	3.30	36.52	0.05	1.86	3.34	4.64	6.93
W	0.56	0.00	0.07	0.09	0.10	0.18	10.83	0.02	1.22	1.52	1.85	3.31
Mo	0.60	0.01	0.05	0.08	0.10	0.14	17.71	0.14	1.47	2.06	2.54	3.77
As	10.58	0.00	0.60	1.09	1.51	1.96	37.32	0.04	1.77	2.82	3.67	6.37
Pb	82.48	0.92	11.58	20.92	27.14	34.92	16.75	0.08	1.60	2.54	3.24	5.20

Note: The unit for element content is μg/g.

It can be observed from the single-element anomaly planar graph (Figure 9) that all six elements exhibit well-defined concentration centers, and the third-level anomaly zones are distinctly delineated. Consequently, in conjunction with the geological conditions of the study area, an analysis of the elemental abnormal characteristics was conducted.

The anomaly of element U is situated centrally within the study area, covering an area of 0.15 km^2 with a peak value of 31.74×10^{-6}. The anomalous region primarily corresponds to Quaternary sediments. This area is traversed by three northeast-oriented faults and one northwest-oriented fault. Notably, the concentrated centers in the eastern portion of the anomaly exhibit a beaded distribution pattern, which aligns well with the inferred orientation of the faults. As the primary ore-forming element, U undoubtedly serves as a reliable indicator of uranium ore bodies. However, considering the regional geochemical characteristics, it is evident that the western part of the study area comprises the Yuechengling granite body, where U contents in the uranium-bearing layer are typically lower than those in moderately acidic granite bodies [44]. Consequently, due to the influence of high-uranium granite bodies, the indicative significance of the U anomaly in this study area may not be singular.

The anomalous distributions of the As and Pb elements exhibit similarities and extend in a north-to-south orientation. The main body of the anomaly is situated in the central part of the study area. The anomalous region of As covers an area of 0.13 km^2, with a peak value of 37.32×10^{-6}. The concentration center is located at the intersection of inferred NE-trending and NW-trending faults, and the anomaly remains open to both the east and west regions of the study area. As migrates in hydrothermal fluids along faults in the form of As-S complexes (e.g., $AsS_3{}^{3-}$) accompanied by U^{6+}. Upon encountering reducing barriers, such as organic matter and pyrite, within carbon-silica mudstones, U^{6+} is reduced to U^{4+}, resulting in the precipitation of uranium minerals. $AsS_3{}^{3-}$ either combines with Fe to form arsenopyrite or adsorbs onto the surface of pyrite. Common uranium ore assemblages in the study area include pitchblende-arsenopyrite-galena-pyrite associations. Consequently, the anomalous distribution of As serves as an effective indicator of uranium mineralization. The anomalous region of Pb spans an area of 0.13 km^2, with a peak value of 16.75×10^{-6}. The long axis of the anomaly trends in a north-northeast direction, and its concentration center is inferred to align along northeast-distributed faults in a beaded pattern. Pb functions as a comprehensive tracer of hydrothermal activities, uranium migration and enrichment, and mineralization ages [45]. The geochemical behavior of deposits is closely linked to the activation-precipitation process of uranium. A significant positive correlation exists between Pb and U in ores, making Pb a key associated element for uranium exploration [46].

The Th element anomaly is situated at the intersection of the central structure within the study area. Its long axis extends in the northwest direction. The anomaly covers an area of 0.11 km^2, with a peak value of 36.52×10^{-6}. Measurement lines L01 and L04 traverse the inner, middle, and outer zones of the anomalies induced by high localized values. As a decay product of the U element, the Th element anomaly primarily corresponds to moderately acidic magmatic rock masses and alkaline rock bodies associated with multi-stage uranium-bearing activities [47]. Furthermore, granite and carbon-silicone-mudstone uranium mineralization exhibits a close relationship with the spatial distribution of the Th element anomaly.

The W element anomaly is predominantly concentrated in survey lines L01 and L04. Extending along the northwest direction, faults are distributed in a linear pattern, with an anomalous area of 0.06 km^2 and a peak anomaly value of 10.83×10^{-6}. Based on the known section of line L00, lens-shaped scheelite occurs within the uranium-bearing layer of the study area. These scheelite deposits exhibit a close genetic and spatial relationship with uranium orebodies. According to previous studies, pitchblende has been found to contain relatively high concentrations of W. Under microscopic observation, white tungsten ore primarily occurs as net-like veins interspersed within pitchblende. Its strong depletion of heavy rare earth elements (HREEs) aligns with the characteristics of the northern Dushiling white tungsten deposit; however, the occurrence of uranium mineralization is significantly later than that of regional tungsten mineralization. These features suggest that tungsten source layers that were formed early may provide some ore-forming materials for subsequent uranium mineralization [2].

The Mo element anomaly is predominantly situated in the northern and central regions of the study area, exhibiting a band-like and bead-like distribution along structural trends. The anomalous area spans 0.11 km^2, with a peak value of 17.71×10^{-6}. For carbon-silicon-mudstone uranium deposits, the Mo element often demonstrates superior indicative properties compared to the U element. In shallow continental shelf sediments, abundant organic matter, phosphorus, iron, and manganese oxides are present, which can reduce multivalent elements such as molybdenum, uranium, and vanadium to lower-valence ions through reactions with hydrogen sulfide, thereby enriching them [42,48]. Consequently, when exploring concealed carbon-silicon-mudstone uranium deposits within the study area and utilizing multi-element combination anomalies such as U-Th-W-Mo-As-Pb while paying particular attention to the nesting sites of each element, this approach proves to be more effective than relying solely on the single-element anomaly of U.

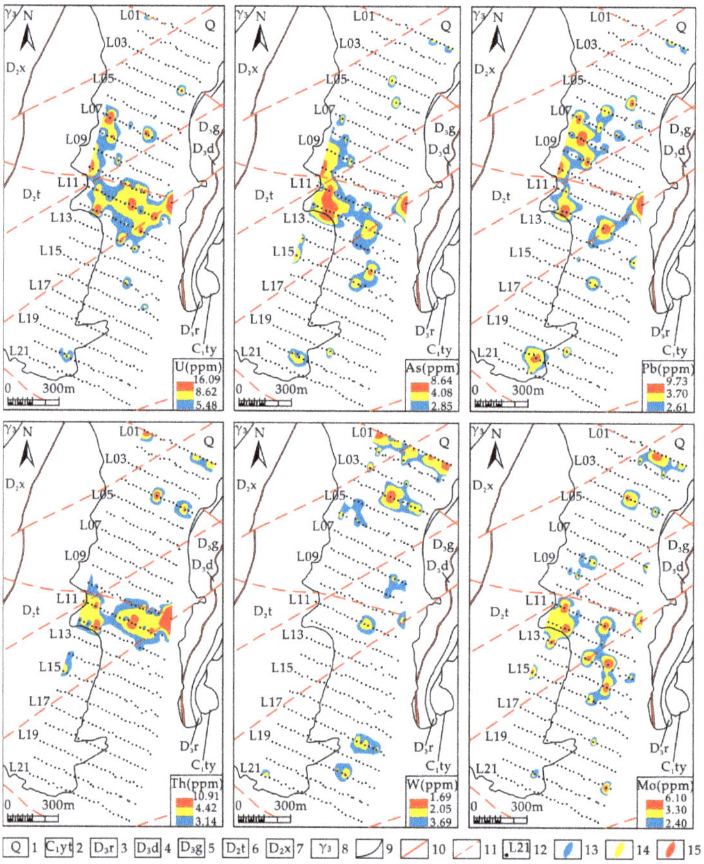

Figure 9. Single-element anomaly map of geoelectrical extraction in the study area. 1—Quaternary; 2—Yingtang Formation, Lower Carboniferous; 3—Rongxian Formation, Upper Devonian; 4—Dongcun Formation, Upper Devonian; 5—Guilin Formation, Upper Devonian; 6—Tangjiawan Formation, Middle Devonian; 7—Xindu Formation, Middle Devonian; 8—Caledonian granite; 9—stratigraphic boundary; 10—fault; 11—inferred fault; 12—sampling location and exploration line number; 13—outer anomaly zone; 14—middle anomaly zone; 15—inner anomaly zone.

5.3. Abnormal Characteristics of Soil Ionic Conductivity and Thermally Released Soil Mercury

The anomalies in soil ionic conductivity and thermally released soil mercury were distributed across the study area from the north to the south. The anomaly's zoning was distinct, exhibiting single-point, beaded, band-like, and irregular planar distributions (Figure 10). The anomalous area of soil ionic conductivity measured 0.16 km^2, with a peak abnormal value of 5.22 μS/cm. These findings exhibited certain discrepancies compared to the anomalies extracted using in situ electrochemistry. At the convergence sites of each element's anomalies, the conductivity anomalies were not pronounced. This could potentially be attributed to the uranium ore being situated near the geochemical barrier within the oxidation-reduction transition zone [49]. Additionally, there is an oxidation-reduction environmental gradient extending from the north to the south in the study area. In the deep oxidation zone, the low mineralization degree of interlayer water results in high soil ionic conductivity, thereby causing the observed conductivity anomalies. Conversely, in the deep reduction zone, the interlayer water exhibits a high mineralization degree while the soil ionic conductivity remains low, resulting in no apparent anomalies. This phenomenon arises due to variations in the types of ions contributing to conductivity between the deep interlayer water and surface soil water [29].

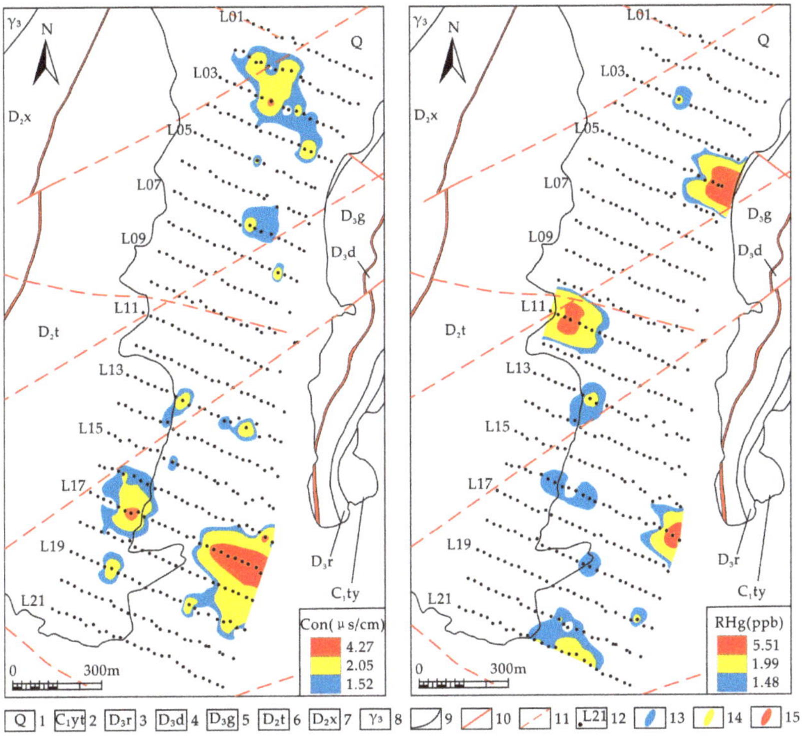

Figure 10. Anomaly map of soil ionic conductivity and thermally released mercury in the study area. 1—Quaternary; 2—Yingtang Formation, Lower Carboniferous; 3—Rongxian Formation, Upper Devonian; 4—Dongcun Formation, Upper Devonian; 5—Guilin Formation, Upper Devonian; 6—Tangjiawan Formation, Middle Devonian; 7—Xindu Formation, Middle Devonian; 8—Caledonian granite; 9—stratigraphic boundary; 10—fault; 11—inferred fault; 12—sampling location and exploration line number; 13—outer anomaly zone; 14—middle anomaly zone; 15—inner anomaly zone.

The anomalous area of thermally released soil mercury is 0.11 km^2, with a peak value of 49.39×10^{-9}. The anomaly is relatively large at the intersection of the L11 line's structure, and it matches well with the anomalies of various elements extracted via geo-electrochemistry. Due to the strong penetrating power of mercury, the tendency of mercury vapor to rise along the structural fracture zone, and the control of uranium deposits by interlayer fractures zones [34,50], it is inferred that the two open anomalies located at L05 and L16 lines, as well as the string-of-beads anomalies distributed along the southeast direction at L13–21 lines, may indicate the existence of concealed fault fracture zones, which also have good potential for mineral exploration.

5.4. Target Area Delineation and Engineering Validation

5.4.1. Demarcation of the Target Area

Based on the metallogenic geological conditions and the coincidence of anomalies, including primary metallogenic elements, associated elemental anomalies, soil thermometric mercury, and ionic conductivity anomalies in the study area, four integrated anomaly zones and one priority anomaly verification area were delineated. These consist of one Class A integrated anomaly zone, two Class B integrated anomaly zones, and one Class C integrated anomaly zone (Figure 11).

The A-1 integrated anomaly zone, situated in the central part of the study area, comprises mono-elemental anomalies (U, As, Pb, Th, W, and Mo) and soil thermometric mercury anomalies, covering a total anomalous area of 0.4 km^2. Geo-electrochemical extraction reveals the large-scale, high-intensity, and well-zoned anomalies of primary metallogenic and associated elements. Characterized by favorable metallogenic geological conditions, this anomaly zone is jointly controlled by two NE-trending inferred faults and one NW-trending fault. It incorporates high-temperature to low-temperature elemental associations, which are indicative of multiphase and multistage mineralization processes in the region. As the most prospective target for concealed uranium mineralization, the No. One priority anomaly verification area was delineated within the A-1 zone and prioritized for drilling verification.

The B-1 integrated anomaly zone, situated in the northern sector of the study area, consists of mono-elemental anomalies (As, Pb, Th, W, and Mo), with a total anomalous coverage of 0.08 km^2. The associated elemental anomalies exhibit moderate spatial extent and relatively high intensities, demonstrating relatively complete zonation patterns. Notably, the tungsten (W) anomaly displays a belt-shaped distribution along an inferred NW-trending fault, establishing this zone as a preferred target area for scheelite exploration. Comprehensive analysis suggests modest mineral exploration potential in this zone.

The B-2 integrated anomaly zone, situated in the northern sector of the study area, incorporates mono-elemental anomalies (U, As, Pb, Th, W, and Mo) coupled with soil ionic conductivity and thermometric mercury anomalies, exhibiting a composite anomalous area of 0.08 km^2. The associated elemental anomalies demonstrate moderate spatial extent with relatively high intensity, showing relatively complete zonation patterns. The ionic conductivity of the soil and the thermally released soil mercury in this region exhibit significant anomalies. It is hypothesized that concealed fractures and fragmentation zones may exist within the concealed fault fracture zones. In conjunction with the B-1 comprehensive anomaly area, utilizing a combination of trenching, drilling, and other exposure techniques is recommended to investigate continuous high-value-point areas and high-value connection zones between adjacent sampling lines. This approach will aid in the detailed analysis and evaluation of the two comprehensive anomaly regions.

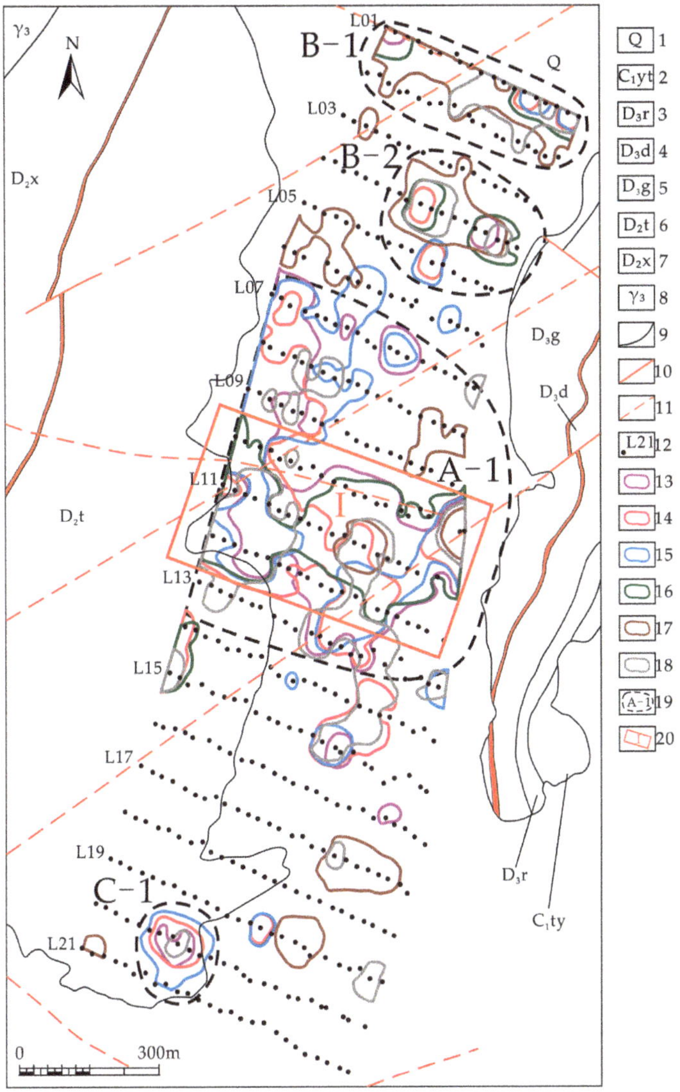

Figure 11. Comprehensive anomaly map of the study area. 1—Quaternary; 2—Yingtang Formation, Lower Carboniferous; 3—Rongxian Formation, Upper Devonian; 4—Dongcun Formation, Upper Devonian; 5—Guilin Formation, Upper Devonian; 6—Tangjiawan Formation, Middle Devonian; 7—Xindu Formation, Middle Devonian; 8—Caledonian granite; 9—stratigraphic boundary; 10—fault; 11—inferred fault; 12—sampling location and exploration line number; 13—U element anomaly; 14—As element anomaly; 15—Pb element anomaly; 16—Th element anomaly; 17—W element anomaly; 18—Mo element anomaly; 19—comprehensive anomaly area and number; 20—key anomaly verification area and number.

The C-1 integrated anomaly zone, situated in the southern sector of the study area, comprises mono-elemental anomalies (U, As, Pb, and Mo), with a composite anomalous area of 0.03 km^2. The primary metallogenic and associated elemental anomalies exhibit limited spatial extent and moderate continuity, demonstrating poor spatial correlations with soil ionic conductivity and thermometric mercury anomalies. No surface expressions of fault structures or ore-hosting strata outcrops were observed. However, the U-As-Pb-Mo element combination anomaly demonstrates good consistency in this region. Targeted geological surveys are recommended for conducting an initial assessment of the anomalies and mineralization-related alterations. Favorable locations for mineralization can then be identified for engineering deployment.

5.4.2. Engineering Validation

Through the geo-electrochemical extraction of element anomalies and assessments of mineralization potential, drilling verification was conducted in the No. One priority anomaly verification area. Drilling projects were successively arranged at points 2, 4, 8, and 12 on the L11 line, where each element showed a single-peak anomaly area, namely, ZK192-11, ZK190-7, ZK192-1, and ZK192-17 (Table 4). An industrial uranium orebody was discovered within the F1 interlayer compressional fracture zone. The orebody exhibits a lenticular morphology, with local occurrences of graphitization, hematitization, and pyritization. The ore body is controlled by drilling a length of 225 m, with a thickness ranging from 0.81 to 9.96 m, an average thickness of 6.67 m, a grade ranging from 0.05% to 1.32%, and an average grade of 0.19%. The ore types can be classified as fractured brecciated dolomite and argillaceous siltstone. The dolomite-type ore exhibits a dark gray color and massive structure, with well-developed hematitization and pyritization alterations. Calcite veinlets were observed to be distributed along cleavage fractures (Figure 12a). Petrographically, the ore is primarily composed of dolomite and calcite. Microscopic examination under alizarin red staining reveals dolomite grains ranging from 0.03 to 0.06 mm in size, accounting for 70–75% of the rock volume, while calcite and other minerals constitute 25–30%. A porphyroclastic to cataclastic texture is evident, characterized by mineral clasts set within a finer-grained matrix. The clasts show no significant displacement and display mutually adapted boundaries. Fractures are filled with comminuted host rock material and calcite (Figure 12c). Based on the backscattered electron images of the dolomite-type ores (not collected during this study but from the same Guangzitian area), uraninite occurs as cryptocrystalline aggregates displaying disseminated and veinlet textures. The veinlets range from 0.02 to 0.15 mm in width. Constituent particles and devitrified microcrystals range from tens to hundreds of nanometers in scale. Vein-type uraninite is closely associated with colloform pyrite and galena (Figure 12e). The siltstone-type ore exhibits grayish-black to grayish-green coloration (Figure 12b) and a fine silt texture. It is primarily composed of quartz clasts within an argillaceous cement. Quartz clasts range from 0.005 to 0.01 mm in size, are subrounded, well-rounded, and moderately sorted. The clay matrix is relatively minor in abundance (Figure 12d). According to the backscattered electron images of the siltstone-type ores (not collected in this study but from the same Guangzitian area), it was revealed that uraninite occurs as disseminated and colloform masses within intergranular spaces, with a particle size of 0.02–0.06 mm Galena is relatively abundant in the ore, primarily exhibiting colloform or veinlet textures in close association with uraninite. Certain galena grains occur as inclusion-like forms enveloped by uraninite (Figure 12f).

Based on the above analysis, in key anomaly verification area No. One, the ore body corresponds well with the anomaly (Figure 13), and it has considerable potential for mineral exploration. Drilling projects can be arranged from the west to the east based on the characteristics of the geo-electrochemical extraction profile and planar anomalies to control the deep ore body. Similarly, the A-1, B-1, and B-2 comprehensive anomaly zones also have good prospects for mineral exploration. Conducting more in-depth research and engineering arrangements for these areas is recommended.

Table 4. Borehole parameters for Line L11.

Parameter	ZK192-11	ZK190-7	ZK192-1	ZK192-17
Elevation (m)	258.10	246.65	231.44	228.32
Drilling Depth (m)	458.92	514.68	643.87	781.45
Initial Borehole Inclination (°)	85.7	81.4	82.4	81.8
Depth of Ore Encounter (m)	442.08	498.89	636.71	360.80
True Thickness of Ore Body (m)	6.73	9.17	0.81	9.96
Ore Body Grade Range (%)	0.05–1.32			
Average Ore Body Grade (%)	0.186			

Figure 12. Images of the uranium ore hand specimen from borehole ZK192-17, identification photomicrographs of thin-polished sections, and in situ ore microscopic features. (**a**)—Hand specimen of grey cataclastic-micritic calcareous dolomite ore; (**b**)—hand specimen of fine-grained silty sandstone ore; (**c**)—cataclastic texture of the dolomite ore; (**d**)—fine-grained silt texture of the silty sandstone ore; (**e**)—veinlet pitchblende encrusting galena and pyrite within dolomite ore (reflected light) [24]; (**f**)—colloidal pitchblende coexisting with scheelite and galena in silty sandstone ore (backscattered electron image) [24]. Ptc: pitchblende; Py: pyrite; Gn: galena; Sh: scheelite.

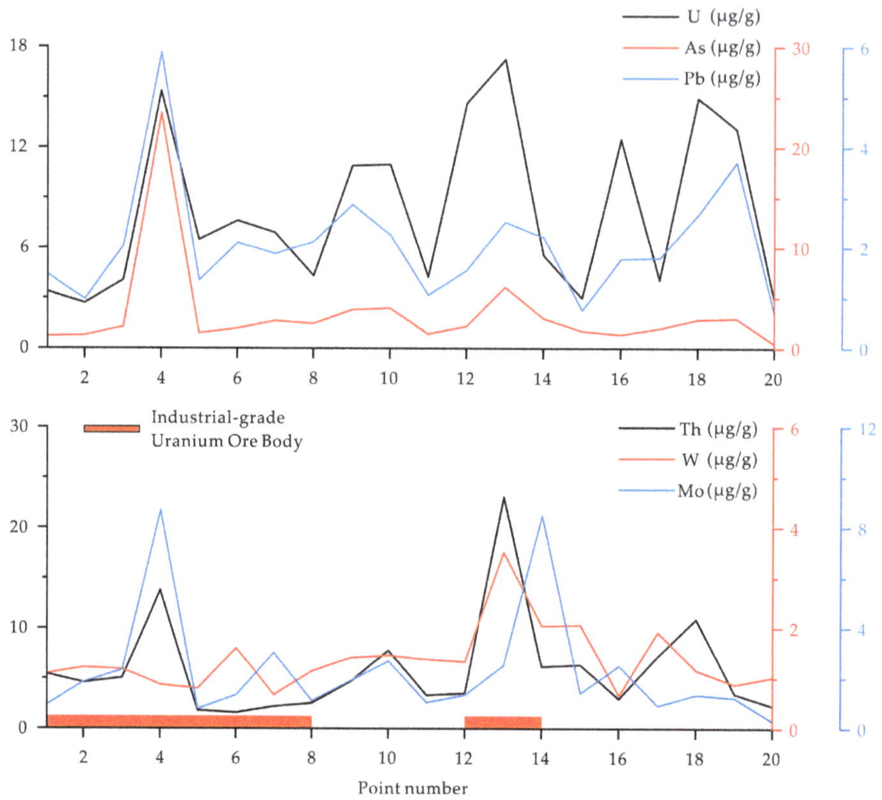

Figure 13. The comprehensive geological geodesic anomaly profile of Line 11.

6. Conclusions

(1) The research results regarding the effectiveness of geo-electrochemical integrated technology in mineral exploration along the L00 line demonstrate that this technology can accurately identify the locations of deep ore bodies. It is evident that a technical approach primarily based on geo-electrochemical extraction and measurement—complemented by soil ionic conductivity and thermally released soil mercury, is both feasible and effective in identifying concealed carbon-silicon-mudstone uranium minerals within the study area;

(2) Based on the geological characteristics, geochemical features, and genetic analysis of the deposit in the study area, the formation of carbonate rocks and siliceous rocks, interlayer compression fracture zones, cross-cutting fault zones, and surrounding rock alteration are identified as key geological indicators in prospecting. Abnormalities in primary ore-forming elements such as U, Th, and Mo serve as direct prospecting indicators, while W, As, and Pb function as indirect indicators. Additionally, soil ionic conductivity and thermally released soil mercury are considered significant prospecting indicators;

(3) Based on the convergence sites of U-Th-W-Mo-As-Pb anomalies and in conjunction with metallogenic geological conditions, one key anomaly verification area, one Class A comprehensive anomaly area, two Class B comprehensive anomaly areas, and one Class C comprehensive anomaly area were delineated. With respect to the Guangzitian deposit, this provides crucial technical support and a theoretical foundation for the principles of "exploring the edges and identifying the bottom, delving deep and uncovering blind spots" and "connecting points to form a surface" in the uranium mining areas in northern Guangxi.

Author Contributions: Writing—original draft preparation, X.Z. and Y.M.; Writing—review and editing, M.W. and Q.L.; Collecting the samples, X.Z., Y.M., Y.J. (Yuheng Jiang), W.Y., J.Z. and M.W.; Software and data processing, X.Z. and Y.J. (Yuxiong Jiang); Funding acquisition, M.W. and Q.L. All authors have read and agreed to the published version of the manuscript.

Funding: This research was funded by the National Natural Science Foundation of China (Grant No. 42163004) and the Guangxi Natural Science Foundation (2025GXNSFAA069927).

Institutional Review Board Statement: Not applicable.

Informed Consent Statement: Not applicable.

Data Availability Statement: Dataset available on request from the authors.

Conflicts of Interest: The authors declare no conflicts of interest.

References

1. Xu, D.R.; Chi, G.X. Diversity of uranium deposits in China—An introduction to the special issue. *Ore Geol. Rev.* **2021**, *129*, 103944. [CrossRef]
2. Xiao, W.; Fan, H.H. Genesis of the Guangzitian uranium—Tungsten deposit in northern Guangxi—Geochemical evidences from in-situ U-Pb dating of pitchblende and mineral geochemistry. *Geol. Rev.* **2022**, *68*, 831–844.
3. Cheng, Q. Singularity theory and methods for mapping geochemical anomalies caused by buried sources and for predicting undiscovered mineral deposits in covered areas. *J. Geochem. Explor.* **2012**, *122*, 55–70. [CrossRef]
4. Liang, M.; Gao, W. Soil geochemical characteristics and ore prospecting prediction in Gaojianzi area, northern Hebei province. *Geoscience* **2023**, *37*, 1567–1579.
5. Liu, P.F.; Wen, M.L. The application of geo-electrochemical integrated technology to the prospecting in the Xiyi Pb-Zn deposit, Yunnan province. *Geophys. Geochem. Explor.* **2016**, *40*, 655–660.
6. Liu, P.F.; Luo, X.R. Geo-electrochemical anomaly prospecting for uranium deposits in southeastern China. *Appl. Geochem.* **2018**, *97*, 226–237. [CrossRef]
7. Qiu, W.; Luo, X.R. Comparative study between the geo-electrochemical method and other geo-chemical exploration methods. *Gold Sci. Technol.* **2011**, *19*, 7–11.
8. Gao, Y.; Luo, X.R. Geo-electrochemical extraction method for concealed gold deposits prospecting in high-altitude permafrost regions: A case study from the Dagou area, Golmud, Qinghai province, China. *Guilin Univ. Technol.* **2017**, *37*, 274–279.
9. Li, S.; Sun, B.B. 1:50 000 electro-geochemical survey in the Luokedun lead-zinc polymetallic deposit, Inner Mongolia. *Geophys. Geochem. Explor.* **2020**, *44*, 514–522.
10. Kang, M.; Luo, X.R. Improvement and applied results of geoelectrical chemistry methods. *Geol. Explor.* **2003**, *5*, 63–66.
11. Luo, X.R.; Yang, X. Research on the detection of concealed ore deposits via geo-electrochemical measurements and preliminary surveys for mineral exploration. *Geol. Explor.* **1989**, *12*, 43–51.
12. Liu, P.F.; Luo, X.R. Main study progress and prospect of geo-electrochemistry technology in China. *Met. Mine* **2022**, *11*, 10–22.
13. Hou, D.M.; Luo, X.R. A comparative study of the prospecting for hidden uranium deposits by applying geo-electrochemical method in China and Australia. *Geol. Bull. China* **2012**, *31*, 126–135.
14. Man, R.H.; Luo, X.R. Application of the geo-electrochemical method on prospecting for concealed uranium deposits in the Dongsheng area of the Ordos Basin. *Bull. Mineral. Petrol. Geochem.* **2015**, *34*, 1007–1013.
15. Huang, N.N. The Research of Using Geo-Eletrochemical Integration Technology to Concealed Uranium Deposits in Dachengliang Area of Inner Mongolia. Master's Thesis, Guilin University of Technology, Guilin, China, April 2017.
16. Wen, M.L.; Luo, X.R. Electro-geochemical method in search of concealed uranium deposits in a area of Jiangxi province. *Mineral. Explor.* **2011**, *2*, 404–408.
17. Ouyang, F.; Luo, X.R. Characteristics of geo-electric extraction anomalies and prospecting prediction in Lejia survey area, Xiangshan orefield, Jiangxi province. *Met. Mine* **2022**, *11*, 154–166.
18. Wen, X.Q.; Luo, X.R. New method study in search of uranium deposit. *Guilin Univ. Technol.* **2003**, *1*, 26–30.
19. Zhang, J.; Wen, M.L. Prospecting prediction using geoelectric extraction in the periphery of the No.374 deposit in Rongshui, Guangxi. *Geoscience* **2024**, *38*, 214–223.
20. Zhao, X.Y.; Ouyang, F. Application of geo-electrochemical methods to prospecting in the concealed uranium: A case study of Lujing area, Zhuguang. *Met. Mine* **2022**, *11*, 167–178.
21. Tang, B. Research on Mineralization Mineralogical Characteristics and Metallogenic Mechanism of Guangzitian Deposits. Master's Thesis, East China University of Technology, Nanchang, China, June 2017.

22. He, Y.K.; Zhang, D.H. Fluid geochemistry and its mineralization significance of the uranium ore deposit in Quanzhou, Guangxi province. *Geol. Explor.* **2010**, *46*, 670–680.
23. Zhang, L.; Wang, F.Y. Metallogenesis of uranium deposits in China: Perspectives from uraninite chemistry. *Ore Geol. Rev.* **2024**, *173*, 106251. [CrossRef]
24. Xiao, W. Mineralization of Uranium Deposits in the Quanzhou Ore Field in Guangxi Province. Master's Thesis, China University of Geosciences, Beijing, China, May 2015.
25. Luo, X.R. The Mechanism of Electro-Geochemical Halo-Formation and the Application of Electrochemical Method to Exploration of Metallic Ore Deposits. Ph.D. Thesis, Hefei University of Technology, Hefei, China, May 2005.
26. Liu, P.F.; Wen, M.L. Geo-electrochemical extraction technology test and the prospecting prediction in cold plateau area. *Geol. China* **2015**, *42*, 2000–2007.
27. Luo, X.R. Research and application effectiveness of novel methods in concealed deposit exploration. *Geol. Explor.* **1995**, *1*, 44–49.
28. Wen, M.L.; Ouyang, F. Anomaly characteristics, components of soil ionic conductivity and prospecting prediction of tin deposits in Pinggui area, Guangxi. *Geol. Bull. China* **2013**, *32*, 784–789.
29. Hou, L.H.; Tan, K.R. Experimental study on soil ionic conductivity measurement in prospecting and exploration of in-situ leachable sandstone-type uranium deposits. *Geotecton. Metallog.* **2002**, *4*, 448–451.
30. Luo, X.R.; Zhou, T.F. Ionic conductivity anomaly characteristics and ionic components of soil in gold deposits in the Daxin'anlin forest overburden region. *Geol. Explor.* **2005**, *2*, 46–50.
31. Zheng, C.J.; Luo, X.R. Comparison of conventional-warming RHg mercury release of soil with ladder-warming RHg mercury release, and ore prediction in Jielagepu au deposit, southern Tibet. *Mineral. Petrol.* **2018**, *38*, 99–110.
32. Wei, J.S. Application and discussion of geochemical mercury measurement methods. *Geophys. Geochem. Explor.* **1999**, *1*, 22–28.
33. Su, H.Y.; Luo, X.R. Application of soil heat-released mercury survey in prospecting prediction of copper-nickel polymetallic deposits in western Mianluening, Shaanxi province. *Min. Explor.* **2021**, *12*, 338–347.
34. Zhou, Z.Y.; Luo, X.R. Measurement of heat releasing hg of soil and prospecting prediction in Shilongshan polymetallic gold deposit, Luoning county of Henan province. *Mineral. Petrol.* **2018**, *38*, 49–58.
35. Liu, P.F.; Luo, X.R. Retrospect and prospect for geo-electrochemical technology research in the past three decades of China. *J. Guilin Univ. Technol.* **2018**, *38*, 47–55.
36. Tang, R.; Ouyang, F. The prediction of electrochemical prospecting in Youfang area of the Xiangshan orefield. *Geophys. Geochem. Explor.* **2021**, *45*, 1425–1438.
37. Shi, Y.H.; Yang, Z.P. Determination of trace elements in electroadsorption mineral exploration foam plastic samples by ICP-MS. *Spectrosc. Spectr. Anal.* **2009**, *29*, 1687–1690.
38. Wang, D. Study on Searching for Concealed Uranium Deposits by Geo-Electrochemical Survey in Lejia Area, Xiangshan, Jiangxi Province. Master's Thesis, Guilin University of Technology, Guilin, China, March 2020.
39. Lu, C.Y.; Ouyang, F. Application of geo-electrochemical extraction technology in Chuandong survey area of Zhugongtang lead-zinc mine in Guizhou [J/OL]. *Min. Explor.* **2024**, 1–17.
40. Chen, J.Y.; Bai, Q. Discussion on the prospecting directions of carbonaceous-siliceous-argillaceous type uranium deposit in Hongliuhe-Fangshankou area, Gansu. *Uranium Geol.* **2020**, *36*, 500–509.
41. Zhong, F.J.; Pan, J.Y. Exploration targeting model for carbonaceous-siliceous-argillaceous (C-SA) type uranium deposits in China. *Acta Mineral. Sin.* **2015**, *35* (Suppl. S1), 372–373.
42. Fu, J.; Zhao, N.B. Indicator characteristics of elementary geochemistry and anomaly model for Chinese carbonaceous-siliceous-argillaceous rock type uranium deposit. *Uranium Geol.* **2014**, *30*, 298–304.
43. Wen, M.L.; Luo, X.R. A study of the ore-prospecting work based on multiple geosciences information in the Jinchuan Cu-Ni deposit. *Geol. China* **2013**, *40*, 594–601.
44. Fu, J.; Zhao, N.B. Characteristics of indicator elements combination and genesis of four types of uranium mineralization in China. *Geophys. Geochem. Explor.* **2015**, *39*, 217–221.
45. Song, H.; Xu, Z.Q. Geochemistry and LA-ICP-MS zircon U-Pb geochronological dating of diabase dykes and their relationship with mineralization of the carbonate-siliceous-pelitic rock type uranium deposits in Daxin-Qinjia, western Guangxi. *Acta Petrol. Sin.* **2019**, *35*, 2845–2863.
46. Ye, Y.Q.; Chen, Y.L. Vertical zoning characteristics of trace element in gangue mineral of uranium deposit 510-1 in Zoige. *Uranium Geol.* **2015**, *31*, 510–516.
47. Fu, J.; Zhao, N.B. Geochemical characteristics of uranium, thorium and potassium anomalies in China in relation to uranium mineralization. *Geophys. Geochem. Explor.* **2014**, *38*, 200–204.
48. Fu, J.; Zhao, N.B. Geochemical anomaly characteristics of molybdenum in China and their indicative role for uranium mineralization. In Proceedings of the 2013 Annual Academic Conference of the Chinese Nuclear Society, Harbin, China, 11 September 2013.

49. Zhao, N.B.; Fu, J. Study on the anomaly extraction of soil metal active state in sandstone type uranium deposit. *Bull. Sci. Technol.* **2019**, *35*, 33–39.
50. Zou, D.F.; Yao, T. Experimental study on the application of soil thermorelease mercury methodin a gold mine area of Jiangxi province. *World Nonferr. Met.* **2020**, *18*, 65–67.

applied sciences

MDPI

Article

Occurrence State and Extraction of Lithium from Jinyinshan Clay-Type Lithium Deposit, Southern Hubei: Novel Blank Roasting–Acid Leaching Processes

Hao Zhang [1,2], Peng Li [1,2,*], Wensheng Zhang [3,*], Jiankang Li [1,2], Zhenyu Chen [2], Jin Yin [3], Yong Fang [2], Shuang Liu [4], Jian Kang [4] and Dan Zhu [4]

1 State Key Laboratory of Deep Earth and Mineral Exploration, Institute of Mineral Resources, Chinese Academy of Geological Sciences, Beijing 100037, China; zhanghao6576@126.com (H.Z.)
2 MNR Key Laboratory of Metallogeny and Mineral Resource Assessment, Institute of Mineral Resources, Chinese Academy of Geological Sciences, Beijing 100037, China; fangyong202110@163.com (Y.F.)
3 Fourth Geological Team of Hubei Geological Bureau, Xianning 437100, China
4 Hubei Geological Research Laboratory, Wuhan 430034, China
* Correspondence: lipeng031111@163.com (P.L.); zhangdizhi@126.com (W.Z.)

Abstract

Addressing the technological bottlenecks in the efficient utilization of clay-type Li deposits in China, this study systematically investigates Li occurrence states and develops clean extraction processes using the Jinyinshan clay-type Li deposit in southern Hubei as a case study. The research aims to provide technical guidance for subsequent geological exploration and development of such deposits. Analytical techniques, including AMICS, EPMA, and LA-ICP-MS, reveal that Li primarily occurs in structurally bound forms within cookeite (82.55% of total Li), illite (6.65%), and rectorite (5.20%), with mineral particle sizes concentrated in fine-grained fractions (<45 μm). Leveraging process mineralogical insights, two industrially adaptable blank roasting–acid leaching processes were innovatively developed. Process I employs a full flow of blank roasting–hydrochloric acid leaching–Li-Al separation–Ca/Mg removal–concentration for Li precipitation–three-stage counter-current washing. Optimizing roasting temperature (600 °C), hydrochloric acid concentration (18 wt%), and leaching parameters achieved a 92.37% Li leaching rate. Multi-step purification yielded lithium carbonate with >99% Li_2CO_3 purity and an overall Li recovery of 73.89%. Process II follows blank roasting–sulfuric acid leaching–Al removal via alum precipitation–Al/Fe removal–freeze crystallization for sodium sulfate removal–Ca/Mg removal–concentration for Li precipitation–three-stage counter-current washing. Parameter optimization and freezing impurity removal achieved an 89.11% Li leaching rate, producing lithium carbonate with >98.85% Li_2CO_3 content alongside by-products like crude sodium chloride and ammonium alum. Both processes enable resource utilization of Al-rich residues, with the hydrochloric acid-based method excelling in stability and the sulfuric acid-based approach offering superior by-product valorization potential. This low-energy, high-yield clean extraction system provides critical theoretical and technical foundations for scaling clay-type Li deposit utilization, advancing green Li extraction and industrial chain development.

Keywords: clay-type Li deposit; occurrence state; cookeite; blank roasting–acid leaching; Li extraction

1. Introduction

As a critical mineral resource for strategic emerging industries, lithium (Li) holds significant strategic importance and research value in fields such as medicine, agriculture, new energy, electronics, information technology, aerospace, and national defense [1,2]. Currently, China's Li resources are primarily concentrated in granite-type, pegmatite-type, salt-lake-type, and clay-type deposits [1,3,4]. Among these, salt-lake-type Li deposits are predominantly sulfate-rich and typically exhibit high Mg/Li ratios, resulting in complex and costly extraction processes. Traditional granite-type Li deposits often face challenges such as low grades and high development costs. Although pegmatite-type Li deposits remain the primary source of China's Li supply [1,3,5,6], their limited mineralization depth has constrained growth in new reserves in recent years. Meanwhile, intensified exploration efforts have led to the discovery of increasingly abundant clay-type Li deposits with high Li content and potential for independent development [7–11], positioning sedimentary clay-type Li deposits as a growing focal point.

Current extraction technologies for clay-type Li deposits primarily include direct leaching, roasting–leaching, additive-assisted roasting–leaching, and chlorination–sulfidation methods, with variations in techniques depending on genetic deposit types [12–16]. For instance, volcanic clay-type deposits often contain silicon (Si)–magnesium (Mg) clay minerals such as smectite, characterized by layered structures comprising silica tetrahedral and magnesia octahedral sheets. Li predominantly occurs as structurally bound substitutions within octahedral layers, complicating extraction processes [17]. Although direct leaching, additive-assisted roasting, and chlorination–sulfidation have been applied to such deposits, additive-assisted roasting remains the most prevalent method [18,19]. In carbonate clay-type deposits, Li is mainly adsorbed in smectite interlayers. Medium-temperature roasting alters the smectite structure, rendering Li exchangeable for leaching via hydrogen ions (sulfuric acid solution) or ferric ions. For this deposit type, previous studies employed inorganic trivalent ferric salt solutions to leach medium-temperature-roasted clay ore, achieving a Li leaching rate exceeding 80% under optimized conditions: roasting at 600 °C, 15 wt% $FeCl_3$ solution, liquid–solid (L/S) ratio of 5 mL/g, leaching temperature of 80 °C, and reaction duration of 240 min [19].

When utilizing clay-type Li resources to produce lithium carbonate, beyond leaching Li from the ore, it is essential to remove impurity ions from the Li-bearing leachate to ensure Li^+ concentration and purity meet production requirements. Current research on clay-type Li deposits remains limited, particularly regarding impurity removal and precipitation processes for Li-bearing leachates, with no documented literature on specialized methods. Consequently, fundamental principles and techniques largely draw from mainstream Li ore processing approaches. Mature industrial methods with relevance to clay-type Li deposits include the sulfuric acid method, sulfate method, and limestone roasting method [20–23]. For the sulfuric acid and sulfate methods, Li-bearing leachates undergo impurity removal via pH adjustment with lime, followed by further purification with sodium carbonate. The purified solution is then evaporated and concentrated before saturated sodium carbonate solution is added to precipitate qualified Li carbonate. These processes carry significant environmental burdens: sulfuric acid digestion of Li-bearing ore releases SO_2/SO_3 emissions (acid rain precursors), while sulfate roasting of lepidolite emits toxic HF/SiF_4 gases [14,19]. Both methods produce acidic residues containing leachable heavy metals (e.g., As, Cd), which significantly contribute to groundwater contamination risks. The limestone roasting method, the oldest technique for Li salt production, involves high-temperature roasting of Li-bearing minerals like spodumene or lepidolite with limestone to form alkali metal aluminates. Subsequent leaching, concentration, and purification yield Li carbonate. Despite its simple flow and raw materials, this method suffers from excessive limestone consumption, high waste residue generation, significant energy demands, and low Li recovery rates [7].

The ore properties of Li resources in the Jinyinshan deposit, southern Hubei, exhibit distinct differences from other domestic clay-type Li deposits [24,25]. This uniqueness provides a natural laboratory for comprehensive mineralogical studies and mineral processing-metallurgical extraction experiments on clay-type Li ores, offering critical insights for China's Li exploration and development. Through detailed mineralogical analysis of ore samples from the deposit, this study elucidates the mineral composition, chemical characteristics, and Li occurrence states in siltstone and slate, establishing a foundation for subsequent beneficiation and extraction. By optimizing leaching parameters and extraction pathways, efficient recovery and comprehensive utilization of clay-type Li ore were achieved. The two full-scale industrially adaptable blank roasting–acid leaching processes developed herein not only produce high-purity lithium carbonate but also address environmental impacts by minimizing energy consumption and pollution. This research provides novel pathways for effective resource utilization of clay-type Li deposits while contributing valuable references for environmental protection and sustainable development.

2. Regional Geological Setting

Jinyinshan deposit is situated in the northern Jiangnan Uplift Belt at the junction of the Yangtze and Cathaysia plates, belonging to the Puqi-Tongshan Cu-Au-Sb polymetallic metallogenic belt (Figure 1a). This deposit is located approximately 30 km at 220° azimuth from Chibi City and administratively falls under Zhaoliqiao Town. Hydrogeologically, it is classified as a simple-to-moderate fissure-watered type. Engineering geological conditions are moderate-to-good, and the geo-environmental quality is rated as medium [26]. Regionally exposed strata consist predominantly of the Mesoproterozoic Qingbaikou System Lengjiaxi Group to the Mesozoic Cretaceous System, exhibiting essentially continuous outcrops. The basement comprises regionally shallow metamorphic rocks, while the cover sequence includes carbonate rocks, siliceous rocks, and clastic rocks (Figure 1b). Major regional dike rocks include lamprophyre, spessartite, minette, and quartz veins. South of the study area lies the Mesozoic Yanshanian Mufushan batholith, primarily composed of biotite monzogranite and two-mica monzogranite, representing typical S-type granites [27–29]. This batholith exhibits distinct specialization for rare metal mineralization (Li, Be, Nb, Ta).

The region has undergone multiple phases of tectonic movement and deformation, resulting in complex structural features (Figure 1b). Fangshan overturned anticline extends along a NEE-SWW trend with a "7"-shaped configuration, plunging eastward. Its core exposes shallow metamorphic rocks of the Lengjiaxi Group, while the limbs consist of Sinian-Cambrian strata. The northern limb exhibits overturned bedding, dipping 170–180° with steep dips of 60–70°; the southeastern limb maintains normal bedding, dipping 130–140° at angles of 30–50°. The nearly E-W trending F9 fault extends ~9 km from Chongyang Huangdinao in the east, through Zheping, to Wanjialing at the Hunan border in the west, though surface geomorphological expressions are subtle. The fault generally dips 166–185° at 48–69°, transitioning westward in the Wanjialing segment to dips of 185–222° at 25–48°. Its fracture zone ranges from 0.10 to 2.25 m in width, displaying an overall tight, pinch-and-swell structure with localized narrowing. Distinctly developed bleaching halos of variable thickness flank both sides. Within the fracture zone, tectonic breccias—predominantly cataclasite and fault breccia—exhibit silicification, bleaching alteration, and intermittent stibnite mineralization [30]. This structure initially acted as a compressional-shear fault, later transitioning to extensional right-lateral strike-slip characteristics, serving as a large-scale ore-conducting and ore-hosting structure.

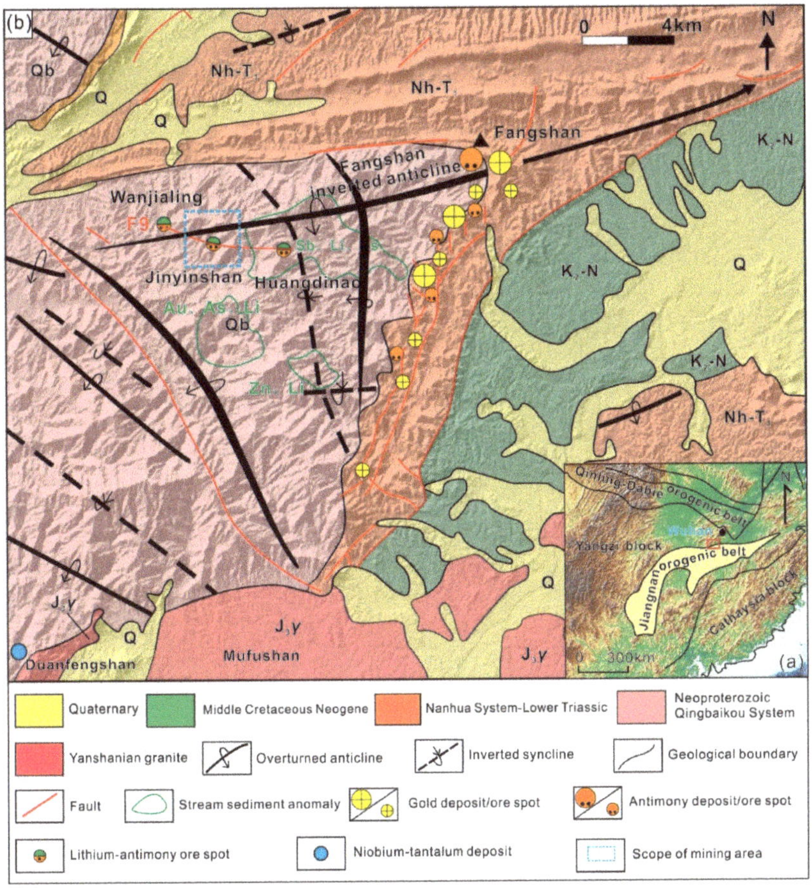

Figure 1. The geotectonic location (**a**) and the regional geological and mineral map (**b**) of Jinyinshan Li deposit in Hubei province [25].

The Li ore bodies occur within the nearly E-W trending F9 fault fracture zone and adjacent wall rocks, exhibiting a stratiform distribution. The mineralized body generally dips southward at approximately 60°, extending ~1.2 km in length with an average thickness of 5.36 m and an average Li_2O grade of 0.24 wt%. The thickness remains continuous along the strike from west to east and relatively stable down-dip. Ore grades vary locally along the strike (enriched in some sections, depleted in others) and display a trend of decreasing enrichment from shallow to deep levels down-dip. The mineralized lithologies comprise silicified cataclasite, fractured silty slate, fractured argillaceous slate, metamorphic siltstone, and lamprophyre [25]. The dominant ore textures are fractured and cataclastic, while primary structures include massive, stockwork, and disseminated types. Major ore minerals consist of clay minerals, sericite, quartz, and iron oxides. Clay minerals are predominantly chlorite, fibrous hydromica, and kaolinite, with most exhibiting silicification [24,25]. Locally, minor relict clastic occurrences are observed.

3. Sample Collection and Their Mineralogical and Chemical Characteristics

3.1. Collection and Preparation

Ore samples for testing were selected based on a comprehensive evaluation of the current exploration layout and ore body occurrence characteristics. Systematic sampling targeted mineralized bodies in six core profiles (ZK2001, ZK2301, ZK2601, ZK2903, ZK3001, and ZK3002) from the Jinyinshan mining area, southern Hubei. Sampled lithologies include cataclastic rock, silty slate, metamorphic siltstone, and lamprophyre. A combination of oriented channel sampling from trenches and split-core sampling from drill holes yielded a cumulative sample weight of 391.5 kg. This sampling strategy ensured complete coverage of orebody characteristics across varying elevations through three-dimensional integrated surface and subsurface sampling, guaranteeing spatially representative distribution within the deposit.

Sample preprocessing sequentially involved standardized crushing, multi-stage sieving, and splitting operations, ultimately categorizing samples into three categories: test specimens, chemical analysis samples, and archival reference samples. The chemical analysis samples were divided into 3 equal parts for component analysis. Preliminary analysis data indicated a Li_2O content of 0.265 wt% in the samples (Table 1), consistent with the theoretical average grade (0.24 wt%) calculated from the deposit's geostatistical model [25], thereby confirming the samples' representativeness.

Table 1. Analysis results of chemical composition for samples.

Composition	**Li_2O**	**Al_2O_3**	**SiO_2**	**K_2O**	**Na_2O**	**MgO**
Content (wt%)	0.265	16.19	66.39	2.62	0.45	0.59
Composition	$^{T}Fe_2O_3$	TiO_2	CaO	MnO	P_2O_5	Total
Content (wt%)	7.06	0.89	0.20	0.10	0.21	94.97
Composition	Nb	Ta	Ga	REE	Sb	Total
Content (μg/g)	16.7	1.20	25.1	302	336	681

The AMICS test sample was prepared through particle size fractionation. The bulk sample was first crushed to <74 μm using a jaw crusher and roller mill, then homogenized by cone-and-quartering. This processed material was sieved into three fractions: >74 μm, 45–74 μm, and <45 μm, with respective yields of 60.17%, 11.25%, and 28.58%. Test samples were prepared from each fraction, and the final AMICS results represent a comprehensive analysis integrating data from all three size fractions. Separately, representative Li ore samples were selected for probe thin sections preparation.

3.2. Analysis Method and Mineralogical Characteristics

To determine in detail the mineral composition, content, and the occurrence state and concentration of Li within the samples, this study utilized multiple analytical techniques including the automated mineral identification and characterization system (AMICS), electron probe microanalyzer (EPMA), and laser ablation inductively coupled plasma mass spectrometry (LA-ICP-MS) to conduct a comprehensive mineralogical investigation.

3.2.1. AMICS, EPMA and LA-ICP-MS

The AMICS system of Hubei Provincial Geological Experimental Testing Center integrates a ZEISS Sigma 300 high-resolution field emission scanning electron microscope (FESEM), a Bruker XFlash 6/60 energy-dispersive X-ray spectrometer (EDS), and a specialized software suite comprising AMICSTool, Investigator, MineralSTDManager, and AMICSProcess. Operational parameters are configured as follows: 20 kV accelerating voltage, high vacuum environment, 8.5 mm working distance, high-current mode activation, backscattered electron detection using HDBSD, and 60 μm objective lens grating aperture.

Major element compositional analyses were conducted at the Electron Probe Laboratory of the Chinese Academy of Geological Sciences (Beijing) utilizing a JEOL JXA-8230 electron probe microanalyzer (EPMA). Sample preparation involved carbon coating to achieve a conductive layer approximately 20 nm thick. Analytical conditions included an accelerating voltage of 15 kV, beam current of 20 nA, and beam diameter of 2 μm. The measurement times were 10 s per element and 5 s for background acquisition. Natural mineral and synthetic oxide standards were used, and the ZAF correction method was applied to all analytical data.

In situ laser ablation inductively coupled plasma mass spectrometry (LA-ICP-MS) was performed at the National Geological Experimental Testing Center. The test employed the American ASIJ-200 343 nm femtosecond laser and the German Thermo X-Series inductively coupled plasma mass spectrometer on-line. In situ point analyses were performed using a laser ablation spot size of 24 μm, a repetition rate of 6 Hz, and NIST SRM 610 synthetic silicate glass as the primary external calibration standard. To ensure analytical precision, standardized reference materials were systematically interspersed within every batch of 5–7 sample mounts. Each measurement cycle incorporated a 20 s background signal acquisition (gas blank), followed by 50 s of sample ablation, and concluded with a 20 s washout period to minimize cross-contamination. Raw data were reduced using ICPMS DataCal 10.7 software, applying background subtraction, and drift correction.

3.2.2. Chemical Composition

The Li_2O content in the samples from the Jinyinshan deposit is 0.265 wt%, with major chemical constituents including SiO_2 (66.39 wt%), Al_2O_3 (16.19 wt%), and Fe_2O_3 (7.06 wt%). Concentrations of rare and dispersed metals such as REE, Ga, Nb, Ta, and Sb are relatively low. Detailed ore composition data are provided in Table 1.

3.2.3. Mineral Composition

Integrated analysis using AMICS and SEM-EDS reveals that quartz is the dominant mineral in the samples (55.07%), followed by clay minerals including illite ($(K_{<1}(Al,R^{2+})_2[(Si,Al)Si_3O_{10}](OH)_2 \cdot nH_2O$, 13.55%), cookeite ($LiAl_4(Si_3Al)O_{10}(OH)_8$, 7.25%), and rectorite ($(Na,Ca)Al_4(Si,Al)_8O_{20}(OH)_4 \cdot 2H_2O$, 4.86%). The dominant metallic minerals are iron oxides/hydroxides (3.20%, commonly referred to as "limonite" in field descriptions; Figure 2e,f). Accessory minerals are diverse and include florencite, apatite, anatase, and zircon. Detailed mineralogical composition is provided in Appendix A.

3.2.4. Li Occurrence State

The test samples were primarily composed of quartz and clay minerals (cookeite, illite, rectorite), with no traditional Li-bearing minerals such as spodumene observed (Figure 2). Given that SEM-EDS cannot provide quantitative Li characterization, this study employed EPMA and LA-ICP-MS for in situ microanalysis of major clay minerals (detailed data provided in Appendix B). Integrated with chemical phase analysis, this approach comprehensively determined Li occurrence states and distribution patterns.

Cookeite exhibits significant Li enrichment (11,308–16,048 ppm), with an average content of 14,079.5 ppm. The structure of chlorite-group minerals is composed of alternating stacking of talc layers (T-O-T) and brucite layers (octahedral sheet, O). This brucite layer is an octahedral sheet sandwiched between the T-O-T layers. Its crystal chemical structure indicates that Li^+ mainly replaces the cations (e.g., Al^{3+}) in the interlayer octahedron sheet with its structural characteristics [31].

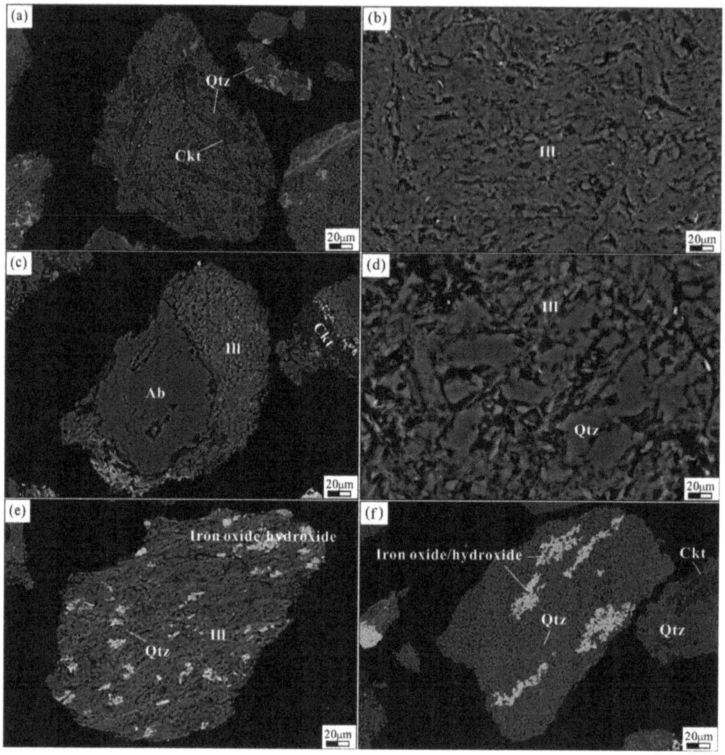

Figure 2. BSE photos of ore minerals in Jinyinshan deposit. (**a**) quartz and cookeite association; (**b**) fine scaly illite; (**c**) illite and albite association; (**d**) illite and quartz association; (**e**) iron oxides/hydroxides sporadic distribution and illite, quartz multiphase association; (**f**) quartz and cookeite, iron oxides/hydroxides association. Ab: Albite; Ckt: Cookeite; Ill: Illite; Qtz: Quartz.

Illite shows lower Li concentrations (555–685 ppm), averaging 607 ppm. Structural charge imbalance requires interlayer cations (Li^+, K^+, Na^+) to compensate for positive charge deficits from Al^{3+} replacing Si^{4+} in tetrahedral sites ($Li^+ + Al^{3+} \rightarrow Si^{4+}$) [32]. Simultaneously, Mg^{2+} substitution for Al^{3+} in octahedral sites generates localized negative charges, compensated by interlayer Li^+ through inner-/outer-sphere complexation ($Li^+ + Mg^{2+} \rightarrow Al^{3+}$) [33–35]. Therefore, Li^+ mainly exists in the interlayer space as interlayer cations in illite. However, illite's limited interlayer space restricts its capacity to accommodate H_2O and cations, resulting in constrained Li uptake [36].

Rectorite also demonstrates high Li content (864–1969 ppm, average 1322.2 ppm) with substantial variation. As a regular interlayer mineral (mica layer/smectite layer), Li^+ in rectorite primarily resides in the interlayer space of the mica layer (analogous to illite), and partially occupies exchangeable cation sites within the smectite layer [37,38]. Crucially, rectorite's expandible interlayer domains enable extensive adsorption of cations (e.g., Li^+, K^+), thus conferring substantially higher Li enrichment capacity than illite.

The content of each mineral was provided by AMICS analysis (Appendix A), which was used to calculate the occurrence ratio of Li in the main mineral phases. The following formula was applied: Total metallic Li content per mineral = mineral content × average Li concentration (Table 2). Li was predominantly hosted in cookeite (82.55% of total Li), followed by illite (6.65%) and rectorite (5.20%). Collectively, these clay minerals accounted for 94.40% of Li, confirming cookeite as the dominant Li host in the deposit.

Table 2. The distribution table of Li in various minerals.

Mineral	Content (%)	Li Content (ppm)	Li Metal Content (ppm)	Li Partition (%)
Cookeite	7.25	14,079.50	1020.76	82.55
Illite	13.55	607.00	82.25	6.65
Rectorite	4.86	1322.20	64.26	5.20
Total	25.66	16,008.7	1167.27	94.40
	Li grade		1236.58 *	100.00

Note: * Conversion according to the results of chemical analysis of Li_2O in Table 1.

3.2.5. Minerals Particle Size Characteristics

Particle size distributions were quantified using the Equivalent Circular Diameter method based on SEM-BSE images. Mineral projections were analyzed in ImageJ (v1.53) with $n \geq 300$ particles per sample to ensure statistical significance. Whole-rock mineral and individual phase size distributions (Figure 3) reveal that Li-bearing minerals (illite, cookeite, and rectorite) display submicroscopic flaky morphology with predominantly fine-grained sizes (over 90% of particles < 45 μm). These minerals commonly intergrow with quartz and primarily occur as aggregates (Figure 2). Quartz contains a significant coarse-grained fraction (11.45% > 150 μm). As the predominant metallic mineral, limonite shows a similar size distribution to Li-bearing clay minerals, with most particles (<45 μm) in the fine fraction.

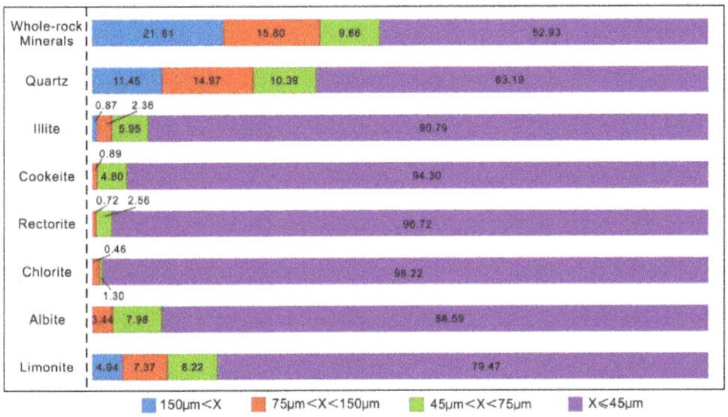

Figure 3. Cumulative diagram of mineral particle size distribution.

4. Experimental Methods and Results

Current mainstream Li extraction processes for traditional Li ores such as lepidolite ($KLi_2Al(Si_4O_{10})(F,OH)_2$) and spodumene ($LiAlSi_2O_6$) in the industrial sector primarily employ the sulfuric acid method, sulfate method, and limestone roasting method [14,20–23,39,40]. Conversely, clay-type Li deposits present greater extraction challenges due to their complex mineral compositions and diverse Li occurrence states. Existing extraction technologies mainly include direct leaching, roasting–leaching, and additive-assisted roasting–leaching methods [12–16].

Based on systematic process mineralogy and exploratory test characterization, this study confirms that Li is predominantly hosted in clay minerals including cookeite and rectorite. Upon roasting, these minerals undergo significant phase transformations where Al-Si bonds in silicates are activated, thereby enhancing the leaching rate for Li and other metallic elements [41–43]. Through iterative experimentation, two clean extraction processes—blank roasting–hydrochloric acid leaching and blank roasting–sulfuric acid leaching—were ultimately identified, both offering high Li leaching efficiencies with notable economic and environmental benefits.

4.1. Process Flow I: Blank Roasting–Hydrochloric Acid Leaching Method

4.1.1. Optimization of Li Leaching Conditions

Condition optimization tests investigated grinding fineness, roasting temperature, roasting time, hydrochloric acid concentration, L/S ratio, leaching temperature, and leaching time. Results indicate that grinding fineness, L/S ratio, and leaching time exerted negligible effects on Li leaching rate (Figure 4a–c). In contrast, roasting temperature, roasting time, hydrochloric acid concentration, and leaching temperature exerted substantial regulatory effects (Figure 4d–g).

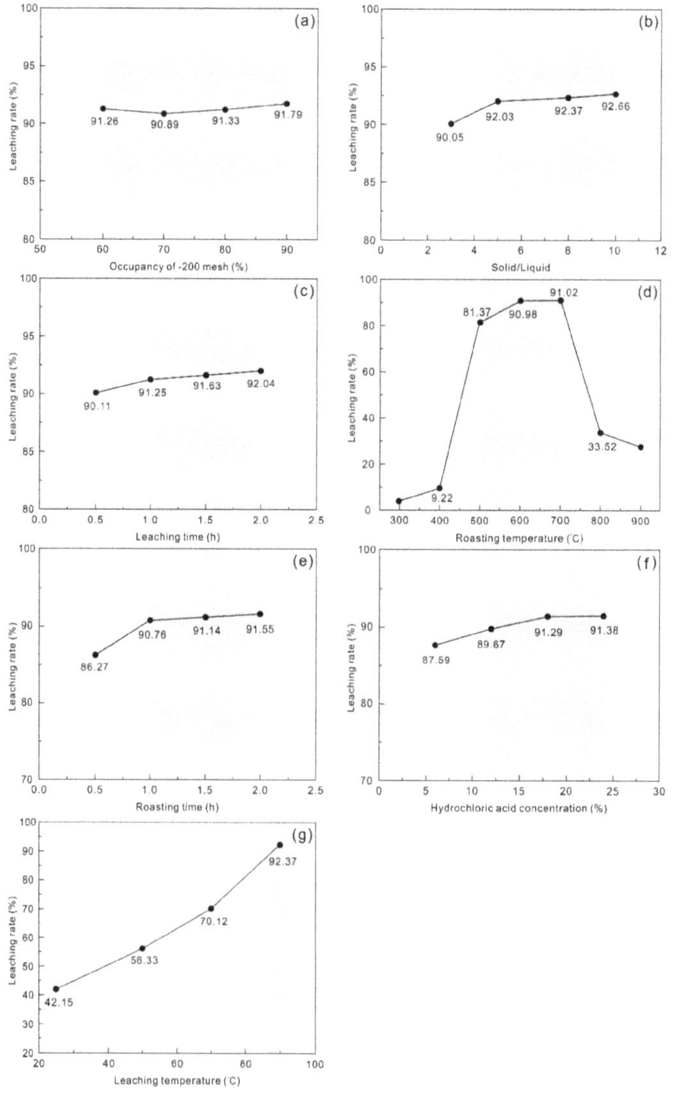

Figure 4. Effects of grinding fineness (**a**), liquid–solid ratio (**b**), leaching time (**c**), roasting temperature (**d**), roasting time (**e**), hydrochloric acid concentration (**f**) and leaching temperature (**g**) on lithium leaching rate in blank roasting–hydrochloric acid leaching process. The shadow range represents the suitable range after the parameter conditions are optimized.

Roasting Temperature

As the roasting temperature increased from 300 °C to 900 °C, the Li leaching rate exhibited a unimodal trend (Figure 4d). At 500 °C, the leaching rate abruptly increased to 81.37%. Further increases to 600 °C and 700 °C yielded rates of 90.98% and 91.02%, respectively. At 800 °C, however, these sharply decreased to 33.52%.

Roasting Time

The Li leaching rate generally increased with prolonged roasting time (Figure 4e). At 0.5 h, the rate reached 86.27%, rising to 90.76% at 1 h. Extending time beyond 1.5 h stabilized above 91%, confirming sufficient thermal activation as a prerequisite for liberating structural Li.

Hydrochloric Acid Concentration

Increasing hydrochloric acid concentrations progressively enhanced the leaching rate (Figure 4f). At 6 wt% hydrochloric acid, the rate was 87.59%, approaching 90% at 12 wt%. Continued concentration increases pushed the rate above 91%. Balancing cost and efficacy, 18 wt% hydrochloric acid was laboratory-optimized, though industrial implementation may adopt a gradient range of 12–18 wt%.

Leaching Temperature

Increasing leaching temperature from 25 °C to 90 °C enhanced the Li recovery rate from 42.15% to 92.37%, demonstrating a gradual upward trend (Figure 4g). Thus, 90 °C is optimal.

Tests on grinding fineness, L/S ratio, and leaching time confirmed their negligible influence (Figure 4a–c). Consequently, the optimized parameters were: grinding fineness of 200 mesh accounts for 60%, roasting temperature of 600 °C, roasting duration of 1.5 h, hydrochloric acid concentration of 18 wt%, an L/S ratio of 5, leaching temperature of 90 °C, and leaching time of 1 h.

4.1.2. Leachate Purification and Li Precipitation Test

Compared to spodumene and lepidolite concentrates, clay-type Li ore features complex composition and polymetallic associations. After blank roasting–acid leaching, significant coexisting metal ions (Al^{3+}, Fe^{3+}, Ca^{2+}, Mg^{2+}) dissolve alongside target Li, substantially complicating Li separation and purification. Initial leaching results indicated Al and Fe concentrations exceeding 7.5 g/L in both hydrogen chloride and hydrogen sulfate systems, followed by Ca and Mg. Addressing the characteristic low-Li and high-Al leachate, this study innovatively implements a process flow of leachate roasting–water leaching for Li-Al separation–alkali precipitation for Ca and Mg removal–concentration for Li precipitation (Figure 5). This approach utilizes the hydrolysis properties of $AlCl_3$ in the leachate to convert Al to amorphous alumina, achieving Li-Al separation via water leaching (roasting temperature 600 °C, roasting time 2 h, liquid–liquid ratio of 3, leaching temperature of 90 °C, and leaching time of 1 h.). Concurrently, volatilized HCl from calcination is captured via closed-loop absorption, significantly reducing acid consumption in the leaching circuit.

The secondary leachate primarily contains Ca^{2+} and Mg^{2+} impurities. Although mature separation techniques exist for Li-Ca-Mg systems, conventional pH adjustment with lime introduces new Ca^{2+} contamination, while NaOH—despite higher alkalinity and lower dosage—may cause Li adsorption losses [44,45]. Therefore, this study employed widely used Na_2CO_3 as a precipitant. At 1.5 times the stoichiometric ratio, Li loss was limited to 0.13%, demonstrating both high efficiency and cost-effectiveness.

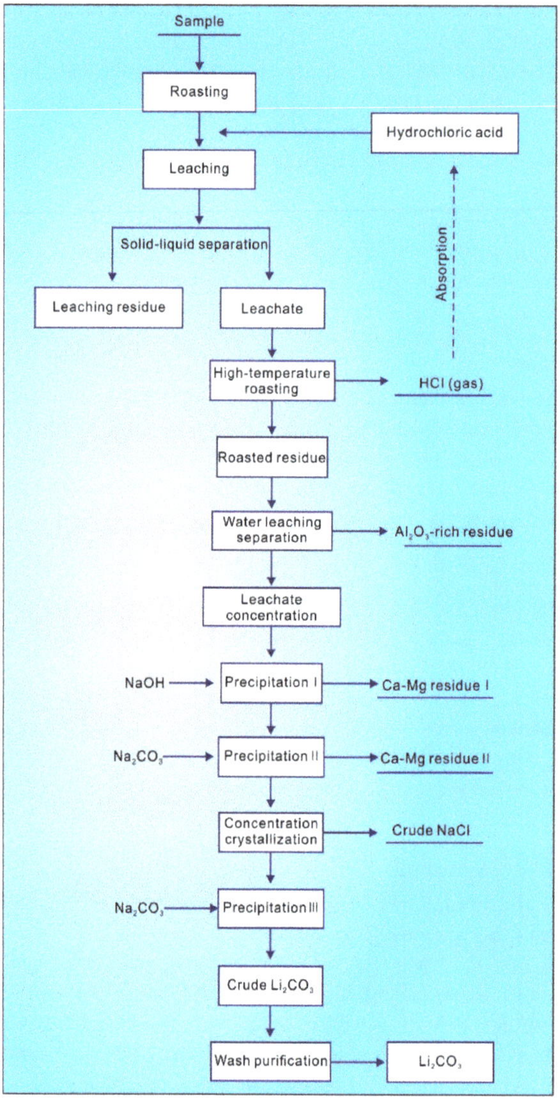

Figure 5. Process Flow I: Blank roasting–hydrochloric acid leaching process.

4.1.3. Purified Solution Concentration and Li Precipitation

The optimal Li^+ concentration for carbonate precipitation is 20–25 g/L [46]. The Ca/Mg-removed solution was further concentrated to crystallize crude sodium chloride, adjusting the final purified solution to 20–25 g/L Li^+.

The concentrated solution was heated to 90 °C, and 400 g/L Na_2CO_3 solution was added dropwise at a constant rate of 10 mL/min (1.1 times stoichiometric requirement) under agitation. After 30 min of precipitation, immediate filtration yielded crude lithium carbonate. X-ray diffraction (XRD) analytical results showed 85.73% Li_2CO_3 content, with Na_2CO_3 and K_2CO_3 as primary impurities, achieving 90.05% Li precipitation rate. In practical production, Li-rich filtrate can be recycled to subsequent cycles, increasing the overall rate beyond 98% at equilibrium.

4.1.4. Crude Lithium Carbonate Washing and Purification

Preliminary purification of crude lithium carbonate involved three-stage cross-flow washing and three-stage counter-current washing (L/S ratio: 2; temperature: 90 °C; time: 15 min; results in Table 3).

Table 3. Washing test results of blank calcination-hydrochloric acid leaching process.

Washing Method	Washing Stage	Na^+ Concentration (g/L)	K^+ Concentration (g/L)	Li^+ Concentration (g/L)	Li Loss Rate (%)
Cross-flow washing	1	22.08	5.73	2.22	8.37
	2	0.61	0.10	2.31	
	3	0.12	0.02	2.40	
Counter-current washing	1	22.39	5.92	2.15	2.77
	2	0.69	0.12	2.09	
	3	0.17	0.03	2.13	

Counter-current washing demonstrated superior performance. Leveraging Li_2CO_3 solubility characteristics, this method minimized Li losses. By the third stage, Na^+ and K^+ concentrations in wash liquor fell below 0.2 g/L, confirming an approximately 99% washing rate. Purified concentrate analysis indicated 99.03% Li_2CO_3 content, with impurities of 0.14% Na, 0.005% K, and 0.07% Ca; Fe and Mg below detection limits.

4.1.5. Full Flow Process I

Following detailed experimental investigations, Process I established a comprehensive Li extraction procedure comprising blank roasting–hydrochloric acid leaching–Li-Al separation from leachate–Ca/Mg removal–concentration for Li precipitation–three-stage counter-current washing (Figure 5). Full-process test parameters indicate that under conditions of 600 °C roasting temperature, a 1.5 h roasting time, 18 wt% hydrochloric acid concentration, 90 °C leaching temperature, 1 h leaching time, and L/S ratio of 5, a Li leaching rate of 92.37% and an Al leaching rate of 47.62% were achieved.

Roasting of the leachate followed by water leaching achieved Li-Al separation with a 93.91% Li leaching rate, while Al and Fe were completely converted into Al-rich residue. The concentrated leachate was treated with NaOH and Na_2CO_3 to remove Mg and Ca, yielding a purified solution with Li loss rates of 2.59% and 0.13%, respectively. Subsequent evaporation and concentration crystallized crude sodium chloride salt. Filtration produced a Li-rich purified solution, from which sodium carbonate precipitation generated a crude lithium carbonate product containing 85.73% Li_2CO_3, with a Li precipitation rate of 90.05% (in actual production, filtrate recycling could further enhance precipitation rate). Three-stage counter-current washing of the crude lithium carbonate yielded a final product with 99.03% Li_2CO_3 content (increased washing cycles may further improve purity).

Final outputs of Process I include: lithium carbonate product (0.48% yield, 99.03% Li_2CO_3, 73.89% Li total recovery), alumina-rich residue (34% yield, 46.88% Al_2O_3, 47.62% Al recovery), and crude sodium chloride product (1% yield, 99.13% NaCl).

4.2. Process Flow II: Blank Roasting–Sulfuric Acid Leaching Test

4.2.1. Optimization of Li Leaching Conditions

Process II similarly conducted condition optimization tests for grinding fineness, roasting temperature, roasting time, sulfuric acid concentration, L/S ratio, leaching temperature, and leaching time. Results indicate that grinding fineness and L/S ratio exerted limited influence on the Li leaching rate (Figure 6a,b), while roasting temperature, roasting time, sulfuric acid concentration, leaching temperature, and leaching time significantly affected outcomes (Figure 6c–g). Specifically, the Li leaching rate initially rose and then declined with prolonged leaching time, peaking at 89.11% at 2 h before decreasing to 83.79% at 4 h (Figure 6c). At a 400 °C roasting temperature, the leaching rate was only 2.80%, rapidly increasing to 79.55% at 500 °C and peaking at 84.89% at 600 °C, above which it rapidly increased (Figure 6d). Both prolonged roasting time and increased sulfuric acid concentration showed positive correlations, achieving an 84.91% rate at 2 h roasting and 87.16% at 30% sulfuric acid concentration (Figure 6e,f). Leaching temperature demonstrated the most pronounced effect, with the rate increasing by 147.5% when raised from 25 °C to 90 °C (Figure 6g).

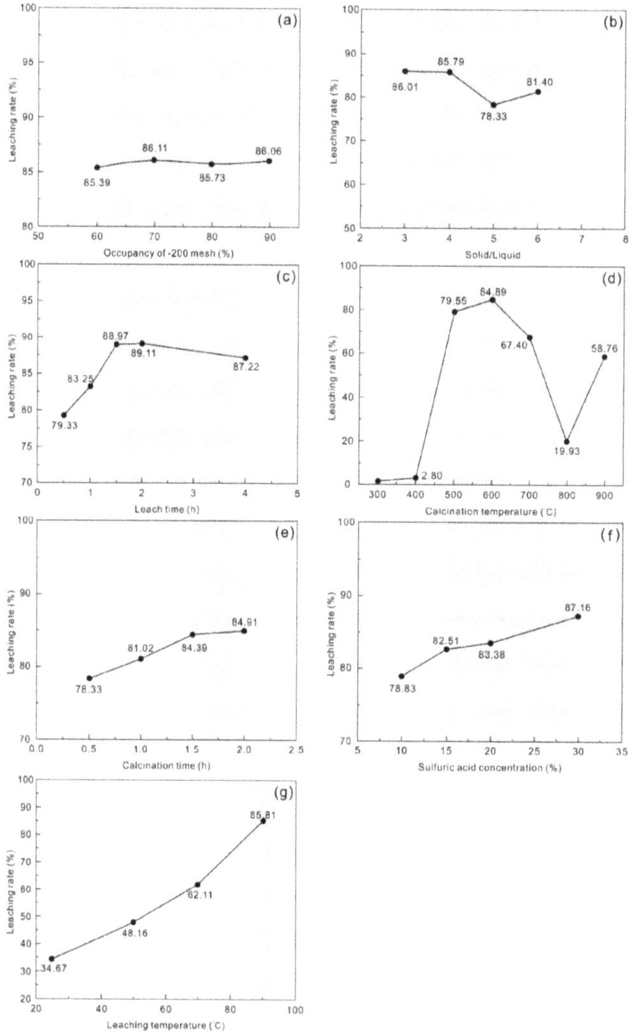

Figure 6. Effects of grinding fineness (**a**), liquid–solid ratio (**b**), leaching time (**c**), calcination temperature (**d**), calcination time (**e**), hydrochloric acid concentration (**f**) and leaching temperature (**g**) on lithium leaching rate in blank calcination–sulfuric acid leaching process. The shadow range represents the suitable range after the parameter conditions are optimized.

4.2.2. Leachate Purification and Li Precipitation Test

From a technical perspective, the key challenge in processing sulfuric acid leachate remains Li-Al separation, with alum precipitation being the mainstream method. This involves adding appropriate amounts of $(NH_4)_2SO_4$ or K_2SO_4 to form alum salts with $Al_2(SO_4)_3$ in the leachate for Al removal. Based on the characteristics of blank roasting–sulfuric acid leachate, this study adopted a process flow of Al removal via alum precipitation–alkali precipitation for Fe/Al removal–alkali precipitation for Ca/Mg removal–freezing crystallization for sodium sulfate removal–concentration for Li precipitation.

Since cyclic leaching tests were not conducted, the leachate was evaporated to double concentration prior to alum precipitation to ensure the Al crystallization rate. Ammonium alum crystallization, a well-established Al removal method, is affected by metal ions like Fe in the leachate, which consume ammonium sulfate. Consequently, ammonium sulfate dosage was optimized. Results (Figure 7) demonstrate that increasing the ammonium aluminum molar ratio positively enhances Al removal. Li loss during alum crystallization primarily stems from crystal entrapment and is of negligible correlation with ammonium sulfate dosage, maintaining low loss rates. Optimal conditions (1.1 × stoichiometric $(NH_4)_2SO_4$, 150 rpm agitation, 4 °C crystallization temperature, 60 min crystallization time) achieved 85.33% Al removal (residual Al 3.89 g/L) with 1.73% Li loss.

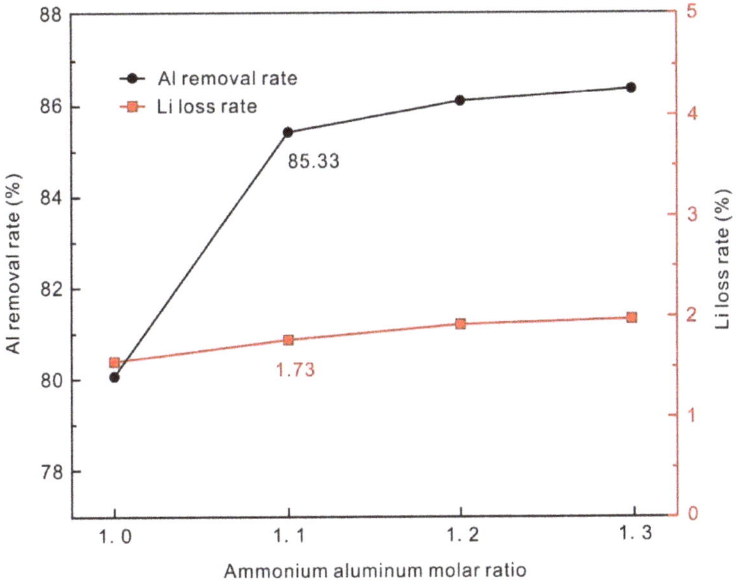

Figure 7. Experimental results of ammonium aluminum molar ratio on Al removal rate.

4.2.3. Alkali Precipitation for Fe/Al and Ca/Ma Removal

The single-pass Al removal efficiency is inherently limited by ammonium alum solubility, necessitating deep removal of residual Al/Fe. Alkali precipitation to form $Al(OH)_3$ and $Fe(OH)_3$ is the most common method, but $Al(OH)_3$ exhibits strong Li^+ adsorption [47]. Precise alkali addition is critical to prevent the formation of highly adsorbent colloids. Furthermore, as Al is amphoteric, pH must be strictly controlled near 7 to avoid aluminate formation at elevated pH, which complicates complete Al removal. Building on the previous step, the solution was first concentrated to a Li^+ concentration of approximately 5 g/L. At this stage, the concentrations of Ca^{2+} and Mg^{2+} were approximately 557 mg/L and

3.41 g/L, respectively. Subsequently, 300 g/L simultaneous NaOH and Na_2CO_3 addition was used for Ca/Mg removal, outperforming stepwise addition (Table 4).

Table 4. Test results of Ca and Mg removal process.

Charge Method	Mg^{2+} Concentration (mg/L)	Ca^{2+} Concentration (mg/L)	Li Loss Rate (%)
Split-step	No detection	9.58	1.38
Single step	No detection	10.33	1.32

4.2.4. Purified Solution Concentration and Li Precipitation

The purified solution was concentrated to 20% of its original volume to obtain an Li-rich solution. Based on prior tests, the solution was heated to 90 °C, and 400 g/L Na_2CO_3 solution (1.1 × stoichiometric) was added dropwise at 10 mL/min under agitation for 30 min. Immediate filtration produced crude lithium carbonate assayed at 89.52% Li_2CO_3 with 87.35% Li precipitation efficiency.

4.2.5. Crude Lithium Carbonate Washing and Purification

For preliminary purification, three-stage counter-current washing (L/S ratio 2; 90 °C; 15 min) was conducted. By Stage 3, Na^+ and K^+ concentrations in wash liquor decreased to 0.1 g/L, with 2.71% Li loss, indicating effective purification. Washed product analysis showed 98.85% Li_2CO_3 content, with impurities at 0.11% Na, 0.005% K, and 0.07% Ca; Fe and Mg were below detection limits (Table 5). Product quality could be further enhanced with additional washing cycles.

Table 5. Washing test results of blank calcination–sulfuric acid leaching process.

Washing Method	Washing Series	Na^+ Concentration (g/L)	K^+ Concentration (g/L)	Li^+ Concentration (g/L)	Li Loss Rate (%)
Counter-current washing	1	17.62	2.56	2.03	
	2	0.45	0.08	2.09	
	3	0.09	0.03	2.11	2.71

4.2.6. Full Process Flow II

Following detailed experimental investigations, Process II was established: blank roasting–sulfuric acid leaching–Al removal via alum precipitation–Fe/Al removal–freezing crystallization for sodium sulfate removal–Ca/Mg removal–concentration for Li precipitation–three-stage counter-current washing (Figure 8). Key performance metrics under optimized conditions (roasting: 600 °C, 2 h; leaching: 15 wt% sulfuric acid, 90 °C, 2 h, L/S ratio 3) include a Li leaching rate of 89.11% and an Al leaching rate of 44.33%. Al removal via ammonium sulfate freeze crystallization achieved an 85.33% rate (crystallization rate: 85.33%) with 1.73% Li loss. Subsequent Fe/Al removal via pH adjustment (NaOH) incurred 1.91% Li loss. After concentration and freezing to remove sodium sulfate, NaOH and Na_2CO_3 were added to eliminate Ca/Mg (3.85% Li loss). The purified solution underwent further concentration/freezing for sodium sulfate removal, yielding a Li-rich solution. Sodium carbonate precipitation produced crude lithium carbonate (89.52% Li_2CO_3), which was upgraded to 98.85% Li_2CO_3 purity via three-stage counter-current washing.

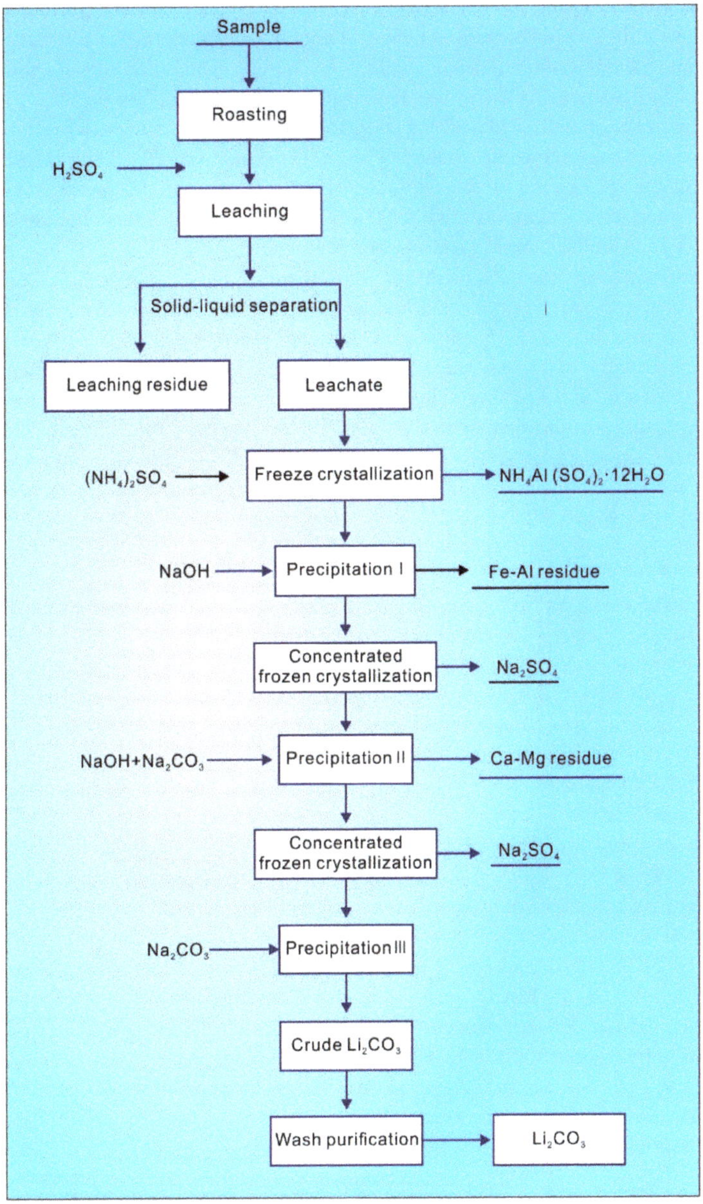

Figure 8. Process II: Blank calcination–sulfuric acid leaching process.

Final outputs of Process II include, lithium carbonate product (0.46% yield, 98.85% Li_2CO_3, 69.89% Li total recovery), crude ammonium alum (29.48% yield, 37.83% Al recovery), and crude sodium sulfate (43% yield).

5. Discussion

5.1. Significance of Mineralogy for Li Extraction

Elucidating the occurrence state of Li and mineralogical characteristics in the Jinyinshan clay-type Li deposit provides a critical mineralogical foundation for understanding Li extraction behavior and designing efficient processes. The study reveals that Li primarily resides in structural sites within cookeite (71.95% of total Li), followed by illite (6.30%) and rectorite (4.71%). As the dominant Li-host mineral, cookeite's crystal chemistry dictates that structural Li release requires disrupting its silicate framework [30]. This explains the typically low efficiency of direct acid leaching and necessitates blank roasting as a pretreatment step. During roasting, cookeite undergoes lattice expansion and disruption [41,42], transforming structurally bound Li into acid-accessible forms. In contrast, Li within rectorite and illite interlayers is more readily leachable, yet its significantly lower abundance compared to cookeite limits its overall contribution to leaching efficiency.

Li-host minerals (illite, cookeite, rectorite) predominantly exhibit fine particle sizes, with over 90% below 45 μm (Figure 3), and frequently occur as intimate intergrowths or aggregates with quartz (Figure 2). This fine-grained intergrowth necessitates fine grinding (−200 mesh accounting for 60%) to sufficiently liberate mineral surfaces for subsequent roasting activation and acid leaching. However, it concurrently leads to significantly elevated concentrations of impurity ions (Al^{3+}, Fe^{3+}, Ca^{2+}, Mg^{2+}) in the leachate. The substantial specific surface area of sub-45 μm particles promotes the dissolution of impurity elements from associated minerals, directly necessitating the complex and crucial leachate purification steps developed in this study.

5.2. Li Extraction Advantages of the Two Processes

Process I established a method comprising blank roasting–hydrochloric acid leaching–Li-Al separation from leachate–Ca/Mg removal–concentration for Li precipitation–three-stage counter-current washing. Through optimization of leaching parameters (600 °C roasting for 1.5 h, 18 wt% hydrochloric acid leaching at 90 °C for 2 h, L/S ratio 5), a Li leaching rate of 92.37% was achieved. Subsequent purification yielded Li carbonate with 0.48% yield, 99.03% Li_2CO_3 purity, and 73.89% Li total recovery. The product complies with GB/T 11075-2013 standards while generating alumina-rich residue and crude sodium chloride as by-products [48].

Process II adopted blank roasting–sulfuric acid leaching–Al removal via alum precipitation–Al/Fe removal–concentration/freezing for sodium sulfate removal–Ca/Mg removal–concentration for Li precipitation–three-stage counter-current washing. Under conditions of 600 °C roasting for 2 h, 15 wt% sulfuric acid leaching at 90 °C for 2 h, and an L/S ratio of 3, an 89.11% Li leaching rate was attained. The final Li carbonate product exhibited 0.46% yield, 98.85% Li_2CO_3 purity, and 69.89% Li recovery, meeting national standards while producing crude ammonium alum and sodium sulfate by-products [48].

Compared to traditional additive-assisted roasting, both developed blank roasting–acid leaching clean processes significantly reduce reagent consumption and avoid introducing new impurity ions. However, they exhibit distinct emphases in acid system selection, core impurity removal technologies, and by-product profiles.

Leaching efficiency and stability: Under optimized conditions, Process I achieved marginally higher Li leaching efficiency (92.37% vs. 89.11%) and total Li recovery (73.89% vs. 69.89%). This enhanced performance is potentially attributable to hydrochloric acid's superior dissolution capability for certain silicates or the more favorable dissolution kinetics of the activated calcine in hydrochloric acid. Process I's innovative roasting–water leaching step for Li-Al separation achieved high efficiency (93.91% Li recovery), while concurrently recovering acid via closed-loop absorption. This contributes to potentially superior operational stability and process controllability for the hydrochloric acid system.

Impurity removal and product purity: Both processes yielded high-purity lithium carbonate (>98.85%) meeting national standards [48]. Process I's final product purity (99.03%) slightly exceeded Process II (98.85%), likely due to its unique Li-Al separation mechanism, which more thoroughly removes aluminum (fixed primarily in the residue) and its relatively simpler subsequent Ca/Mg removal step using Na_2CO_3. In contrast, Process II requires multiple purification stages (alum precipitation for Al, alkali precipitation for Fe/Al and Ca/Mg), increasing cumulative Li loss risks.

By-products and economics: Process I generates an Al-rich residue (valuable but requiring specific utilization pathways like metallurgy or construction materials) and crude sodium chloride (lower value). Process II's core advantage lies in diversified by-product valorization: Crude ammonium alum finds applications in water treatment, papermaking, and tanning, while crude sodium sulfate is a significant chemical feedstock (glass, detergents, pulp). Strategic by-product sales can substantially offset production costs and enhance overall project economics.

6. Conclusions

(1) Detailed mineralogical studies confirm Li predominantly occurs in cookeite (82.55% of total Li), followed by illite (6.65%) and rectorite (5.20%). These Li-bearing minerals exhibit microscopic flaky morphologies with fine particle sizes (<45 μm), predominantly occurring as aggregates. Cookeite shows particularly high Li enrichment (11,308–16,048 ppm, average 14,079.5 ppm), providing critical mineralogical foundations for extraction process design.

(2) By optimizing leaching conditions and refining purification steps, two industrially adaptable flowsheets were innovatively developed: Process I (blank roasting–hydrochloric acid leaching–Li-Al separation–Ca/Mg removal–concentration precipitation–three-stage counter-current washing) and Process II (blank roasting–sulfuric acid leaching–alum precipitation for Al removal–Al/Fe removal–concentration/freezing for sodium sulfate removal–Ca/Mg removal–concentration precipitation–three-stage counter-current washing).

(3) Post-optimization, the blank roasting–acid leaching processes stably yielded high-purity Li carbonate ($\geq$99% Li_2CO_3) compliant with national standards, achieving $\geq$70% overall Li recovery. This validates the technical feasibility and effectiveness for deposit development.

(4) Beyond Li carbonate, the processes generate valuable by-products including alumina-rich residue, crude sodium chloride, and ammonium alum. Strategic utilization pathways for these materials will maximize resource rate and enhance economic returns while reducing environmental footprints.

Author Contributions: Writing—original draft preparation, H.Z.; Writing—review and editing, P.L., W.Z. and J.L.; Collecting the samples, J.Y.; Software and data processing, Z.C., Y.F., S.L., J.K. and D.Z.; Funding acquisition, P.L. All authors have read and agreed to the published version of the manuscript.

Funding: This work is financially supported by the National Key Research and Development Program (2024YFC2909300), the National Natural Science Foundation of China (42330806, U2444204, and 42472111), and Geological Survey Project grants from the China Geological Survey (DD20230289).

Institutional Review Board Statement: Not applicable.

Informed Consent Statement: Not applicable.

Data Availability Statement: The original contributions presented in this study are included in the article. Further inquiries can be directed to the corresponding author.

Conflicts of Interest: The authors declare no conflicts of interest.

Appendix A. AMICS Analysis Results of Mineral Composition of Samples

Mineral	Content (%)	Area (%)	Mineral Phases Number	Particle Number
Quartz	55.07	54.74	36,096	1,078,978
Illite	13.55	12.50	13,323	266,225
Cookeite	7.25	7.65	14,224	91,276
K-feldspar	5.36	5.39	22,239	76,738
Rectorite	4.86	4.57	12,168	111,464
Unknown	3.87	4.99	23,482	107,222
Iron oxide/hydroxide	3.20	2.20	4809	95,553
Albite	1.91	1.9	3455	40,380
Chlorite	1.31	1.14	6146	53,273
Biotite	0.94	0.82	5176	53,972
Anatase	0.20	0.12	1652	4510
Apatite	0.11	0.09	181	711
Scorodite	0.09	0.08	78	515
Pyrite	0.05	0.03	37	250
Siderite	0.05	0.03	188	1050
Florencite	0.03	0.02	572	910
Dolomite	0.02	0.01	23	33
Zircon	<0.01	<0.01	89	132

Note: The mineral phases number reflects the diversity of association phases of minerals. Particle number is the total number of entities of independent mineral particles. Area Percentage: (Number of mineral particles × Average particle area)/Total mineral area = Area percentage of the mineral.

Appendix B. EPMA and LA-ICP-MS Analysis Results of Cookeite, Illite and Rectorite

Mineral	Sample	Li ppm	Na_2O wt%	MgO wt%	Al_2O_3 wt%	SiO_2 wt%	K_2O wt%	CaO wt%	TiO_2 wt%	^{T}FeO wt%	Total wt%
Cookeite	A-01	15,057.00	0.00	0.04	42.90	38.39	0.06	0.00	0.00	0.11	81.50
	A-02	13,895.00	0.01	0.07	43.50	38.70	0.07	0.06	0.01	0.45	82.86
	A-03	13,298.00	0.08	0.02	43.20	38.00	0.06	0.06	0.01	0.52	81.95
	A-04	11,308.00	0.08	0.02	42.30	37.92	0.07	0.09	0.10	0.61	82.19
	A-05	14,871.00	0.05	0.02	43.32	38.50	0.06	0.03	0.01	0.35	82.34
	A-06	16,048.00	0.01	0.05	42.55	38.84	0.08	0.05	0.06	0.46	82.09
Illite	B-01	612.00	0.32	1.50	33.20	52.10	8.40	0.50	0.72	2.00	98.74
	B-02	585.00	0.18	1.80	34.50	50.30	7.60	0.40	0.65	3.10	98.53
	B-03	700.00	0.45	0.80	31.00	54.00	8.20	0.30	0.50	1.50	96.75
	B-04	550.00	0.25	1.20	32.80	51.50	7.90	0.60	0.68	2.50	97.43
	B-05	555.00	0.12	2.00	30.70	53.80	7.30	0.20	0.45	3.60	98.17
Rectorite	C-01	1046.00	2.83	0.51	36.61	52.32	0.29	0.45	0.08	0.23	93.32
	C-02	864.00	2.58	0.64	36.14	53.01	0.19	0.46	0.07	0.42	93.51
	C-03	1969.00	2.69	0.54	37.14	51.88	0.20	0.52	0.09	0.25	93.31
	C-04	1422.00	2.95	0.77	36.28	52.32	0.17	0.25	0.08	0.28	93.10
	C-05	1310.00	2.76	0.26	36.54	52.22	0.37	0.39	0.10	0.11	92.75

References

1. Li, J.K.; Liu, X.F.; Wang, D.H. The metallogenetic regularity of lithium deposit in China. *Acta Geol. Sin.* **2014**, *88*, 2269–2283. (In Chinese with English abstract)
2. Wang, D.H.; Sun, Y.; Liu, X.F.; Tian, S.H.; Dai, J.J.; Liu, L.J.; Ma, S.C. Deep exploration technology and prospecting direction for lithium energy metal. *Geol. Surv. China* **2018**, *5*, 1–9. (In Chinese with English abstract)
3. Liu, L.J.; Wang, D.H.; Liu, X.F.; Li, J.K.; Dai, H.Z.; Yan, W.D. The main types, distribution features and present situation of exploration and development for domestic and foreign lithium min. *Geol. China* **2017**, *44*, 263–278. (In Chinese with English abstract)
4. Xu, X.W.; Li, H.L.; Shi, F.P.; Yao, F.J.; Chen, J.Z.; Yang, Z.Q.; Hong, T.; Qiang, K. Metallogenic characteristics and prospecting of granitic pegmatite-type rare metal deposits in the Tugeman area, middle part of Altyn Tagh. *Acta Petrol. Sin.* **2019**, *35*, 3303–3316. (In Chinese with English abstract)

5. Li, J.K.; Fu, X.F.; Zou, T.R.; Wang, D.H. Prospecting breakthrough of lithium deposit in Jiajika, Sichuan and analysis of prospecting prospects in the periphery. *Miner. Depos.* **2014**, *S1*, 1175–1176. (In Chinese with English abstract)
6. Fan, J. *Research on Exploitation and Industry Development Strategy of Lithium Mineral Resources in China*; China University of Geosciences: Beijing, China, 2016; pp. 1–110. (In Chinese with English abstract)
7. Kesler, S.E.; Gruber, P.W.; Medina, P.A.; Keoleian, G.A.; Everson, M.P.; Wallington, T.J. Global lithium resources: Relative importance of pegmatite, brine and other deposits. *Ore Geol. Rev.* **2012**, *48*, 55–69. [CrossRef]
8. Verley, G.G.; Vidal, M.F.; MacNeill, E. Report on the Sonora Lithium Project. In *Technical Report on the Feasibility Study for the Sonora Lithium Project, Mexico*; Bacanora Lithium: Sonora, Mexico, 2012.
9. Benson, T.R.; Coble, M.A.; Rytuba, J.J.; Mahood, G.A. Lithium enrichment in intracontinental rhyolite magmas leads to Li deposits in caldera basins. *Nat. Commun.* **2017**, *8*, 270. [CrossRef]
10. Xie, R.Q.; Zhao, Z.H.; Tong, X.; Xie, X.; Song, Q.; Fan, P.Q. Review of the research on the development and utilization of clay-type lithium resources. *Particuology* **2024**, *87*, 46–53. [CrossRef]
11. Li, Q.Z.; Wang, J.B.; Liu, D.W.; Jia, C.S. Exploration and development status of global lithium resources and overview of typical foreign lithium deposits. *Geol. Explor.* **2025**, *61*, 0636–0654. (In Chinese with English abstract)
12. Wen, H.J.; Luo, C.G.; Du, S.J.; Yu, W.X.; Gu, H.N.; Ling, K.Y.; Cui, Y.; Li, Y.; Yang, J.H. Carbonate-hosted clay-type lithium deposit and its prospecting significance. *Chin. Sci. Bull.* **2020**, *65*, 53–59. (In Chinese with English abstract) [CrossRef]
13. Xiao, H. Study on Recovery and Reuse of Valuable Metals from Waste Ferrous Lithium Phosphate Cathode Materials by Roasting Leaching Method. Master's Thesis, Jiangxi University of Science and Technology, Ganzhou, China, 2023; pp. 1–54.
14. Li, P.H.; Zeng, Y.; Yu, X.D.; Huang, Y. Current research status on lithium extraction technologies and occurrence characteristics of clay-type lithium deposits. *Ind. Miner. Process.* **2025**, 1–13. (In Chinese with English abstract)
15. Bai, Z.; He, J.H.; Wang, M.X.; Han, Y.X.; Sun, Y.S.; Yuan, S.; Jin, J.P. Pilot-scale case study on vanadium extraction from vanadium-bearing shale using suspension oxidation roasting-curing-leaching process. *Trans. Nonferrous Met. Soc. China* **2025**, 1–30. (In Chinese with English abstract)
16. Ni, C.Q.; Liu, C.; Wang, J.Y.; Song, Y.B.; Liang, Y.T.; Xie, W.Q.; Zhong, H.; He, Z.G. Highly efficient lithium leaching from α-spodumene via binary composite salts low-temperature roasting process. *Powder Technol.* **2025**, *449*, 120404. [CrossRef]
17. Cui, Y. Occurrence and Enrichment Mechanisms of Lithium in Clay-Type Li Deposits: Case Studies from Central Yunnan, Central Guizhou, and McDermitt, USA. Ph.D. Thesis, East China University of Technology, Nanchang, China, 2024; pp. 1–132. (In Chinese with English abstract)
18. Crocker, L.; Lien, R.H. *Lithium and Its Recovery from Low-Grade Nevada Clays*; U.S. Department of the Interior: Washington, DC, USA, 1987.
19. Zhu, L.; Gu, H.N.; Yang, Y.Q.; Wen, H.J.; Wang, N.; Luo, C.G. Research progress of lithium extraction from clay-type lithium ore resources. *Light Met.* **2020**, *12*, 8–13.
20. Fosu, A.Y.; Kanari, N.; Vaughan, J.; Chagnes, A. Literature review and thermodynamic modelling of roasting processes for lithium extraction from spodumene. *Metals* **2020**, *10*, 1312. [CrossRef]
21. Han, S.; Sagzhanov, D.; Pan, J.; Vaziri Hassas, B.; Rezaee, M.; Akbari, H.; Mensah-Biney, R. Direct extraction of lithium from α-spodumene by salt roasting-leaching process. *ACS Sustain. Chem. Eng.* **2022**, *10*, 13495–13504. [CrossRef]
22. Zhang, Y.Q.; Ma, B.Z.; Lv, Y.W.; Wang, C.Y.; Chen, Y.Q. An effective method for directly extracting lithium from α-spodumene by activated roasting and sulfuric acid leaching. *J. Ind. Eng. Chem.* **2023**, *122*, 540–550. [CrossRef]
23. Zhao, H.; Wang, Y.; Chen, H.F. Recent advances in lithium extraction from lithium-bearing clay minerals. *Hydrometallurgy* **2023**, *217*, 106025. [CrossRef]
24. Guo, W.M.; Wang, D.H.; Li, P.; Li, J.K.; Shu, Z.M.; Zhang, W.S.; Li, C. Sb-Li assemblages: Rare assemblages of mineralization and the occurrence of lithium. *Acta Geol. Sin.* **2019**, *93*, 1296–1308. (In Chinese with English abstract)
25. Yin, J.; Wu, J.H.; Li, Y.; Wang, J.; Zhong, W.; Zeng, T.F.; Yang, R.; Chen, W. The occurrence status and genesis of the Jinyinshan lithium deposit in Chibi, Hubei. *Geolgical Rev.* **2024**, *70*, 1558–1570. (In Chinese with English abstract)
26. Li, P.; Liu, X.; Li, J.K.; Huang, Z.B.; Zhou, F.C.; Zhang, L.P. Petrographic and geochemical characteristics of Renli-Chuanziyuan No. 5 pegmatite, NE Hunan, and its metallogenic age. *Acta Geol. Sin.* **2019**, *93*, 1374–1391. (In Chinese with English abstract)
27. Zhang, A.; Shao, C.S.; Wang, C.; Yang, Y.L.; Lu, T. Dataset of the Environmental Geological Survey on a Scale of 1:50000 in the Puqi County (Present Chibi City) Map Sheet, Hubei Province. *Geol. China* **2019**, *46*, 67–79. (In Chinese with English abstract)
28. Jiang, P.F.; Li, P.; Li, J.K.; He, X.M.; Leng, S.L. Zircon U-Pb geochronology and Hf isotopic composition of Be-pegmatites in Maiguo deposit, eastern Mufushan, and their geological implications. *Miner. Depos.* **2021**, *40*, 723–739. (In Chinese with English abstract)
29. Zhou, F.C.; Li, J.K.; Liu, X.; Li, P.; Huang, Z.B.; Shi, W.K.; Su, J.N.; Chen, H.; Huang, X.Q. Geochemical characteristics and genetic significance of ore bodies in Renli Nb-Ta deposit, Hunan Province. *Acta Geol. Sin.* **2019**, *93*, 1392–1404.
30. Shu, Z.M.; Wu, W.X. *Summary of General Survey of Gold (Antimony) Ore in Jinyinshan Area*; The Fourth Geological Team of Hubei Geological Bureau: Xianning, China, 2018; pp. 1–85. (In Chinese)
31. Brigatti, M.F.; Galan, E.; Theng, B.K.G. Structure and mineralogy of clay minerals. *Dev. Clay Sci.* **2006**, *1*, 19–86.

32. Velde, B. (Ed.) *Origin and Mineralogy of Clays: Clays and the Environment*; Springer Science & Business Media: Berlin/Heidelberg, Germany, 2013.
33. Howard, J.J. Lithium and potassium saturation of illite/smectite clays from interlaminated shales and sandstones. *Clays Clay Miner.* **1981**, *29*, 136–142. [CrossRef]
34. Tournassat, C.; Gailhanou, H.; Crouzet, C.; Braibant, G.; Gautier, A.; Gaucher, E.C. Cation exchange selectivity coefficient values on smectite and mixed-layer illite/smectite minerals. *Soil Sci. Soc. Am. J.* **2009**, *73*, 928–942. [CrossRef]
35. Williams, L.B.; Hervig, R.L. Lithium and boron isotopes in illite-smectite: The importance of crystal size. *Geochim. Cosmochim. Acta* **2005**, *69*, 5705–5716. [CrossRef]
36. Whitney, G. Role of water in the smectite-to-illite reaction. *Clays Clay Miner.* **1990**, *38*, 343–350. [CrossRef]
37. Moore, D.M.; Reynolds, R.C. X-ray diffraction and the identification and analysis of clay minerals. *Clays Clay Miner.* **1989**, *38*, 448.
38. Decarreau, A.; Vigier, N.; Pálková, H.; Petit, S.; Vieillard, P.; Fontaine, C. Partitioning of lithium between smectite and solution: An experimental approach. *Geochim. Cosmochim. Acta.* **2012**, *85*, 314–325. [CrossRef]
39. Shi, G.M.; Zhou, Y.C.; Chen, H.J.; Guan, K.C.; Chen, Y.J. Experiment Study on Lithium Extraction with Roasting and Acid Leaching Process for a Sedimentary Clay-type Lithium Ore in Central Yunnan Province. *Met. Mine* **2023**, *01*, 199–203. (In Chinese with English abstract)
40. Wang, Z.; Ma, W.Q.; Shao, X.Q.; Xia, Y.Z.; He, L.H.; Huan, Y.D. Optimization of Lithium Extraction Process by Fluidized Roasting. *Chin. J. Rare Met.* **2025**, *49*, 618–624. (In Chinese with English abstract)
41. Al-Ajeel, A.W.A.; Al-Sindy, S.I. Alumina recovery from Iraqi kaolinitic clay by hydrochloric acid route. *Iraqi Bull. Geol. Min.* **2006**, *2*, 67–76.
42. Al-Zahrani, A.A.; Abdul-Majid, M.H. Extraction of alumina from local clays by hydrochloric acid process. *J. King Abdulaziz Univ. Eng. Sci.* **2009**, *20*, 29–41. [CrossRef]
43. Erdemoğlu, M.; Birinci, M.; Uysal, T. Thermal behavior of pyrophyllite ore during calcination for thermal activation for aluminum extraction by acid leaching. *Clays Clay Miner.* **2020**, *68*, 89–99. [CrossRef]
44. Meshram, P.; Pandey, B.D.; Mankhand, T.R. Extraction of lithium from primary and secondary sources by pre-treatment, leaching and separation: A comprehensive review. *Hydrometallurgy* **2014**, *150*, 192–208. [CrossRef]
45. Han, Y.; Yang, Y.; Ma, Y.L.; Liang, D.X.; Wen, L.P.; Ma, J.; Wang, W. Selective ion channel adsorbents facilitate efficient and low environmental impact extraction of liquid lithium resources. *J. Hazard. Mater.* **2024**, *480*, 136335. [CrossRef] [PubMed]
46. Shu, Q.Y. *Study of Leaching Lithium from Lepidolite and Preparation of Lithium Carbonate*; Nanchang University: Nanchang, China, 2012; pp. 1–60. (In Chinese with English abstract)
47. Dong, Q.; Li, Y.J.; Piao, X.L.; Zhu, S.L. Recovery of lithium from salt-lake bittern using aluminum salt adsorbent. *Chin. J. Rare Met.* **2007**, *03*, 357–361. (In Chinese with English abstract)
48. *GB/T 11075-2013*; Lithium Carbonate. General Administration of Quality Supervision, Inspection and Quarantine, Standardization Administration of China. Standards Press of China: Beijing, China, 2013.